LINEAR ALGEBRA AND GEOMETRY

ALGEBRA, LOGIC AND APPLICATIONS

A Series edited by

R. Göbel
Universitat Gesamthochschule, Essen, FRG
A. Macintyre
The Mathematical Institute, University of Oxford, UK

Volume 1
Linear Algebra and Geometry
A. I. Kostrikin and Yu. I. Manin

Additional volumes

Volume 2
Model Theoretic Algebra
with particular emphasis on Fields, Rings, Modules
Christian U. Jensen and Helmut Lenzing

LINEAR ALGEBRA AND GEOMETRY

By A. I. Kostrikin and Yu. I. Manin

Steklov Institute of Mathematics,
USSR Academy of Sciences, Moscow

Translated from the Second Russian Edition by
M. E. Alferieff

CRC Press
Taylor & Francis Group
Boca Raton London New York

CRC Press is an imprint of the
Taylor & Francis Group, an informa business

First published 1989 by OPA (Amsterdam) B.V.

Published 2021 by CRC Press
Taylor & Francis Group
6000 Broken Sound Parkway NW, Suite 300
Boca Raton, FL 33487-2742

ISBN-13: 978-2-88124-683-8 (hbk)

Visit the Taylor & Francis Web site at
http://www.taylorandfrancis.com

and the CRC Press Web site at
http://www.crcpress.com

Originally published in Russian in 1981 as Линейная Алгеьра и Геометрия
(Lineinaya algebra i geometriya) by Moscow University Press
Издательство московского Университета
The Second Russian Edition was published in 1986
by Nauka Publishers, Moscow Издательство Наука Москва

Library of Congress Cataloging-in-Publication Data

Kostrikin, A. I. (Alekseĭ Ivanovich)
[Lineĭnaiа algebra i geometriiа. English]
Linear algebra and geometry / A. I. Kostrikin and Yu. I. Manin ;
translated from the Russian by M. E. Alferieff.
p. cm.
Translation of: Lineĭnaiа algebra i geometriiа.
Bibliography: p. ix.
Includes index.
ISBN 2-88124-683-4
1. Algebras, Linear. 2. Geometry. I. Manin, IU. I. II. Title.
QA184.K6913 1988
512'.5--dc19

88-39834
CIP

Publisher's Note
The publisher has gone to great lengths to ensure the quality of this reprint but points out that some imperfections in the original may be apparent.

Contents

Preface

There is a different viewpoint for everything; the subject of this book is no exception. To students of mechanics and mathematics at Moscow State University, linear algebra is the material taught to a freshman in the second semester. To the professional algebraist, trained in the spirit of Bourbaki, linear algebra is the theory of algebraic structures of a particular form, namely, linear spaces and linear mappings, or, in the more modern style, the theory of linear categories.

From a more general viewpoint, linear algebra is the careful study of the mathematical language for expressing one of the most general ideas in the natural sciences – the idea of linearity. The most important special case of this idea could quite possibly be the principle of linearity of small increments: almost any natural process is linear in the small almost everywhere. This principle lies at the foundation of all of mathematical analysis and its applications. The vector algebra of three-dimensional physical space, which has historically become the cornerstone of linear algebra, dates back to the same source: after Einstein, we now know that physical space is also approximately linear only in a small neighbourhood of the observer. Fortunately, this small neighbourhood is quite large.

Twentieth-century physics has markedly and unexpectedly expanded the sphere of application of linear algebra, adding to the principle of linearity of small increments the principle of superposition of state vectors. Roughly speaking, the state space of any quantum system is a linear space over the field of complex numbers. As a result, almost all constructions of complex linear algebra have been transformed into a tool for formulating the fundamental laws of nature: from the theory of linear duality, which explains Bohr's quantum principle of complementarity, to the theory of representations of groups, which explains Mendeleev's table, the zoology of elementary particles, and even the structure of space-time.

The selection of the material for this course was determined by our desire not only to present the foundations of the body of knowledge which was essentially completed by the beginning of the twentieth century, but also to give an idea of its applications, which are usually relegated to other disciplines. Traditional teaching dissects the live body of mathematics into isolated organs, whose vitality must be maintained artificially. This particularly concerns the "critical periods" in the history of our science, which are characterised by their attention to the logical structure and detailed study of the foundations. During the last half-century the language and fundamental concepts were reformulated in set-theoretic language; the unity of mathematics came to be viewed largely in terms of the unity of its logical principles. We wanted to reflect in this book, without ignoring the remarkable achievements of this period, the emerging trend toward the synthesis of mathematics as a tool for understanding the outside world.

(Unfortunately, we had to ignore the theory of the computational aspects of linear algebra, which has now developed into an independent science.)

Based on these considerations, this book, just as the "Introduction to Algebra" written by one of us, includes not only material for a course of lectures, but also sections for independent reading, which can also be used for seminars. There is no strict division here. Nevertheless, in accordance with the standard programme, a lecture course (one semester with two lectures per week) should include the basic material in Sections 1-9 of Chapter 1; Sections 2–8 of Chapter 2; Sections 1, 3, 5 and 6 of Chapter 3 and Sections 1 and 3–6 of Chapter 4. By basic material we mean not the proof of difficult theorems (of which there are only a few in linear algebra), but rather the system of concepts which should be mastered. Accordingly, many theorems from these sections can be presented in a simpler version or omitted entirely; due to the lack of time, such abridgement is unavoidable. It is up to the instructor to determine how to prevent the lectures from becoming a tedious listing of definitions. We hope that the remaining sections of the course will be of some help in this task.

A. I. Kostrikin
Yu. I. Manin

Bibliography

1 Kostrikin, A. I., (1977) Vvedenie v algebru (Introduction to Algebra), Nauka, Moscow
2 Lang, S., (1971) Algebra, Addison-Wesley, Reading, MA
3 Gel'fand, I. M., (1961) Lectures on Linear Algebra, Interscience Publishers, Inc., New York
4 Halmos, P. R., (1958) Finite-Dimensional Vector Spaces, D. Van Nostrand Company, Inc., New York
5 Artin, E., (1957) Geometric Algebra, Interscience Publishers, Inc., New York
6 Galzman, I. M. and Ju. I. Ljubic, (1974) Finite-Dimensional Linear Analysis: A Systemic Presentation in Problems Form, MIT Press, Cambridge, MA
7 Voevodin, V. V., (1977) Vychislitel'nye osnovy lineinoi algebry (Computational Foundations of Linear Algebra), Nauka, Moscow

CHAPTER 1

Linear Spaces and Linear Mappings

§1. Linear Spaces

1.1. Vectors, whose starting points are located at a fixed point in space, can be multiplied by a number and added by the parallelogram rule. This is the classical model of the laws of addition of displacements, velocities, and forces in mechanics. In the general definition of a vector or a linear space, the real numbers are replaced by an arbitrary field and the simplest properties of addition and multiplication of vectors are postulated as an axiom. No traces of the "three-dimensionality" of physical space remain in the definition. The concept of dimensionality is introduced and studied separately.

Analytic geometry in two- and three-dimensional space furnishes many examples of the geometric interpretation of algebraic relations between two or three variables. However, as expressed by N. Bourbaki, "... the restriction to geometric language, conforming to a space of only three dimensions, would be just as inconvenient a yoke for modern mathematics as the yoke that prevented the Greeks from extending the concept of numbers to relations between incommensurate quantities ...".

1.2. Definition *A set is said to be a linear (or vector) space L over a field K if it is equipped with a binary operation $L \times L \to L$, usually denoted as addition $(l_1, l_2) \to l_1 + l_2$, and an external binary operation $K \times L \to L$, usually denoted as multiplication $(a, l) \to al$, which satisfy the following axioms:*

a) *Addition of the elements of L, or vectors, transforms L into a commutative (abelian)group. Its zero element is usually denoted by 0; the element inverse to l is usually denoted by $-l$.*

b) *Multiplication of vectors by elements in the field K, or scalars, is unitary, i.e., $1l = l$ for all l, and is associative, i.e., $a(bl) = (ab)l$ for all $a, b \in K$ and $l \in L$.*

c) *Addition and multiplication satisfy the distributivity laws, i.e.*

$$a(l_1 + l_2) = al_1 + al_2, \ (a_1 + a_2)l = a_1 l + a_2 l$$

for all $a, a_1, a_2 \in K$ and $l, l_1, l_2 \in L$.

1.3. Here are some very simple consequences of this definition.

a) $0l = a0 = 0$ for all $a \in K$ and $l \in L$. Indeed $0l + 0l = (0+0)l = 0l$, whence according to the property of contraction in an abelian group, $0l = 0$. Analogously, $a0 + a0 = a(0+0) = a0$, that is, $a0 = 0$.

b) $(-1)l = -l$. Indeed, $l + (-1)l = 1l + (-1)l = (1 + (-l))l = 0l = 0$, so that the vector $(-1)l$ is the inverse of l.

c) If $al = 0$, then either $a = 0$ or $l = 0$. Indeed, if $a \neq 0$, then $0 = a^{-1}(al) = (a^{-1}a)l = 1l = l$.

d) The expression $a_1 l_1 + \ldots + a_n l_n = \sum a_i l_i$ is uniquely defined for any $a_1, \ldots, a_n \in K$ and $l_1, \ldots, l_n \in L$: because of the associativity of addition in an abelian group it is not necessary to insert parentheses indicating order for the calculation of double sums. Analogously, the expression $a_1 a_2 \ldots a_n l$ is uniquely defined.

An expression of the form $\sum_{i=1}^n a_i l_i$ is called a *linear combination* of vectors $l_1, \ldots, l_n$; the scalars a_i are called the coefficients of this linear combination.

The following examples of linear spaces will be encountered often in what follows.

1.4. Zero-dimensional space. This is the abelian group $L = \{0\}$, which consists of one zero. The only possible law is multiplication by a scalar: $a0 = 0$ for all $a \in K$ (verify the validity of the axioms !).

Caution: zero-dimensional spaces over *different fields* are *different spaces*: the field K is specified in the definition of the linear space.

1.5. The basic field K as a one-dimensional coordinate space. Here $L = K$; addition is addition in K and multiplication by scalars is multiplication in K. The validity of the axioms of the linear space follows from the axioms of the field.

More generally, for any field K and a subfield $\mathbf{K}$ of it, K can be interpreted as a linear space over $\mathbf{K}$. For example, the field of complex numbers $\mathbf{C}$ is a linear space over the field of real numbers $\mathbf{R}$, which in its turn is a linear space over the field of rational numbers $\mathbf{Q}$.

1.6. n-dimensional coordinate space. Let $L = K^n = K \times \ldots \times K$ (Cartesian product of $n \geq 1$ factors). The elements of L can be written in the form of rows of length n $(a_1, \ldots, a_n)$, $a_i \in K$ or columns of height n. Addition and multiplication by a scalar is defined by the formulas:

$$(a_1, \ldots, a_n) + (b_1, \ldots, b_n) = (a_1 + b_1, \ldots, a_n + b_n),$$
$$a(a_1, \ldots, a_n) = (aa_1, \ldots, aa_n).$$

The preceding example is obtained by setting $n = 1$. One-dimensional spaces over K are called straight lines or K-lines; two-dimensional spaces are called K-planes.

1.7. Function spaces. Let S be an arbitrary set and let $F(S)$ be the set of functions on S with values in K or mappings of S into K. As usual, if $f : S \to K$ is such a function, then $f(s)$ denotes the value of f on the element $s \in S$.

Addition and multiplication of functions by a scalar are defined pointwise:

$$(f+g)(s) = f(s) + g(s) \quad \text{for all} \quad s \in S,$$

$$(af)(s) = a(f(s)) \quad \text{for all} \quad a \in K, s \in S.$$

If $S = \{1, \ldots, n\}$, then $F(S)$ can be identified with K^n: the function f is associated with the "vector" formed by all of its values $(f(1), \ldots, f(n))$. The addition and multiplication rules are consistent with respect to this identification.

Every element $s \in S$ can be associated with the important "delta function δ_s centred on $\{s\}$", which is defined as $\delta_s(s) = 1$ and $\delta_s(t) = 0$, if $t \neq s$. If $S = (1, \ldots, n)$, then δ_{ik}, *the Kronecker delta*, is written instead of $\delta_i(k)$.

If the set S is finite, then any function from $F(S)$ can be represented uniquely by a linear combination of delta functions: $f = \sum_{s \in S} f(s)\delta_s$. Indeed, this equality follows from the fact that the left side equals the right side at every point $s \in S$. Conversely, if $f = \sum_{s \in S} a_s \delta_s$, then taking the value at the point s we obtain $f(s) = a_s$.

If the set S is infinite, then this result is incorrect. More precisely, it cannot be formulated on the basis of our definitions: sums of an infinite number of vectors in a general linear space are not defined ! Some infinite sums can be defined in linear spaces which are equipped with the concept of a limit or a topology (see Chapter 10). Such spaces form the basic subject of functional analysis.

In the case $S = \{1, \ldots, n\}$, the function δ_i is represented by the vector $e_i = (0, \ldots, 0, 1, 0, \ldots, 0)$ (1 at the ith place and 0 elsewhere) and the equality $f = \sum_{s \in S} f(s)\delta_s$ transforms into the equality

$$(a_1, \ldots, a_n) = \sum_{i=1}^{n} a_i e_i.$$

1.8. Linear conditions and linear subspaces. In analysis, primarily real-valued functions defined over all $\mathbf{R}$ or on intervals $(a, b) \subset \mathbf{R}$ are studied. For most applications, however, the space of all such functions is too large: it is useful to study continuous or differentiable functions. After the appropriate definitions are introduced, it is usually proved that the sum of continuous functions is continuous and that the product of a continuous function by a scalar is continuous; the same assertions are also proved for differentiability.

This means that the continuous or differentiable functions themselves form a linear space.

More generally, let L be a linear space over the field K and let $M \subset L$ be a subset of L, which is a subgroup and which transforms into itself under multiplication by a scalar. Then M together with the operations induced by the operations in L (in other words, the restrictions of the operations defined in L to M) is called a *linear subspace* of L, and the conditions which an arbitrary vector in L must satisfy in order to belong to M are called *linear conditions.*

Here is an example of linear conditions in the coordinate space K^n. We fix scalars $a_1, \ldots, a_n \in K$ and define $M \subset L$:

$$(x_1, \ldots, x_n) \in M \Leftrightarrow \sum_{i=1}^{n} a_i x_i = 0. \tag{1}$$

A combination of any number of linear conditions is also a linear condition. In other words, the intersection of any number of linear subspaces is also a linear subspace (check this !). We shall prove later that any subspace in K^n is described by a finite number of conditions of the form (1).

An important example of a linear condition is the following construction.

1.9. The dual linear space. Let L be a linear space over K. We shall first study the linear space $F(L)$ of *all* functions on L with values in K. We shall now say that a function $f \in F(L)$ is *linear* (or, as is sometimes said, a *"linear functional"*), if it satisfies the conditions

$$f(l_1 + l_2) = f(l_1) + f(l_2), \ f(al) = af(l)$$

for all $l, l_1, l_2 \in L$ and $a \in K$. From here, by induction on the number of terms, we find that

$$f(\sum_{i=1}^{n} a_i l_i) = \sum_{i=1}^{n} a_i f(l_i).$$

We assert that *linear functions form a linear subspace of* $F(L)$ or "*the condition of linearity is a linear condition*". Indeed, if f, f_1 and f_2 are linear, then

$$(f_1 + f_2)(l_1 + l_2) = f_1(l_1 + l_2) + f_2(l_1 + l_2) =$$

$$= f_1(l_1) + f_1(l_2) + f_2(l_1) + f_2(l_2) = (f_1 + f_2)(l_1) + (f_1 + f_2)(l_2).$$

(Here the following are used successively: the rule for adding functions, the linearity of f_1 and f_2, the commutativity and associativity of addition in a field, and again the rule of addition of functions.) Analogously,

$$(af)(l_1 + l_2) = a[f(l_1 + l_2)] = a[f(l_1) + f(l_2)] =$$

$$= a[f(l_1)] + a[f(l_2)] = (af)(l_1) + (af)(l_2).$$

Thus $f_1 + f_2$ and af are also linear.

The space of linear functions on a linear space L is called a dual space or the space conjugate to L and is denoted by L^*.

In what follows we shall encounter many other constructions of linear spaces.

1.10. Remarks regarding notation. It is very convenient, but not entirely consistent, to denote the zero element and addition in K and L by the same symbols. All formulas of ordinary high-school algebra, which can be imagined in this situation, are correct: refer to the examples in §1.3.

Here are two examples of cases when a different notation is preferable.

a) Let $L = \{x \in \mathbf{R} | x > 0\}$. We regard L as an abelian group *with respect to multiplication* and we introduce in L multiplication by a scalar from $\mathbf{R}$ according to the formula $(a, x) \to x^a$. It is easy to verify that all conditions of Definition 1.2 are satisfied, though in the usual notation they assume a different form: the zero vector in L is 1; $1l = l$ is replaced by $x^1 = x$; $a(bl) = (ab)l$ is replaced by the identity $(x^b)^a = x^{ba}$; $(a + b)l = al + bl$ is replaced by the identity $x^{a+b} = x^a x^b$; etc.

b) Let L be a vector space over the field of complex numbers $\mathbf{C}$. We define a new vector space $\bar{L}$ with the same additive group L, but a different law of multiplication by a scalar:

$$(a, l) \mapsto \bar{a}l,$$

where $\bar{a}$ is the complex conjugate of a. From the formulas $\overline{a + b} = \bar{a} + \bar{b}$ and $\overline{ab} = \bar{a}\bar{b}$ it follows without difficulty that $\bar{L}$ is a vector space. If in some situation L and $\bar{L}$ must be studied at the same time, then it may be convenient to write $a * l$ or $a \circ l$ instead of $\bar{a}l$.

1.11. Remarks regarding diagrams and graphic representations. Many general concepts and theorems of linear algebra are conveniently illustrated by diagrams and pictures. We want to warn the reader immediately about the dangers of such illustrations.

a) *Low dimensionality.* We live in a three-dimensional space and our diagrams usually portray two- or three-dimensional images. In linear algebra we work with space of any finite number of dimensions and in functional analysis we work with infinite-dimensional spaces. Our "low-dimensional" intuition can be greatly developed, but it must be developed systematically.

Here is a simple example: how are we to imagine the general arrangement of two planes in four-dimensional space ? Imagine two planes in $\mathbf{R}^3$ intersecting along a straight line which splay out everywhere along this straight line except at the origin, vanishing into the fourth dimension.

b) *Real field.* The physical space $\mathbf{R}^3$ is linear over a real field. The unfamiliarity of the geometry of a linear space over K could be associated with the properties of this field.

For example, let $K = \mathbf{C}$ (a very important case for quantum mechanics). A straight line over $\mathbf{C}$ is a one-dimensional coordinate space $\mathbf{C}^1$. We have become accustomed to the fact that multiplication of points on the straight line $\mathbf{R}^1$ by a real number a represents an a-fold stretching (for $a > 1$), an a^{-1}-fold compression (for $0 < a < 1$), or their combination with an inversion of the straight line (for $a < 0$).

It is, however, natural to imagine multiplication by a complex number a, acting on $\mathbf{C}^1$, in a geometric representation of $\mathbf{C}^1$ in terms of $\mathbf{R}^2$ ("Argand plane" or the "complex plane" — not to be confused with $\mathbf{C}^2$!). The point $(x, y) \in \mathbf{R}^2$ is then the image of the point $z = x + iy \in \mathbf{C}^1$ and multiplication by $a \neq 0$ corresponds to stretching by a factor of $|a|$ and counterclockwise rotation by the angle $\arg a$. In particular, for $a = -1$ the real "inversion" of the straight line $\mathbf{R}^1$ is the restriction of the rotation of $\mathbf{C}^1$ by 180° to $\mathbf{R}^1$.

In general, it is often useful to think of an n-dimensional complex space $\mathbf{C}^n$ as a $2n$-dimensional real space $\mathbf{R}^{2n}$ (compare §12 on complexification and decomplexification).

Finite fields K, in particular the field consisting of two elements $F_2 = \{0, 1\}$, which is important in encoding theory, are another important example. Here finite-dimensional coordinate spaces are finite, and it is sometimes useful to associate discrete images with a linear geometry over K. For example, F_2^n is often identified with the vertices of an n-dimensional unit cube in $\mathbf{R}^n$ - the set of points $(\epsilon_1, \ldots, \epsilon_n)$, where $\epsilon_i = 0$ or 1. Coordinatewise addition in F_2^n is a Boolean operation: $1 + 0 = 0+1 = 1$; $0+0 = 1+1 = 0$. The subspace consisting of points with $\epsilon_1 + \ldots + \epsilon_n = 0$ defines the simplest code with error detection. If it is stipulated that the points $(\epsilon_1, \ldots, \epsilon_n)$ encode a message only if $\epsilon_1 + \ldots + \epsilon_n = 0$, then in the case when a signal $(\epsilon_1', \ldots, \epsilon_n')$ with $\sum_{i=1}^n \epsilon_i' \neq 0$ is received we can be sure that interference in transmission has led to erroneous reception.

c) *Physical space is Euclidean.* This means that not only are addition of vectors and multiplication by a scalar defined in this space, but the lengths of vectors, the angles between vectors, the areas and volumes of figures, and so on are also defined. Our diagrams carry compelling information about these "metric" properties and we perceive them automatically, though they are in no way reflected in the general axiomatics of linear spaces. It is impossible to imagine that one vector is shorter than another or that a pair of vectors forms a right angle unless the space is equipped with a special additional structure, for example, an abstract inner product. Chapter 2 of this book is devoted to such structures.

EXERCISES

1. Do the following sets of real numbers form a linear space over $\mathbf{Q}$?

a) the positive real numbers;

b) the negative real numbers;

c) the integers;

d) the rational numbers with a denominator $\leq N$;

e) all numbers of the form $a+b\pi$, where a and b are arbitrary rational numbers.

2. Let S be some set and let $F(S)$ be the space of functions with values in the field K. Which of the following conditions are linear ?

a) f vanishes at a given point in S;

b) f assumes the value l at a given point of S;

c) f vanishes at all points in a subset $S_0 \subset S$;

d) f vanishes at at least one point of a subset $S_0 \subset S$.

Below $S = \mathbf{R}$ and $K = \mathbf{R}$:

e) $f(x) \to 0$ as $|x| \to \infty$;

f) $f(x) \to 1$ as $|x| \to \infty$;

g) f has not more than a finite number of points of discontinuity.

3. Let L be the linear space of continuous real functions on the segment $[-1,1]$. Which of the functionals on L are linear functionals ?

a) $f \mapsto \int_{-1}^{1} f(x)dx$;

b) $f \mapsto \int_{-1}^{1} f^2(x)dx$;

c) $f \mapsto f(0)$ (this is the *Dirac delta-function*);

d) $f \mapsto \int_{-1}^{1} f(x)g(x)dx$, where g is a fixed continuous function on $[-1,1]$.

4. Let $L = K^n$. Which of the following conditions on $(x_1,\ldots,x_n) \in L$ are linear :

a) $\sum_{i=1}^{n} a_i x_i = 1; a_1,\ldots,a_n \in K$;

b) $\sum_{i=1}^{n} x_i^2 = 0$ (examine the following cases separately: $K = \mathbf{R}$, $K = \mathbf{C}$, and K is a field with two elements or, more generally, a field whose characteristic equals two);

c) $x_3 = 2x_4$.

5. Let K be a finite field consisting of q elements. How many elements are there in the linear space K^n ? How many solutions does the equation $\sum_{i-1}^{n} a_i x_i = 0$ have ?

6. Let K^∞ be the space of infinite sequences $(a_1,a_2,a_3,\ldots)$, $a_i \in K$, with coordinatewise addition and multiplication. Which of the following conditions on vectors from K^∞ are linear ?

a) only a finite number of the coordinates a_i differs from zero;

b) only a finite number of coordinates a_i vanishes;

c) no coordinate a_i is equal to 1.

Below $K = \mathbf{R}$ or $\mathbf{C}$;

d) Cauchy's condition: for every $\epsilon > 0$ there exists a number $N > 0$ such that $|a_m - a_n| < \epsilon$ for $m, n > N$;

e) Hilbert's condition: the series $\sum_{n=1}^{\infty} |a_n|^2$ converges;

f) (a_i) form a bounded sequence, i.e., there exists a constant c, depending on (a_i), such that $|a_i| < c$ for all i.

7. Let S be a finite set. Prove that every linear functional on $F(S)$ is determined uniquely by the set of elements $\{a_s | s \in S\}$ of the field K: the scalar $\sum_{s\in S} a_s f(s)$ is associated with the function f.

If n is the number of elements of S and $a_s = 1/n$ for all s, we obtain the functional $f \mapsto \frac{1}{n}\sum_{s\in S} f(s)$ — the average arithmetic value of the function.

If $K = \mathbf{R}$ and $a_s \geq 0$, $\sum_{s\in S} a_s = 1$, the functional $\sum_{s\in S} a_s f(s)$ is called the *weighted mean* of the function f (with weights a_s).

§2. Basis and Dimension

2.1. Definition. *A set of vectors $\{e_1, \ldots, e_n\}$ in a linear space L is said to be a (finite)basis of L if every vector in L can be uniquely represented as a linear combination $l = \sum_{i=1}^{n} a_i e_i$, $a_i \in K$. The coefficients a_i are called the coordinates of the vector l with respect to the basis $\{e_i\}$.*

2.2. Examples. a) The vectors $e_i = (0, \ldots, 1, \ldots, 0)$, $1 \leq i \leq n$, in K^n form a basis of K. b) If the set S is finite, the functions $\delta_s \in F(S)$ form a basis of $F(S)$. Both of these assertions were checked in §1.

If a basis consisting of n vectors is chosen in L and every vector is specified in terms of its coordinates with respect to this basis, then addition and multiplication by a scalar are performed coordinatewise:

$$\sum_{i=1}^{n} a_i e_i + \sum_{i=1}^{n} b_i e_i = \sum_{i=1}^{n} (a_i + b_i) e_i, \quad a\sum_{i=1}^{n} a_i e_i = \sum_{i=1}^{n} a a_i e_i.$$

The selection of a basis is therefore equivalent to the identification of L with the coordinate vector space. The notation $l = [a_i, \ldots, a_n]$ or $l = \vec{a}$ is sometimes used instead of the equality $l = \sum_{i=1}^{n} a_i e_i$; in this notation the basis is not indicated explicitly. Here $[a_1, \ldots, a_n]$ stands for the column vector

$$[a_1, \ldots, a_n] = \begin{pmatrix} a_1 \\ \vdots \\ a_n \end{pmatrix} = \vec{a}$$

2.3. Definition. *A space L is said to be finite-dimensional if it is either zero-dimensional* (see §1.4) *or has a finite basis. Otherwise it is said to be infinite-dimensional.*

It is convenient to regard the basis of a zero-dimensional space as an empty set of vectors. Since all of our assertions become trivial for zero-dimensional spaces, we shall usually restrict our attention to non-empty bases.

2.4. Theorem. *In a finite-dimensional space the number of elements in the basis does not depend on the basis.*

This number is called the dimension of the space L and is denoted by $\dim L$ or $\dim_K L$. If $\dim L = n$, then the space L is said to be an n-dimensional space. In the infinite-dimensional case, we write $\dim L = \infty$.

Proof. Let $\{e_1, \ldots, e_n\}$ be a basis of L. We shall prove that no family of vectors $\{e'_1, \ldots, e'_m\}$ with $m > n$ can serve as a basis of L for the following reason: there exists a representation of the zero vector $0 = \sum_{i=1}^{m} x_i e'_i$ such that not all x_i vanish. Hence 0 cannot be uniquely represented as a linear combination of the vectors $\{e'_i\}$: the trivial representation $0 = \sum_{i=1}^{m} 0e'_i$ always exists.

The complete assertion of the theorem already follows from here, since we can now verify that no basis can contain more elements than any other basis.

Let $e'_k = \sum_{i=1}^{n} a_{ik} e_i$, $k = 1, \ldots, m$. For any $x_k \in K$ we have

$$\sum_{k=1}^{m} x_k e'_k = \sum_{k=1}^{m} x_k \sum_{i=1}^{n} a_{ik} e_i = \sum_{i=1}^{n} (\sum_{k=1}^{m} a_{ik} x_k) e_i.$$

Since $\{e_i\}$ form a basis of L, the zero vector can be represented uniquely as a linear combination $\sum_{k=1}^{n} 0e_k$ of $\{e_k\}$. The condition $\sum_{k=1}^{m} x_k e'_k = 0$ is therefore equivalent to a system of homogeneous linear equations for x_k:

$$\sum_{k=1}^{m} a_{ik} x_k = 0, \; i = 1, \ldots, n.$$

Since the number of unknowns m exceeds the number of equations n, this system has a non-zero solution. The theorem is proved.

2.5. Remark. a) Any set of vectors can be a basis if any vector in the space can be uniquely represented as a *finite* linear combination of the elements of this set. In this sense, any linear space has a basis and the basis of an infinite-dimensional space is always infinite. This concept, however, is not very useful. As a rule, infinite-dimensional spaces are equipped with a topology and the possibilities of defining infinite-dimensional linear combinations are included.

b) In general linear spaces, bases are traditionally enumerated by integers from 1 to n (sometimes from 0 to n), but this is not absolutely necessary. The basis $\{\delta_s\}$ in $F(S)$ is naturally enumerated by the elements of the set $s \in S$. A basis in L can also be viewed as simply a subset in L, whose elements are not equipped with any indices (cf. §2.20). Enumeration or rather the order of the elements of a basis is important in the matrix formalism (see §4). In other problems, a different structure on the set of indices enumerating the basis could be important. For example, if S is a finite group, then the manner in which the indices s of the basis (δ_s) are multiplied within S is important, and a random enumeration of S by integers can only confuse the notation.

2.6. Examples. a) The dimension of K^n equals n. b) The dimension of $F(S)$ equals the number of elements in S, if S is finite.

Later we shall learn how to calculate the dimension of linear spaces without constructing their bases. This is very important, because many numerical invariants in mathematics are defined as a dimension (the "Betti number" in topology, the indices of operators in the theory of differential equations); the bases of the corresponding spaces, on the other hand, may be difficult to calculate or they may not have any special significance. For the time being, however, we must work with bases.

The verification of the fact that a given family of vectors $\{e_1, \ldots, e_n\}$ in L forms a basis, according to the definition, consists of two parts. A study of each part separately leads to the following concepts.

2.7. Definition. *The set of all possible linear combinations of a set of vectors in L is called the linear span of the set.*

It is easy to verify that a linear span is a linear subspace of L (see §1.8). The linear span of $\{e_i\}$ is also referred to as the subspace *spanned* or *generated* by the vectors $\{e_i\}$. It can also be defined as the *intersection of all linear subspaces of L containing all e_i* (prove this !). The *dimension of a linear span of a set of vectors* is called the *rank* of the set.

The first characteristic property of a basis is: *its linear span coincides with all of L.*

2.8. Definition. The set of vectors $\{e_i)$ is said to be linearly independent if no non-trivial linear combination of $\{e_i\}$ vanishes, i.e., if $\sum_{i=1}^n a_i e_i = 0$ implies that all the $a_i = 0$. Otherwise, it is said to be linearly dependent.

The fact that the set $\{e_i\}$ is linearly independent indicates that the *zero vector can be represented as a unique linear combination of the elements of the set.* Then any other vector has either a unique representation or no representation. Indeed, comparing the two representations

$$l = \sum_{i=1}^n a_i e_i = \sum_{i=1}^n a_i' e_i,$$

we find that

$$0 = \sum_{i=1}^n (a_i - a_i') e_i,$$

whence $a_i = a_i'$.

From here follows the second characteristic property of a basis: *its elements are linearly independent.*

The combination of these two properties is equivalent to the first definition of a basis.

We note also that *a set of vectors is linearly independent if and only if it forms a basis of its linear span.*

The family $\{e_1, \dots, e_n\}$ is obviously linearly dependent if one of the vectors e_i is the zero vector or two of the vectors e_i are identical (why ?).

More generally, we have the following lemma.

2.9. Lemma. a) The set of vectors $\{e_1, \dots, e_n\}$ is linearly dependent if and only if at least one of the vectors e_j is a linear combination of the others.

b) If the set $\{e_1, \dots, e_n\}$ is linearly independent and the set $\{e_1, \dots, e_n, e_{n+1}\}$ is linearly dependent, then e_{n+1} is a linear combination of $e_1, \dots, e_n$.

Proof. a) If $\sum_{i=1}^n a_i e_i = 0$ and $a_j \neq 0$, then $e_j = \sum_{i=1, i\neq j}^n (-a_j^{-1} a_i) e_i$. Conversely, if $e_j = \sum_{i\neq j} b_i e_i$, then $e_j - \sum_{i \neq j} b_i e_i = 0$.

b) If $\sum_{i=1}^{n+1} a_i e_i = 0$ and not all a_i vanish, then necessarily $a_{n+1} \neq 0$. Otherwise we would obtain a non-trivial linear dependence between $e_1, \dots, e_n$. Therefore, $e_{n+1} = \sum_{i=1}^n (-a_{n+1}^{-1} a_i) e_i$. The lemma is proved.

Let $E = \{e_1, \dots, e_n\}$ be a finite set of vectors in L and let $F = \{e_{i_1}, \dots, e_{i_m}\}$ be a linearly independent subset of E. We shall say that F is *maximal*, if every element in E can be expressed as a linear combination of the elements of F.

2.10. Proposition. *Every linearly independent subset $E' \subset E$ is contained in some maximal linearly independent subset $F \subset E$. The linear spans of F and E coincide with each other.*

Proof. If $E \backslash E'$ contains a vector that cannot be represented as a linear combination of the elements of E', then we add it to E'. According to Lemma 2.9b, the set E'' so obtained will be linearly independent. We apply the same argument to E'', etc. Since E is finite, this process will terminate on the maximal set F. Any element of the linear span of E can evidently be expressed as a linear combination of the vectors in the set F.

In the case $E' = \emptyset$, E'' must be chosen as a non-zero vector from E, if it exists; otherwise, F is empty.

2.11. Remark. This result is also true for infinite sets E. To prove this assertion it is necessary to apply transfinite induction or Zorn's lemma: see §2.18–§2.20.

The maximal subset is not necessarily unique. Let $E = \{(1,0),\ (0,1),\ (1,1)\}$ and $E' = \{(1,0)\}$ in K^2. Then E' is contained in two maximal independent subsets

$\{(1,0), (0,1)\}$ and $\{(1,0), (1,1)\}$. However, the number of elements in the maximal subset is determined uniquely; it equals the dimension of the linear span of E and is called the rank of the set E.

The following theorem is often useful.

2.12. Theorem on the extension of a basis. *Let $E' = \{e_1, \dots, e_m\}$ be a linearly independent set of vectors in a finite-dimensional space L. Then there exists a basis of L that contains E'.*

Proof. Select any basis $\{e_{m+1}, \dots, e_n\}$ of L and set $E = \{e_1, \dots, e_m, e_{m+1}, \dots, e_n\}$. Let F denote a maximal linearly independent subset of E containing E'. This is the basis sought.

Actually, it is only necessary to verify that the linear span of F coincides with L. But, according to Proposition 2.10, it equals the linear span of E, while the latter equals L because E contains a basis of L.

2.13. Corollary (monotonicity of dimension). *Let M be a linear subspace of L. Then $\dim M \leq \dim L$ and if L is finite-dimensional, then $\dim M = \dim L$ implies that $M = L$.*

Proof. If M is infinite-dimensional, then L is also infinite-dimensional. Indeed, we shall first show that M contains arbitrarily large independent sets of vectors. If a set of n linearly independent vectors $\{e_1, \dots, e_n\}$ has already been found, then its linear span $M' \subset M$ cannot coincide with M, for otherwise M would be n-dimensional. Therefore, M contains a vector e_{n+1}, that cannot be expressed as a linear combination of $\{e_1, \dots, e_n\}$ and Lemma 2.9b shows that the set $\{e_1, \dots, e_n, e_{n+1}\}$ is linearly independent. We now assume that M is infinite-dimensional while L is n-dimensional. Then according to the proof of Theorem 2.4, any $n+1$ linear combinations of elements of the basis of L are linearly dependent, which contradicts the infinite-dimensionality of M.

It remains to analyse the case when M and L are finite-dimensional. In this case, according to Theorem 2.12, any basis of M can be extended up to the basis of L, whence it follows that $\dim M \leq \dim L$.

Finally, if $\dim M = \dim L$, then any basis of M must be a basis of L. Otherwise, its extension up to a basis in L would consist of $> \dim L$ elements, which is impossible.

2.14. Bases and flags. One of the standard methods for studying sets S with algebraic structures is to single out sequences of subsets $S_0 \subset S_1 \subset S_2 \dots$ or $S_0 \supset S_1 \supset S_2 \supset \dots$ such that the transition from one subset to the next one is simple in some sense. The general name for such sequences is *filtering* (increasing and decreasing respectively). In the theory of linear spaces, a strictly increasing sequence of subspaces $L_0 \subset L_1 \subset \ldots \subset L_n$ of the space L is called a ***flag***. (This term is motivated by the following correspondence: flag $\{0\ \textit{point}\} \subset \{\textit{straight line}\} \subset$ $\subset \{\textit{plane}\}$ corresponds to "nail", "staff", and "sheet of cloth".)

The number n is called the ***length*** of the flag $L_0 \subset L_1 \subset \ldots \subset L_n$.

The flag $L_0 \subset L_1 \subset \ldots \subset L_n \subset \ldots$ is said to be ***maximal*** if $L_0 = \{0\}$, $\bigcup L_i = L$ and a subspace cannot be inserted between L_i, L_{i+1} (for any i): if $L_i \subset M \subset L_{i+1}$, then either $L_i = M$ or $M = L_{i+1}$.

A flag of length n can be constructed for any basis $\{e_1, \ldots, e_n\}$ of the space L by setting $L_0 = \{0\}$ and $L_i =$ linear span of $\{e_1, \ldots, e_i\}$ (for $i \geq 1$). It will be evident from the proof of the following theorem that this flag is maximal and that our construction gives all maximal flags.

2.15. Theorem, *The dimension of the space L equals the length of any maximal flag of L.*

Proof. Let $L_0 \subset L_1 \subset L_2 \subset \ldots$ be a maximal flag in L. For all $i \geq 1$ we select a vector $e_i \in L_i \backslash L_{i-1}$ and show that $\{e_1, \ldots, e_i\}$ form a basis of the space L_i.

First of all, the linear span of $\{e_1, \ldots, e_{i-1}\}$ is contained in L_{i-1}, and e_i does not lie in L_{i-1}, whence it follows by induction on i (taking into account the fact that $e_1 \neq 0$) that $\{e_1, \ldots, e_i\}$ are linearly independent for all i.

We shall now show by induction on i that $\{e_1, \ldots, e_i\}$ generate L_i. Assume that this is true for $i - 1$ and let M be the linear span of $\{e_1, \ldots, e_i\}$. Then $L_{i-1} \subset M$ according to the induction hypothesis and $L_{i-1} \neq M$ because $e_i \notin L_{i-1}$. The definition of the maximality of a flag now implies that $M = L_i$.

It is now easy to complete the proof of the theorem. If $L_0 \subset L_1 \subset \ldots \subset L_n = L$ is a finite maximal flag in L, then, according to what has been proved the vectors $\{e_1, \ldots, e_n\}$, $e_i \in L_i \backslash L_{i-1}$, form a basis of L so that $n = \dim L$. If L contains an infinite maximal flag, then this construction provides arbitrarily large linearly independent sets of vectors in L, so that L is infinite-dimensional.

2.16. Supplement. Any flag in a finite-dimensional space L can be extended up to the maximal flag, and its length is therefore always $\leq \dim L$. Indeed, we continue to insert intermediate subspaces into the starting flag as long as it is possible to do so. This process cannot continue indefinitely, because the construction of systems of vectors $\{e_1, \ldots, e_i\}$, $e_i \in L_i \backslash L_{i-1}$ for *any* flag gives linearly independent systems (see the beginning of the proof of Theorem 2.15). Therefore, the length of the flag cannot exceed $\dim L$.

2.17. The basic principle for working with infinite-dimensional spaces: Zorn's lemma or transfinite induction. Most theorems in finite-dimensional linear algebra can be easily proved by making use of the existence of finite bases and Theorem 2.12 on the extension of bases; many examples of this will occur in what follows. But the habit of using bases makes it difficult to make the transition to functional analysis. We shall now describe a set-theoretical principle which, in very many cases, eliminates the need for bases.

We recall (see §6 of Ch. 1 in "Introduction to Algebra") that a *partially ordered set* is a set X together with a binary *ordering relation* $\leq$ on X that is reflexive ($x \leq x$), transitive (if $x \leq y$ and $y \leq z$, then $x \leq z$), and antisymmetric (if $x \leq y$ and $y \leq x$, then $x = y$). It is entirely possible that a pair of elements $x, y \in X$ does not satisfy $x \leq y$ or $y \leq x$. If, on the other hand, for any pair either $x \leq y$ or $y \leq x$, then the set is said to be *linearly ordered* or a *chain*.

An *upper bound* of a subset Y in a partially ordered set X is any element $x \in X$ such that $y \leq x$ for all $y \in Y$. An upper bound of a subset may not exist: if $X = \mathbf{R}$ with the usual relation $\leq$ and $Y = \mathbf{Z}$ (integers), then Y does not have an upper bound.

The *greatest element* of the partially ordered set X is an element $n \in X$ such that $x \leq n$ for all $x \in X$; a *maximal element* is an element $m \in X$ for which $m \leq x \in X$ implies that $x = m$. The greatest element is always maximal, but not conversely.

2.18. Example. A typical example of an ordered set X is the set of all subsets $\mathcal{P}(S)$ of the set S, or some part of it, ordered by the relation $\subseteq$. If S has more than two elements, then $\mathcal{P}(S)$ is partially ordered, but it is not linearly ordered (why ?). The element $S \in \mathcal{P}(S)$ is maximal and is even the greatest element in $\mathcal{P}(S)$.

2.19. Zorn's lemma. *Let X be a non-empty partially ordered set, any chain in which has an upper bound in X. Then some chain has an upper bound that is simultaneously the maximal element in X.*

Zorn's lemma can be derived from other, intuitively more plausible, axioms of set theory. But logically it is equivalent to the so-called axiom of choice, if the remaining axioms are accepted. For this reason, it is convenient to add it to the basic axioms which is, in fact, often done.

2.20. Example of the application of Zorn's lemma: existence of a basis in infinite-dimensional linear spaces.

Let L be a linear space over the field K. We denote by $X \subset \mathcal{P}(L)$ *the set of linearly independent subsets of vectors in L, ordered by the relation $\subseteq$.*

In other words, $Y \in X$ if any finite linear combination of vectors in Y that equals zero has zero coefficients. Let us check the conditions of Zorn's lemma: if S is a chain in X, then it has an upper bound in X. Indeed, let $Z = \bigcup_{Y \in S} Y$. Obviously, $Y \subseteq Z$ for any $Y \in S$; in addition, Z forms a linearly independent set of vectors, because any finite set of vectors $\{y_1, \ldots, y_n\}$ from Z is contained in some element $Y \in S$. Actually, let $y_i \in Y_i \in S$; since S is a chain, one of every two elements $Y_i, Y_j \in S$ is a subset of the other; deleting in turn the smallest sets from such pairs, we find that amongst the Y_i there exists a greatest set; this set contains all the $y_1, \ldots, y_n$, which are thus linearly independent.

We shall now make an application of Zorn's lemma. Here, only part of it is required: the existence of a maximal element in X. By definition, this is a linearly independent set of vectors $Y \in X$ such that if any vector $l \in L$ is added to it, then the set $Y \cup \{l\}$ will no longer be linearly independent. Exactly the same argument as in Lemma 2.9b then shows that l is a (finite) linear combination of the elements of Y, i.e., Y forms a basis of L.

EXERCISES

1. Let L be the space of polynomials of x of degree $\leq n-1$ with coefficients in the field K. Verify the following assertions.

a) $1, x, \ldots, x^{n-1}$ form a basis of L. The coordinates of the polynomial f in this basis are its coefficients.

b) $1, x-a, (x-a)^2, \ldots, (x-a)^{n-1}$ form a basis of L. If char $K = p \geq n$, then the coordinates of the polynomial f in this basis are: $\{f(a), f'(a), \frac{f''(a)}{2!}, \ldots, \frac{f^{n-1}(a)}{(n-1)!}\}$.

c) Let $a_1, \ldots, a_n \in K$ be pairwise different elements. Let $g_i(x) = \prod_{j \neq i}(x - -a_j)(a_i - a_j)^{-1}$. The polynomials $g_1(x), \ldots, g_n(x)$ form a basis of L ("interpolation basis"). The coordinates of the polynomial f in this basis are $\{f(a_1), \ldots, f(a_n)\}$.

2. Let L be an n-dimensional space and let $f : L \to K$ be a non-zero linear functional. Prove that $M = \{l \in L | f(l) = 0\}$ is an $(n-1)$-dimensional subspace of L. Prove that all $(n-1)$-dimensional subspaces are obtained by this method.

3. Let L be an n-dimensional space and $M \subset L$ an m-dimensional subspace. Prove that there exist linear functionals $f_1, \ldots, f_{n-m} \in L^*$ such that $M = \{l | f_1(l) = \ldots = f_{n-m}(l) = 0\}$.

4. Calculate the dimensions of the following spaces:

a) the space of polynomials of degree $\leq p$ of n variables;

b) the space of homogeneous polynomials (forms) of degree p of n variables;

c) the space of functions in $F(S)$, $|S| < \infty$ that vanish at all points of the subset $S_0 \subset S$.

5. Let K be a finite field with characteristic p. Prove that the number of elements in this field equals p^n for some $n \geq 1$. (Hint: interpret K as a linear space over a simple subfield consisting of all "sums of ones" in $K : 0, 1, 1+1, \ldots$).

6. In the infinite-dimensional case the concept of a flag is replaced by the concept of a chain of subspaces (ordered with respect to inclusion). Using Zorn's lemma, prove that any chain is contained in a maximal chain.

§3. Linear Mappings

3.1. Definition. *Let L and M be linear spaces over a field K. The mapping $f : L \to M$ is said to be linear if for all $l, l_1, l_2 \in L$ and $a \in K$ we have*

$$f(al) = af(l), \; f(l_1 + l_2) = f(l_1) + f(l_2).$$

A linear mapping is a homomorphism of additive groups. Indeed, $f(0) = 0f(0) = 0$ and $f(-l) = f((-1)l) = -f(l)$. Induction on n shows that

$$f\left(\sum_{i=1}^{n} a_i l_i\right) = \sum_{i=1}^{n} a_i f(l_i)$$

for all $a_i \in K$ and $l_i \in L$.

The linear mappings $f : L \to L$ are also called *linear operators* on L.

3.2. Examples. a) The *null* linear mapping $f : L \to M$, $f(l) = 0$ for all $l \in L$. The identity linear mapping $f : L \to L$, $f(l) = l$ for all $l \in L$. It is denoted by id_L or id (from the English word "identity"). *Multiplication by a scalar $a \in K$* or the *homothety transformation* $f : L \to L$, $f(l) = al$ for all $l \in L$. The null operator is obtained for $a = 0$ and the identity operator is obtained for $a = 1$.

b) The linear mappings $f : L \to K$ are linear functions or functionals on L (see §1.9). Let L be a space with the basis $\{e_1, \ldots, e_n\}$. For any $1 \leq i \leq n$, the mapping $e^i : L \to K$, where $e^i(l)$ is the ith coordinate of l in the basis $\{e_1, \ldots, e_n\}$, is a linear functional.

c) Let $L = \{x \in \mathbf{R} | x > 0\}$ be equipped with the structure of a linear space over $\mathbf{R}$, described in §1.10a, $M = \mathbf{R}^1$. The mapping $\log : L \to M, x \mapsto \log x$ is $\mathbf{R}$ linear.

d) Let $S \subset T$ be two sets. The mapping $F(T) \to F(S)$, which associates to any function on T its restriction to S, is linear. In particular, if $S = \{s\}$, $s \in T$, $f \in F(T)$, then the mapping $f \mapsto$ (value of f at the point s) is linear.

Linear mappings with the required properties are often constructed based on the following result.

3.3. Proposition. *Let L and M be linear spaces over a field K and $\{l_1, \ldots, l_n\} \subset L$, $\{m_1, \ldots, m_n\} \subset M$ two sets of vectors with the same number of elements. Then:*

a) if the linear span of $\{l_1, \ldots, l_n\}$ coincides with L, then there exists not more than one linear mapping $f : L \to M$, for which $f(l_i) = m_i$ for all i;

b) if $\{l_1, \ldots, l_n\}$ are also linearly independent, i.e., they form a basis of L, then such a mapping exists.

Proof. Let f and f' be two mappings for which $f(l_i) = f'(l_i) = m_i$ for all i. We shall study the mapping $g = f - f'$, where $(f - f')(l) = f(l) - f'(l)$. It is easy to verify that it is linear. In addition, it transforms all l_i, and therefore any linear combination of vectors l_i, into zero. This means that f and f' coincide on every vector in L, whence $f' = f$.

Now let $\{l_1, \ldots, l_n\}$ be a basis of L. Since every element of L can be uniquely represented in the form $\sum_{i=1}^{n} a_i l_i$, we can define the set-theoretical mapping $f : L \to M$ by the formula

$$f\left(\sum_{i=1}^{n} a_i l_i\right) = \sum_{i=1}^{n} a_i m_i.$$

It is obviously linear.

In this proof we made use of the difference between two linear mappings $L \to M$. This is a particular case of the following more general construction.

3.4. Let $\mathcal{L}(L, M)$ denote the set of linear mappings from L into M. For $f, g \in \mathcal{L}(L, M)$ and $a \in K$ we define af and $f + g$ by the formulas

$$(af)(l) = a(f(l)), \; (f + g)(l) = f(l) + g(l)$$

for all $l \in L$. Just as in §1.9, we verify that af and $f + g$ are linear, so that $\mathcal{L}(L, M)$ is a linear space.

3.5. Let $f \in \mathcal{L}(L, M)$ and $g \in \mathcal{L}(M, N)$. The set-theoretical composition $g \circ f = gf : L \to N$ is a linear mapping. Indeed,

$$(gf)(l_1 + l_2) = g[f(l_1 + l_2)] = g[f(l_1) + f(l_2)] = g[f(l_1)] + g[f(l_2)] = gf(l_1) + gf(l_2)$$

and, analogously, $(gf)(al) = a(gf(l))$.

Obviously, $\mathrm{id}_M \circ f = f \circ \mathrm{id}_L = f$. In addition, $h(gf) = (hg)f$ when both parts are defined, so that the parentheses can be dropped; this is the general property of associativity of set-theoretical mappings. Finally, the composition (gf) is linear with respect to each of the arguments with the other argument held fixed: for example, $g \circ (af_1 + bf_2) = a(g \circ f_1) + b(g \circ f_2)$.

3.6. Let $f \in \mathcal{L}(L, M)$ be a bijective mapping. Then it has a set-theoretic inverse mapping $f^{-1} : M \to L$. We assert that f^{-1} is automatically linear. For this, we must verify that

$$f^{-1}(m_1 + m_2) = f^{-1}(m_1) + f^{-1}(m_2), \; f^{-1}(am_1) = af^{-1}(m_1)$$

for all $m_1, m_2 \in M$ and $a \in K$. Since f is bijective, there exist uniquely defined vectors $l_1, l_2 \in L$ such that $m_i = f(l_i)$. Writing the formulas

$$f(l_1) + f(l_2) = f(l_1 + l_2), \quad af(l_1) = f(al_1),$$

applying f^{-1} to both sides of each formula, and replacing in the result l_i by $f^{-1}(m_i)$, we obtain the required result.

Bijective linear mappings $f : L \to M$ are called *isomorphisms*. The spaces L and M are said to be *isomorphic* if an isomorphism exists between them.

The following theorem shows that the dimension of a space completely determines the space up to isomorphism.

3.7. Theorem. *Two finite-dimensional spaces L and M over a field K are isomorphic if and only if they have the same dimension.*

Proof. The isomorphism $l : L \to M$ preserves all properties formulated in terms of linear combinations. In particular, it transforms any basis of L into a basis of M, so that the dimensions of L and M are equal. (It also follows from this argument that a finite-dimensional space cannot be isomorphic to an infinite-dimensional space.)

Conversely, let the dimensions of L and M equal n. We select the bases $\{l_1, \ldots, l_n\}$ and $\{m_1, \ldots, m_n\}$ in L and M respectively. The formula

$$f\left(\sum_{i=1}^{n} a_i l_i\right) = \sum_{i=1}^{n} a_i m_i$$

defines a linear mapping of L into M acccording to Proposition 3.3. It is bijective because the formula

$$f^{-1}\left(\sum_{i=1}^{n} a_i m_i\right) = \sum_{i=1}^{n} a_i l_i$$

defines the inverse linear mapping f^{-1}.

3.8. Warning. Even if an isomorphism between two linear spaces L and M exists, it is defined uniquely only in two cases:

a) $L = M = \{0\}$ and

b) L and M are one-dimensional, while K is a field consisting of two elements (try to prove this !).

In all other cases there exist many (if K is infinite, then infinitely many) isomorphisms. In particular, there exist many isomorphisms of the space L with itself. The results of §3.5 and §3.6 imply that they form a group with respect to set-theoretic composition. This group is called the *general (or full) linear group of the space L*. Later we shall describe it in a more explicit form as the group of non-singular square matrices.

It sometimes happens that an isomorphism, not depending on any arbitrary choices (such as the choice of bases in the spaces L and M in the proof of Theorem 3.7), is defined between two linear spaces. We shall call such isomorphisms *canonical* or *natural* isomorphisms (a precise definition of these terms can be given only in the language of categories; see §13). Natural isomorphisms should be carefully distinguished from "accidental" isomorphisms. We shall present two characteristic examples which are very important for understanding this distinction.

3.9. "Accidental" isomorphism between a space and its dual. Let L be a finite-dimensional space with the basis $\{e_1, \dots, e_n\}$. We denote by $e^i \in L^*$ the linear functional

$$l \mapsto e^i(l), \text{where } e^i(l) \text{ is the ith coordinate of } l \text{ in the basis } \{e_i\}$$

(do not confuse this with the ith power which is not defined in a linear space). We assert that the *functionals* $\{e^1, \dots, e^n\}$ *form a basis of* L^**, the so-called dual basis with respect to the basis* $\{e_1, \dots, e_n\}$. An equivalent description of $\{e^i\}$ is as follows: $e^i(e_k) = \delta_{ik}$ (the Kronecker delta: 1 for $i = k$, 0 for $i \neq k$).

Actually, any linear functional $f : L \to K$ can be represented as a linear combination of $\{e^i\}$:

$$f = \sum_{i=1}^{n} f(e_i)e^i.$$

Indeed, the values of the left and right sides coincide on any linear combination $\sum_{k=1}^{n} a_k e_k$, because $e^i\left(\sum_{k=1}^{m} a_k e_k\right) = a_i$ by definition of e^i.

In addition, $\{e_i\}$ are linearly independent: if $\sum_{i=1}^{n} a_i e^i = 0$, then for all k, $1 \leq$ $\leq k \leq n$, we have $a_k = \left(\sum_{i=1}^{n} a_i e^i\right)(e_k) = 0$.

Therefore, L and L^* have the same dimension n and even the isomorphism $f : L \to L^*$ which transforms e_i into e^i, is defined.

This isomorphism is not, however, canonical: generally speaking it changes if the basis $\{e_1, \dots, e_n\}$ is changed. Thus if L is one-dimensional, then the set $\{e_1\}$ is a basis of L for any non-zero vector $e_1 \in L$. Let $\{e^1\}$ be the basis dual to $\{e_1\}$, $e^1(e_1) = 1$. Then the basis $\{a^{-1}e^1\}$ is the dual of $\{ae_1\}$, $a \in K \setminus \{0\}$. But the linear mappings $f_1 : e_1 \mapsto e^1$ and $f_2 : ae_1 \mapsto a^{-1}e^1$ are different, provided that $a^2 \neq 1$.

3.10. Canonical isomorphism between a space and its second dual. Let L be a linear space, L^* the space of linear functions on it, and $L^{**} = (L^*)^*$ the space of linear functions on L^*, called the "double dual of the space L".

We shall describe the canonical mapping $\epsilon_L : L \to L^{**}$, which is independent of any arbitrary choices. *It associates to every vector* $l \in L$ *a function on* L^**, whose value on the functional* $f \in L^*$ *equals* $f(l)$*; using a shorthand notation:*

$$\epsilon_L : l \mapsto [f \mapsto f(l)].$$

We shall verify the following properties of ϵ_L:

a) *for each* $l \in L$ *the mapping* $\epsilon_L(l) : L^* \to K$ *is linear.* Indeed this means that the *expression for* $f(l)$ *as a function of* f *with fixed* l *is linear with respect to* f. But this follows from the rules for adding functionals and multiplying them by scalars (§1.7).

Therefore, ϵ_L does indeed determine the mapping of L into L^{**}, as asserted.

b) *The mapping* $\epsilon_L : L \to L^{**}$ *is linear.* Indeed, this means that the expression $f(l)$ *as a function of* l *with fixed* f *is linear*, which is so, because $f \in L^*$.

c) If L is *finite-dimensional, then the mapping* $\epsilon_L : L \to L^{**}$ *is an isomorphism.* Indeed, let $\{e_1, \ldots, e_n\}$ be a basis of L, $\{e^1, \ldots, e^n\}$ the basis of the dual space L^*, and $\{e'_1, \ldots, e'_n\}$ the basis of L^{**} dual to $\{e^1, \ldots, e^n\}$.

We shall show that $\epsilon_L(e_i) = e'_i$, whence it will follow that ϵ_L is an isomorphism (in this verification, the use of the basis of L is harmless, because it did not appear in the definition of ϵ_L !).

Actually, $\epsilon_L(e_i)$ is, by definition, a functional on L^*, whose value at e^k is equal to $e^k(e_i) = \delta_{ik}$ (the Kronecker delta). But e'_i is exactly the same functional on L^*, by definition of a dual basis.

We note that if L is infinite-dimensional, then $\epsilon_L : L \to L^{**}$ remains injective, but it is no longer surjective (see Exercise 2). In functional analysis instead of the full space L^*, only the subspaces of linear functionals L' which are *continuous in an appropriate topology* on L and K are usually studied, and then the mapping $L \to L''$ can be defined and is sometimes an isomorphism. Such (topological) spaces are said to be *reflexive*. We have proved that *finite-dimensional spaces* (ignoring the topology) *are reflexive.*

We shall now study the relationship between linear mappings and linear subspaces.

3.11. Definition. *Let* $f : L \to M$ *be a linear mapping. The set* $\ker f = \{l \in L | f(l) = 0\} \subset L$ *is called the kernel of* f *and the set* $\operatorname{Im} f = \{m \in M | \exists l \in L, f(l) = m\} \subset M$ *is called the image of* f.

It is easy to verify that the kernel of f is a linear subspace of L and that the image of f is a linear subspace of M. We shall verify, as an example, the second assertion. Let $m_1, m_2 \in \operatorname{Im} f, a \in K$. Then there exist vectors $l_1, l_2 \in L$ such that $f(l_1) = m_1, f(l_2) = m_2$. Hence $m_1 + m_2 = f(l_1 + l_2)$, $am_1 = f(al_1)$. Therefore $m_1 + m_2 \in \operatorname{Im} f$ and $am_1 \in \operatorname{Im} f$.

The mapping f is injective, if and only if $\ker f = \{0\}$. In fact, if $f(l_1) = f(l_2)$, $l_1 \neq l_2$, then $0 \neq l_1 - l_2 \in \ker f$. Conversely, if $0 \neq l \in \ker f$, then $f(l) = 0 = f(0)$.

3.12. Theorem. *Let* L *be a finite-dimensional linear space and let* $f : L \to M$ *be* *a linear mapping. Then* $\ker f$ *and* $\operatorname{Im} f$ *are finite-dimensional and*

$$\dim \ker f + \dim \operatorname{Im} f = \dim L.$$

Proof, The kernel of f is finite-dimensional as a consequence of §2.13. We shall select a basis $\{e_1, \ldots, e_m\}$ of $\ker f$ and extend it to the basis $\{e_1, \ldots, e_m, e_{m+1}, \ldots, \ldots, e_{m+n}\}$ of the space L, according to Theorem 2.12. We shall show that the vectors $f(e_{m+1}), \ldots, f(e_{m+n})$ form a basis of $\operatorname{Im} f$. The theorem will follow from here.

Any vector in $\operatorname{Im} f$ has the form

$$f\left(\sum_{i=1}^{m+n} a_i e_i\right) = \sum_{i=m+1}^{m+n} a_i f(e_i).$$

Therefore, $f(e_{m+1}), \ldots, f(e_{m+n})$ generate $\operatorname{Im} f$.

Let $\sum_{i=m+1}^{m+n} a_i f(e_i) = 0$. Then $f\left(\sum_{i=m+1}^{m+n} a_i e_i\right) = 0$. This means that

$$\sum_{i=m+1}^{m+n} a_i e_i \in \ker f,$$

that is,

$$\sum_{i=m+1}^{m+n} a_i e_i = \sum_{j=1}^{m} a_j e_j.$$

This is possible only if all coefficients vanish, because $\{e_1, \cdots, e_{m+n}\}$ is a basis of L. Therefore, the vectors $f(e_{m+1}, \cdots, f(e_{m+n})$ are linearly independent. The theorem is proved.

3.13. Corollary. *The following properties of* f *are equivalent (in the case of finite-dimension* L*):*

a) f *is injective,*

b) $\dim L = \dim \operatorname{Im} f$.

Proof. According to the theorem, $\dim L = \dim \operatorname{Im} f$, if and only if $\dim \ker f = 0$ that is, $\ker f = \{0\}$.

EXERCISES

1. Let $f : \mathbf{R}^m \to \mathbf{R}^n$ be a mapping, defined by differentiable functions which, generally speaking, are non-linear and map zero into zero:

$$f(x_1, \ldots, x_m) = (\ldots, f_i(x_1, \ldots, x_m), \ldots),\ i = 1, \ldots, n,$$

$$f_i(0,\dots,0)=0.$$

Associate with it the linear mapping $df_0:\mathbf{R}^m\to\mathbf{R}^n$, called the *differential of f at the point* 0, according to the formula

$$(df_0)(e_j)=\sum_{i=1}^{n}\frac{\partial f_i}{\partial x_j}(0)e'_i=\left(\frac{\partial f_1}{\partial x_j}(0),\dots,\frac{\partial f_n}{\partial x_j}(0)\right),$$

where $\{e_j\},\{e'_i\}$ are standard bases of $\mathbf{R}^m$ and $\mathbf{R}^n$. Show that if the bases of the spaces $\mathbf{R}^m$ and $\mathbf{R}^n$ are changed and df_0 is calculated using the same formulas in the new bases, then the new linear mapping df_0 coincides with the old one.

2. Prove that the space of polynomials $\mathbf{Q}[x]$ is not isomorphic to its dual. (Hint: compare the cardinalities.)

§4. Matrices

4.1. The purpose of this section is to introduce the language of matrices and to establish the basic relations between it and the language of linear spaces and mappings. For further details and examples, we refer the reader to Chapters 2 and 3 of "Introduction to Algebra"; in particular, we shall make use of the theory of determinants developed there without repeating it here. The reader should convince himself that the exposition in these chapters transfers without any changes from the field of real numbers to any scalar field; the only exceptions are cases where specific properties of real numbers such as order and continuity are used.

4.2. Terminology. An $m\times n$ *matrix* A with elements from the set S is a set (a_{ik}) of elements from S which are enumerated by ordered pairs of numbers (i,k), where $1\le i\le m$, $1\le k\le n$. The notation $A=(a_{ik})$, $1\le i\le m$, $1\le k\le n$ is often used; the size need not be indicated.

For fixed i, the set $(a_{i1},\dots,a_{in})$ is called the ith *row* of the matrix A. For fixed k the set $(a_{1k},\dots,a_{mk})$ is called the *kth column* of the matrix A. A $1\times n$ matrix is called a row matrix and an $m\times 1$ matrix is called a column matrix.

If $m=n$, then the matrix A is called a *square* matrix (the terminology "of order n" instead of "of size $n\times n$" is sometimes used),

If A is a square matrix of order n, $S=K$ (field), and $a_{ik}=0$ for $i\ne k$, then the matrix A is called a *diagonal* matrix; it is sometimes written as $\mathrm{diag}(a_{11},\dots,a_{nn})$. In general, the elements (a_{ii}) are called the elements of the *main diagonal.* The elements $(a_{1,k+1};a_{2,k+2};\dots)$ form a diagonal standing above the main diagonal for $k>0$ and the elements $(a_{k+1,1};a_{k+2,2};\dots,)$ for $k>0$ form a diagonal standing below it. If $S = K$ and $a_{ik} = 0$ for $k < i$, the matrix is called an *upper triangular matrix* and if $a_{ik} = 0$ for $k > i$, it is called a *lower triangular matrix*. A diagonal square matrix over K, in which all the elements along the main diagonal are identical, is called a *scalar* matrix. If these elements are equal to unity, the matrix is called the *unit* matrix. The unit matrix of order n is denoted by E_n or simply E if the order is obvious from the context.

All these terms originate from the standard notation for a matrix in the form of a table:

$$A = \begin{pmatrix} a_{11} & a_{12} & \dots & a_{1n} \\ a_{21} & a_{22} & \dots & a_{2n} \\ \dots & \dots & \dots & \dots \\ a_{m1} & a_{m2} & \dots & a_{mn} \end{pmatrix}.$$

The $n \times m$ matrix A^t whose (i,k) element equals a_{ki} is called the transpose of A. (Sometimes the notation $A^t = (a_{ki})$ is ambiguous !)

4.3. Remarks. Most matrices encountered in the theory of linear spaces over a field K have elements from the field itself. However, there are exceptions. For example, we shall sometimes interpret an ordered basis of the space L, $\{e_1, \dots, e_n\}$, as a $1 \times n$ matrix with elements from this space. Another example are *block matrices*, whose elements are also matrices: blocks of the starting matrix. A matrix A is partitioned into blocks by partitioning the row numbers $[1, \dots, m] = I_1 \cup I_2 \cup \dots \cup I_\mu$ and the column numbers $[1, \dots, n] = J_1 \cup \dots \cup J_\nu$ into sequential, pairwise nonintersecting segments

$$\left(\begin{array}{c|c|c|c} A_{11} & A_{12} & \dots & A_{1\nu} \\ \hline \dots & \dots & \dots & \dots \\ \hline A_{\mu 1} & A_{\mu 2} & \dots & A_{\mu\nu} \end{array}\right),$$

where the elements of $A_{\alpha\beta}$ are a_{ik}, $i \in I_\alpha$, $k \in J_\beta$. If $\mu = \nu$, it is possible to define in an obvious manner block-diagonal, block upper-triangular, and block lower-triangular matrices. This same example shows that it is not always convenient to enumerate the columns and rows of a matrix by numbers from 1 to m (or n): often only the order of the rows and columns is significant.

4.4. The matrix of a linear mapping. Let N and M be finite-dimensional linear spaces over K with the distinguished bases $\{e_1, \dots, e_n\}$ and $\{e'_1, \dots, e'_m\}$, respectively. Consider an arbitrary linear mapping $f : N \to M$ and associate with it an $m \times n$ matrix A_f with elements from the field K as follows (note that the dimensions of A_f are the same as those of N, M *in reverse order*). We represent the vectors $f(e_k)$ as linear combinations: $f(e_k) = \sum_{i=1}^m a_{ik} e'_i$. Then by definition $A_f =$ $= (a_{ik})$. In other words, the coefficients of these linear combinations are sequential *columns* of the matrix A_f.

The matrix A_f is called the matrix of the linear mapping f *with respect to the bases* (or *in the bases*) $\{e_k\}$, $\{e'_i\}$.

Proposition 3.3 implies that the linear mapping f is uniquely determined by the images $f(e_k)$, and the latter can be taken as any set of n vectors from the space M. Therefore this correspondence establishes a *bijection* between the set $\mathcal{L}(N, M)$ and the set of $m \times n$ matrices with elements from K (or *over* K). This bijection, however, depends on the choice of bases: see §4.8).

The matrix A_f also allows us to describe a linear mapping f in terms of its action on the coordinates. If the vector l is represented by its coordinates $\vec{x} = [x_1, \ldots, x_n]$ in the basis $\{e_1, \ldots, e_n\}$, that is, $l = \sum_{i=1}^n x_i e_i$ then the vector $f(l)$ is represented by the coordinates $\vec{y} = [y_1, \ldots, y_m]$, where

$$y_i = \sum_{k=1}^n a_{ik} x_k, \ i = 1, \ldots, m.$$

In other words, $\vec{y} = A_f \cdot \vec{x}$ is the usual product of A_f and the column vector $\vec{x}$.

When we talk about the matrix of a *linear operator* $A = (a_{ik})$, it is always assumed that the same basis is chosen in "two replicas" of the space N. The matrix of a linear operator is a square matrix. The matrix of the identity operator is the unit matrix.

According to §3.4, the set $\mathcal{L}(N, M)$ is, in its turn, a linear space over K. When the elements of $\mathcal{L}(N, M)$ are matrices, this structure is described as follows.

4.5. Addition of matrices and multiplication by a scalar. Let $A = (a_{ik})$ and $B = (b_{ik})$ be two matrices of the same size over the field K, $a \in K$. We set

$$A + B = (c_{ik}), \quad \text{where} \quad c_{ik} = a_{ik} + b_{ik},$$

$$aA = (aa_{ik}).$$

These operations define the structure of a linear space on matrices of a given size. It is easy to verify that if $A = A_f$ and $B = A_g$ (in the same bases), then

$$A_f + A_g = A_{f+g}, \ A_{af} = aA_f,$$

so that the indicated correspondence (and it is bijective) is an isomorphism. In particular $\dim \mathcal{L}(N, M) = \dim M \dim N$, so that the space of matrices is isomorphic to K^{mn} (size $m \times n$).

The composition of linear mappings is described in terms of the multiplication of matrices.

4.6. Multiplication of matrices. The product of an $m \times n'$ matrix A over a field K by an $n'' \times p$ matrix B over a field K is defined if and only if $n' = n'' = n$;

AB then has the dimensions $m \times p$ and by definition

$$AB = (c_{ik}), \quad \text{where} \quad c_{ik} = \sum_{j=1}^{n} a_{ij} b_{jk}.$$

It is easy to verify that $(AB)^t = B^t A^t$.

It can happen that AB is defined but BA is not defined (if $m \neq p$), or both matrices AB and BA are defined but have different dimensions (if $m \neq n$) or are defined and have the same dimensions ($m = n = p$) but are not equal. In other words, *matrix multiplication is not commutative*. It is, however, *associative*: *if the matrices AB and BC are defined, then $(AB)C$ and $A(BC)$ are defined and are equal to one another.* Indeed, let $A = (a_{ij})$, $B = (b_{jk})$, and $C = (c_{kl})$. The reader should check that the matrices A and BC are commensurate and that the matrices AB and C are commensurate. Then we can calculate the (il)th element of $(AB)C$ from the formula

$$\sum_k \left(\sum_j a_{ij} b_{jk} \right) c_{kl} = \sum_{j,k} (a_{ij} b_{jk}) c_{kl},$$

and the (il)th element of $A(BC)$ from the formula

$$\sum_j a_{ij} \left(\sum_k b_{jk} c_{kl} \right) = \sum_{j,k} a_{ij} (b_{jk} c_{kl}).$$

Since multiplication in K is associative, these elements are equal to one another. Since we already know that multiplication of matrices defined over K is associative, we can verify that "block multiplication" of block matrices is also associative (see also Exercise 1).

In addition, matrix multiplication is linear with respect to each argument:

$$(aA + bB)C = aAC + bBC; \; A(bB + cC) = bAB + cAC.$$

A very important property of matrix multiplication is that it corresponds to the composition of linear mappings. However, many other situations in linear algebra are also conveniently described by matrix multiplication: this is the main reason for the unifying role of matrix language and the somewhat independent nature of matrix algebra within linear algebra. We shall enumerate some of these situations.

4.7. Matrix of the composition of linear mappings. Let P, N and M be three finite-dimensional linear spaces and let $P \xrightarrow{g} N \xrightarrow{f} M$ be two linear mappings. We choose the bases $\{e''_i\}, \{e'_k\}$, and $\{e_m\}$ in P, N and M respectively and we denote by A_g, A_f, and A_{fg} the matrices of g, f, and fg in these bases. We assert that $A_{fg} = A_f A_g$. Indeed, let $A_f = (a_{ji}), A_g = (b_{ik})$. Then

$$g(e''_k) = \sum_i b_{ik} e'_i,$$

$$fg(e''_k) = \sum_i b_{ik} f(e'_i) = \sum_i b_{ik} \sum_j a_{ji} e_j = \sum_j \left(\sum_i a_{ji} b_{ik} \right) e_k.$$

Therefore the (jk)th element of the matrix A_{fg} equals $\sum_i a_{ji} b_{ik}$, that is, $A_{fg} = A_f A_g$.

According to the results of §4.4–§4.6, after a basis in L is chosen the set of linear operators $\mathcal{L}(L, L)$ can be identified with the set of square matrices $M_n(K)$ of order $n = \dim L$ over the field K. With this identification the structures of linear spaces and rings in both sets are consistent with one another. Bijections, i.e., the linear automorphisms $f : L \to L$, correspond to inverse matrices: if $f \circ f^{-1} = \mathrm{id}_L$, then $A_f A_{f^{-1}} = E_n$, so that $A_{f^{-1}} = A_f^{-1}$. We recall that the matrix A is invertible, or non-singular, if and only if $\det A \neq 0$.

4.8. a) ***Action of a linear mapping on the coordinates.*** In the notation of §4.4, we can represent the vectors of the spaces N and M in terms of coordinates by columns

$$\vec{x} = \begin{pmatrix} x_1 \\ \vdots \\ x_n \end{pmatrix}, \quad \vec{y} = \begin{pmatrix} y_1 \\ \vdots \\ y_m \end{pmatrix}.$$

Then the action of the operator f is expressed in the language of matrix multiplication by the formula

$$\begin{pmatrix} y_1 \\ \vdots \\ y_m \end{pmatrix} = \begin{pmatrix} a_{11} & \dots & a_{1n} \\ \dots & \dots & \dots \\ a_{m1} & \dots & a_{mn} \end{pmatrix} \begin{pmatrix} x_1 \\ \vdots \\ x_n \end{pmatrix},$$

or $\vec{y} = A_f \vec{x}$ (cf. §4.4). It is sometimes convenient to write an analogous formula in terms of the bases $\{e_i\}, \{e'_k\}$ in which it assumes the form

$$f(e_1, \dots, e_n) = (f(e_1), \dots, f(e_n)) = (e'_1, \dots, e'_m) A_f.$$

Here the formalism of matrix multiplication requires that the vectors of M in the expression on the right be multiplied by scalars from the *right* and not from the left; this is harmless, and we shall simply assume that $e'a = ae'$ for any $e' \in M$ and $a \in K$.

In using this notation, we shall sometimes need to verify the associativity or linearity with respect to the arguments of "mixed" products of matrices, some of which have elements in K while others have elements in L, for example

$$((e_1, \dots, e_n)A)B = (e_1, \dots, e_n)(AB)$$

or

$$(e_1 + e'_1, \dots, e_n + e'_n)A = (e_1, \dots, e_n)A + (e'_1, \dots, e'_n)A$$

etc. The formalism in §4.4 and §4.5 transfers automatically to these cases. The same remark holds for the block matrices.

b) *The coordinates of a vector in a new basis.* Let $\{e_i\}$ and $\{e'_i\}$ be two bases in the space L. Any vector $l \in L$ can be represented by its coordinates in these bases: $l = \sum_{i=1}^n x_i e_i = \sum_{k=1}^n x'_k e'_k$. We shall show that there exists a square matrix A of order n, which does not depend on l, such that $\vec{x} = A\vec{x}'$.

Indeed, if $e'_k = \sum_{i=1}^n a_{ik} e_i$, then $A = (a_{ik})$:

$$\sum_{i=1}^n x_i e_i = l = \sum_{k=1}^n x'_k e'_k = \sum_{k=1}^n x'_k \left(\sum_{i=1}^n a_{ik} e_i \right) = \sum_{i=1}^n \left(\sum_{k=1}^n a_{ik} x'_k \right) e_i.$$

The matrix A is called the *matrix of the change of basis* (from the unprimed to the primed basis), or from the primed to the unprimed coordinates. We note that it is invertible: the inverse matrix is the matrix corresponding to a change from the primed to the unprimed basis.

We note that the formula $\vec{x} = A\vec{x}'$ could also have been interpreted as a formula expressing the coordinates of the *new vector* $f(\vec{x}')$ in terms of the coordinates of the vector $\vec{x}$, where f is the linear mapping $L \to L$, described by the matrix A in the basis $\{e_k\}$.

In physics, these two points of view are called "passive" and "active" respectively. In the first case, we describe the *same state of the system* (the vector l) from the point of view of *different observers* (with their own coordinate systems). In the second case, *there is only one observer*, while the state of the system is subjected to transformations consisting, for example, of symmetry transformations of the space of states of this system.

c) *The matrix of a linear mapping in new bases.* In the situation of §4.4, we shall determine how the matrix A_f of the linear mapping changes when we transform from the bases $\{e_k\}, \{e'_i\}$ to the new bases $\{\bar{e}_k\}, \{\bar{e}'_i\}$ of the spaces N and M. Let B be the matrix of the change from the $\{e_k\}$ coordinates to the $\{\bar{e}_k\}$ and C the matrix of the change from the $\{e'_i\}$ coordinates to the $\{\bar{e}'_i\}$ coordinates. We assert that the matrix $\bar{A}_f$ of the mapping f in the bases $\{\bar{e}_k\}, \{\bar{e}'_i\}$ is given by

$$\bar{A}_f = C^{-1} a_f B.$$

Indeed, in terms of the bases we have

$$(\bar{e}_k)\bar{A}_f = f((\bar{e}_k)) = f((e_k)B) = (f(e_k))B = (e'_i)A_f B = (\bar{e}'_i)C^{-1}A_f B.$$

We recommend that the reader perform the analogous calculations in terms of the coordinates.

The particular case $N = M, \{e_i\} = \{e'_i\}, \{\bar{e}_i\} = \{\bar{e}'_i\}, B = C$ is especially important. The matrix of the linear operator f in the new basis equals

$$\bar{A}_f = B^{-1} A_f B.$$

The mapping $M_n(K) \to M_n(K) : A \mapsto B^{-1}AB$ is called a *conjugation* (by means of the non-singular matrix B). Every conjugation is an *automorphism of the matrix algebra* $M_n(K)$:

$$B^{-1}\left(\sum_{i=1}^{n} a_i A_i\right) B = \sum_{i=1}^{n} a_i B^{-1} A_i B, \ a_i \in K;$$

$$B^{-1}(A_1 \dots A_m)B = (B^{-1}A_1B)\dots(B^{-1}A_mB)$$

(in the product on the right, the inner cofactors B^{-1} and B cancel in pairs, because they are adjacent to one another).

Among the elements of $M_n(K)$ the functions which remain unchanged when a matrix is replaced by its conjugate play a special role because with the help of these functions it is possible to construct *invariants of linear operators*: if ϕ is such a function, then setting $\phi(f) = \phi(A_f)$, we obtain a result which depends only on f and not on the basis in which A_f is written. Here are two important examples.

4.9. The determinant and the trace of a linear operator. We set

$$\operatorname{Tr} f = \operatorname{Tr} A_f = \sum_{i=1}^{n} a_{ii}, \quad \text{where} \quad A_f = (a_{ik})$$

(the trace of the matrix A is the sum of the elements of its main diagonal);

$$\det f = \det A_f.$$

The invariance of the determinant relative to conjugation is obvious

$$\det(B^{-1}AB) = (\det B)^{-1} \cdot \det A \cdot \det B = \det A.$$

To establish the invariance of the trace we shall prove a more general fact: *if A and B are matrices such that AB and BA are defined, then* $\operatorname{Tr} AB = \operatorname{Tr} BA$. Indeed,

$$\operatorname{Tr} AB = \sum_i \sum_j a_{ij} b_{ji}, \quad \operatorname{Tr} AB = \sum_j \sum_i b_{ji} a_{ij}.$$

If now B is non-singular, then applying the fact proved above to the matrices $B^{-1}A$ and B, we obtain

$$\operatorname{Tr}(B^{-1}AB) = \operatorname{Tr}(BB^{-1}A) = \operatorname{Tr} A.$$

In §8 we shall introduce the eigenvalues of matrices and operators, symmetric functions of which will furnish other invariant functions.

In concluding this section, we shall present the definitions, names and standard notation for several classes of matrices over the real and complex numbers, which are very important in the theory of Lie groups and Lie algebras and its many applications, in particular, in physics. The first class consists of the so-called *classical groups*: they are actually groups under matrix multiplication. The second class are the *Lie algebras*: they form a linear space and are stable under the *commutation* operation: $[A, B] = AB - BA$. The similarity of the notations for these classes will be explained in §11 and in Exercise 8.

4.10. Classical groups.

a) *The general linear group* $GL(n, K)$. It consists of non-singular $n \times n$ square matrices over the field K.

b) *The special linear group* $SL(n, K)$. It consists of square $n \times n$ matrices over the field K with determinant equal to 1.

In these two cases, K can be any field. Later, we shall restrict ourselves to the fields $K = \mathbf{R}$ or $\mathbf{C}$, though these definitions have been extended to other fields.

c) *The orthogonal group* $O(n, K)$. It consists of $n \times n$ matrices which satisfy the condition $AA^t = E_n$. Such matrices indeed form a group, because

$$E_nE_n^t = E_n, \quad A^{-1}(A^{-1})^t = A^{-1}(A^t)^{-1} = (A^tA)^{-1} = (E_n^t)^{-1} = E_n,$$

and finally,

$$(AB)(AB)^t = ABB^tA^t = AA^t = E_n.$$

In the case that $K = \mathbf{R}$ or $\mathbf{C}$ this group is said to be real or complex, respectively. The elements of the group $O(n, K)$ are called orthogonal matrices. The notation $O(n)$ is usually used instead of $O(n, \mathbf{R})$

d) *The special orthogonal group* $SO(n, K)$. This group consists of orthogonal matrices whose determinant equals unity:

$$SO(n, K) = O(n, K) \cap SL(n, K).$$

The notation $SO(n)$ is usually used instead of $SO(n, \mathbf{R})$.

e) *The unitary group* $U(n)$. It consists of complex $n \times n$ matrices which satisfy the condition $A\bar{A}^t = E_n$, where $\bar{A}$ is a matrix whose elements are the complex conjugates of the corresponding elements of the matrix A: if $A = (a_{ik})$, then $\bar{A} = (\bar{a}_{ik})$. Using the equality $\bar{A}\bar{B} = \overline{AB}$, it is easy to verify that $U(n)$ is a group, as in the preceding example. The elements $U(n)$ are called *unitary matrices*.

The matrix $\bar{A}^t$ is often called the *Hermitian conjugate* of the matrix A; mathematicians usually denote it by A^*, whereas physicists write A^+. We note that the operation of Hermitian conjugation is defined for complex matrices of arbitrary dimensions.

f) *The special unitary group* $SU(n)$. It consists of unitary matrices whose determinant equals unity:

$$SU(n) = U(n) \cap SL(n, \mathbf{C}).$$

It is clear from the definitions that real unitary matrices are orthogonal matrices:

$$O(n) = U(n) \cap GL(n, \mathbf{R}), \quad SO(n) = U(n) \cap SL(n, \mathbf{R}).$$

4.11. Classical Lie algebras. Any additive subgroup of square matrices $M_n(K)$ that is closed under the commutation operation $[A, B] = AB - BA$ is called a (matrix) Lie algebra. (For a general definition, see Exercise 14.) The following sets of matrices form classical Lie algebras; they usually also form linear spaces over K (sometimes over $\mathbf{R}$, though $K = \mathbf{C}$). They are not groups under multiplication !

a) *The algebra* $gl(n, K)$. It consists of all matrices $M^n(K)$.

b) *The algebra* $sl(n, K)$. It consists of all matrices from $M_n(K)$ with zero trace (such matrices are sometimes referred to as "traceless"). Closure under commutation follows from the formula $\mathrm{Tr}[A, B] = 0$, proved in §4.9. We note that Tr is a linear function on spaces of square matrices and linear operators, so that $sl(n, K)$ is a linear space over K.

c) *The algebra* $o(n, K)$. It consists of all matrices in $M_n(K)$ which satisfy the condition $A + A^t = 0$. An equivalent condition is $A = (a_{ik})$, where $a_{ii} = 0$ (if the characteristic of K is not two) and $a_{ik} = -a_{ki}$. Such matrices are called *antisymmetric* or *skew-symmetric* matrices. We note that $\mathrm{Tr}\, A = 0$ for all $A \in$ $\in o(n, K)$.

If $A^t = -A$ and $B^t = -B$, then $[A, B]^t = [B^t, A^t] = [-B, -A] = -[A, B]$ so that $[A, B]$ is skew-symmetric. Such matrices form a linear space over K.

We note in passing that the matrix A is called a *symmetric* matrix if $A^t = A$. The set of such matrices is not closed under commutation, but it is closed under anticommutation $AB + BA$ or Jordan's operation $(AB + BA)/2$.

d) *The algebra* $u(n)$. This algebra consists of complex $n \times n$ matrices which satisfy the condition $A + \bar{A}^t = 0$, or $a_{ik} = -\bar{a}_{ki}$. In particular, the diagonal elements are purely imaginary. Such matrices are called *Hermitian antisymmetric* or *anti-Hermitian* or *skew-Hermitian* matrices. They form a linear space over $\mathbf{R}$, but not over $\mathbf{C}$.

If $A^t = -\bar{A}$ and $B^t = -\bar{B}$, then

$$[A, B]^t = [B^t, A^t] = [-\bar{B}, -\bar{A}] = -\overline{[A, B]},$$

so that $u(n)$ is a Lie algebra.

We note in passing that the matrix A is called a *Hermitian symmetric* or simply Hermitian matrix if $A = \bar{A}^t$, that is, $a_{ki} = \bar{a}_{ik}$. Evidently, real Hermitian matrices are symmetric, while anti-Hermitian matrices are antisymmetric. In particular,

$$o(n, \mathbf{R}) = u(n) \cap sl(n, \mathbf{R}).$$

The matrix A is Hermitian if the matrix $i\bar{A}$ is anti-Hermitian and vice versa.

e) *The algebra* $su(n)$. This is $u(n) \cap sl(n, \mathbf{C})$ – the algebra of traceless anti-Hermitian matrices. They form an $\mathbf{R}$-linear space.

In Chapter 2, while studying linear spaces equipped with Euclidean or Hermitian metrics, we shall clarify the geometric meaning of operators that are represented by matrices from the classes described above, and we shall also enlarge our list.

EXERCISES

1. Formulate precisely and prove the assertion that block matrices over a field can be multiplied block by block, if the dimensions and numbers of blocks are compatible:

$$(A_{ij})(B_{jk}) = \left(\sum_j A_{ij} B_{jk}\right),$$

when the number of columns in the block A_{ij} equals the number of rows in the block B_{jk} and the number of block columns of the matrix A equals the number of block rows of the matrix B.

2. Introduce the concept of an infinite matrix (with an infinite number of rows and/or columns). Find the conditions under which two such matrices defined over a field can be multiplied (examples: finite matrices, i.e., matrices with only a finite number of non-zero elements; matrices in which each column and/or each row has a finite number of non-zero elements). Find the necessary conditions for the existence of triple products.

3. Prove that the equation $XY - YX = E$ cannot be solved in terms of finite square matrices X, Y over a field with a zero characteristic. (Hint: examine the trace of both parts.) Find the solution of this equation in terms of infinite matrices. (Hint: study the linear operators d/dx and multiplication by x on the space of all polynomials of x and use the fact that $\frac{d}{dx}(xf) - x\frac{d}{dx}f = f$.)

4. Give an explicit description of classical groups and classical Lie algebras in the cases $n = 1$ and $n = 2$. Construct an isomorphism of the groups $U(1)$ and $SO(2, \mathbf{R})$.

5. The following matrices over $\mathbf{C}$ are called Pauli matrices:

$$\sigma_0 = \begin{pmatrix} 1 & 0 \\ 0 & 1 \end{pmatrix}, \sigma_1 = \begin{pmatrix} 0 & 1 \\ 1 & 0 \end{pmatrix}, \sigma_2 = \begin{pmatrix} 0 & -i \\ i & 0 \end{pmatrix}, \sigma_3 = \begin{pmatrix} 1 & 0 \\ 0 & -1 \end{pmatrix}.$$

(These matrices were introduced by the well-known German physicist Wolfgang Pauli, one of the creators of quantum mechanics, in his theory of electron spin.) Check their properties:

a) $[\sigma_a, \sigma_b] = 2i\epsilon_{abc}\sigma_c$, where $\{a,b,c\} = \{1,2,3\}$ and ϵ_{abc} is the permutation symbol $\begin{pmatrix} 1 & 2 & 3 \\ a & b & c \end{pmatrix}$.

b) $\sigma_a\sigma_b + \sigma_b\sigma_a = 2\delta_{ab}\sigma_0$ (δ_{ab} is the Kronecker delta).

c) The matrices $i\sigma_1, i\sigma_2, i\sigma_3$ form a basis of $su(2)$ over **R** and a basis of $sl(2)$ over **C**; the matrices $\sigma_0, i\sigma_1, i\sigma_2, i\sigma_3$ form a basis of $u(2)$ over **R** and a basis of $gl(2)$ over **C**.

6. The following matrices over **C** of order 4 are called Dirac matrices (here the σ_a are the Pauli matrices):

$$\gamma_0 = \begin{pmatrix} \sigma_0 & 0 \\ 0 & -\sigma_0 \end{pmatrix}, \quad \gamma_1 = \begin{pmatrix} 0 & \sigma_1 \\ -\sigma_1 & 0 \end{pmatrix}, \quad \gamma_2 = \begin{pmatrix} 0 & \sigma_2 \\ -\sigma_2 & 0 \end{pmatrix}, \quad \gamma_3 = \begin{pmatrix} 0 & \sigma_3 \\ -\sigma_3 & 0 \end{pmatrix}$$

(These matrices were introduced by the well-known English physicist P.A.M. Dirac, one of the creators of quantum mechanics, in his theory of the relativistic electron with spin.) Using the results of Exercises 1 and 5, verify their properties:

a) $\gamma_a\gamma_b + \gamma_b\gamma_a = 2g_{ab}E_4$, where $g_{ab} = 0$ for $a \neq b$ and $g_{00} = 1$, $g_{11} = g_{22} = g_{33} = -1$

b) By definition, $\gamma_5 = i\gamma_1\gamma_2\gamma_3\gamma_0$. Verify that $\gamma_5 = -\begin{pmatrix} 0 & \sigma_0 \\ \sigma_0 & 0 \end{pmatrix}$.

c) $\gamma_a\gamma_5 = -\gamma_5\gamma_a$ for $a = 0,1,2,3$; $\gamma_5^2 = E_4$.

7. Verify the following table of dimensions of the classical Lie algebras (as linear spaces over the corresponding fields):

$gl(n,K)$	$sl(n,K)$	$o(n,K)$	$u(n)$	$su(n)$
n^2	n^2-1	$\frac{n(n-1)}{2}$	n^2	n^2-1

8. Let A be a square matrix of order n, let ϵ be a real variable, and let $\epsilon \to 0$. Show that the matrix $U = E + \epsilon A$ is "unitary up to order ϵ^2" if and only if A is anti-Hermitian:

$$U\bar{U}^t = E + O(\epsilon^2) \Leftrightarrow A + \bar{A}^t = 0.$$

Formulate and prove analogous assertions for other pairs of classical groups and Lie algebras.

9. Let $U = E + \epsilon A$, $V = E + \epsilon B$, where $\epsilon \to 0$. Verify that

$$UVU^{-1}V^{-1} = E + \epsilon^2[A,B] + O(\epsilon^3)$$

(the expression on the left is called the group commutator of the elements U and V).

10. The rank rank A of a matrix over a field is the maximum number of columns which are linearly independent. Prove that rank $A_f = \dim \operatorname{im} f$.

Prove that a square matrix of rank 1 can be represented as the product of a column by a row.

11. Let A and B be $m \times n$ and $m_1 \times n_1$ matrices over a field and choose a fixed enumeration of the ordered pairs (i, j) of row indices $(1 \le i \le m,\ 1 \le j \le m_1)$ (e.g., dictionary order). Similarly, choose a fixed enumeration of the ordered pairs (k, l) of column indices $(1 \le k \le n,\ 1 \le l \le n_1)$. The tensor or Kronecker product $A \otimes B$ is the $mm_1 \times nn_1$ matrix with the element $a_{ik}b_{jl}$ at the location $\alpha\beta$, where α enumerates (i, j) and β enumerates (k, l). Verify the following assertions:

a) $A \otimes B$ is linear with respect to each argument with the other argument held fixed.

b) If $m = n$ and $m_1 = n_1$, then $\det(A \otimes B) = (\det A)^{m_1}(\det B)^m$.

12. How many operations are required in order to multiply two large matrices? Strassen's method, which makes it possible to reduce substantially the number of operations if the matrices are indeed large, is explained in the following series of assertions:

a) The multiplication of two matrices of order N by the usual method requires N^3 multiplications and $N^2(N-1)$ additions.

b)The following multiplication formula, involving 7 multiplications (instead of 8) at the expense of 18 additions (instead of 4) holds for $N = 2$ (it is not assumed that the elements commute):

$$\begin{pmatrix} a & b \\ -c & -d \end{pmatrix} \begin{pmatrix} A & B \\ -C & -D \end{pmatrix} =$$

$$\begin{pmatrix} (a+d)(A+D) - (b+d)(C+D) - d(A-C) - (a-b)D, & (a-b)D - a(D-B) \\ (d-c)A - d(A-C), & (a+d)(A+D) - (a+c)(A+B) - a(D-B) - (d-c)A \end{pmatrix}.$$

c) Applying this method to matrices of order 2^n, partitioned into four $2^{n-1} \times$ $\times 2^{n-1}$ blocks, show that they can be multiplied using 7^n multiplications and $6(7^n - 4^n)$ additions.

d) Extend matrices of order N to the nearest matrix of order 2^n with zeros and show that $O(N^{\log_2 7}) = O(N^{2.81})$ operations are sufficient for multiplying them.

Can you think of something better ?

13. Let $L = M_n(K)$ be the space of square matrices of order n. Prove that for any functional $f \in L^*$ there exists a unique matrix $A \in M_n(K)$ with the property

$$f(X) = \operatorname{Tr}(AX)$$

for all $X \in M_n(K)$. Hence derive the existence of the canonical isomorphism

$$\mathcal{L}(L,L) \to [\mathcal{L}(L,L)]^*$$

for any finite-dimensional space L.

14. A linear space L over K together with a binary operation (commutator) $L \times L \to L$, denoted by $[\,,\,]$ and satisfying the following conditions, is called a Lie algebra over K:

a) the commutator $[l,m]$ is linear with respect to each argument $l,m \in L$ with the second argument fixed;

b) $[l,m] = -[m,l]$ for all l,m;

c) $[l_1,[l_2,l_3]]+[l_3,[l_1,l_2]]+[l_2,[l_3,l_1]] = 0$ (Jacobi's identity) for all $l_1,l_2,l_3 \in L$.

Verify that the classical Lie algebras described in §4.11 are Lie algebras in the sense of this definition.

More generally, prove that the commutator $[X,Y] = XY - YX$ in any associative ring satisfies Jacobi's identity.

§5. Subspaces and Direct Sums

5.1. In this section we shall study some geometric properties of the relative arrangement of subspaces in a finite-dimensional space L. We shall illustrate the first problem with a very simple example. Let $L_1, L_1' \subset L$ be two subspaces. It is natural to assume that they are arranged in the same manner in L if there exists a linear automorphism $f : L \to L$ which transforms L_1 into L_1'. For this, of course, it is necessary that $\dim L_1 = \dim L_1'$, because f preserves all linear relations and, therefore, transforms a basis of L_1 into a basis of L_1'. But this is also sufficient. Indeed, we choose a basis $\{e_1,\ldots,e_m\}$ of L_1 and a basis $\{e_1',\ldots,e_m'\}$ of L_1'. According to Theorem 2.12, they can be extended up to the bases $\{e_1,\ldots,e_m,e_{m+1},\ldots,e_n\}$ and $\{e_1',\ldots,e_m',e_{m+1}',\ldots,e_n'\}$ in the space L. By Definition 3.3 there exists a linear mapping $f : L \to L$ which transforms e_i into e_i' for all i. This mapping is invertible and transforms L_1 onto L_1'.

Thus *all linear subspaces with the same dimension are arranged in the same manner in L.*

Going further, it is natural to study all possible arrangements of (ordered) pairs of the subspaces $L_1, L_2 \subset L$. As above, we shall say that the pairs (L_1,L_2) and (L_1',L_2') are arranged identically if there exists a linear automorphism $f : L \to L$ such that $f(L_1) = L_1'$ and $f(L_2) = L_2'$. Once again, the equalities $\dim L_1 = \dim L_1'$ and $\dim L_2 = \dim L_2'$ are necessary for an arrangement to be the same. Generally speaking, however, these conditions are no longer sufficient. Indeed, if (L_1,L_2)

and (L_1', L_2') are identically arranged, then f maps the subspace $L_1 \cap L_2$ onto $L_1' \cap L_2'$, and for this reason, the condition $\dim(L_1 \cap L_2) = \dim(L_1' \cap L_2')$ is also necessary. If $\dim L_1$ and $\dim L_2$ are fixed, but L_1 and L_2 are otherwise arbitrary, then $\dim(L_1 \cap L_2)$ can assume, generally speaking, a series of values.

To determine these values, we introduce the concept of the sum of linear subspaces.

5.2. Definition. *Let $L_1, \ldots, L_n \subset L$ be linear subspaces of L. The set*

$$\sum_{i=1}^{n} L_i = L_1 + \ldots + L_n = \{\sum_{i=1}^{n} l_i | l_i \in L_i\}.$$

It is easy to verify that the sum is also a linear subspace and that, like the operation of intersection of linear subspaces, this summation operation is associative and commutative. The sum $L_1 + \ldots + L_n$ can also be defined as the *smallest subspace of L which contains all the L_i.*

The following theorem relates the dimension of the sum of two subspaces and their intersection.

5.3. Theorem. *If $L_1, L_2 \subset L$ are finite-dimensional, then $L_1 \cap L_2$ and $L_1 + L_2$ are finite-dimensional and*

$$\dim(L_1 \cap L_2) + \dim(L_1 + L_2) = \dim L_1 + \dim L_2.$$

Proof. $L_1 + L_2$ is the linear span of the union of the bases of L_1 and L_2 and is therefore finite-dimensional; $L_1 \cap L_2$ is contained in the finite-dimensional spaces L_1 and L_2.

Let $m = \dim L_1 \cap L_2$, $n = \dim L_1$, $p = \dim L_2$. Select a basis $\{e_1, \ldots, e_m\}$ of the space $L_1 \cap L_2$. According to Theorem 2.12, it can be extended up to bases of the spaces L_1 and L_2. Let the extended bases be $\{e_1, \ldots, e_m, e'_{m+1}, \ldots, e'_n\}$ and $\{e_1, \ldots, e_m, e''_{m+1}, \ldots, e''_p\}$. We shall say that such a pair of bases in L_1 and L_2 is concordant.

We shall now prove that the set $\{e_1, \ldots, e_m, e'_{m+1}, \ldots, e'_n, e''_{m+1}, \ldots, e''_p\}$ forms a basis of the space $L_1 + L_2$. The assertion of the theorem follows from here:

$$\dim(L_1 + L_2) = p + n - m = \dim L_1 + \dim L_2 - \dim L_1 \cap L_2.$$

Since every vector in $L_1 + L_2$ is a sum of vectors from L_1 and L_2, i.e., a sum of the linear combinations of $\{e_1, \ldots, e_m, e'_{m+1}, \ldots, e'_n\}$ and $\{e_1, \ldots, e_m, e''_{m+1}, \ldots, e''_p\}$, the union of these sets generates $L_1 + L_2$. Therefore, we have only to verify its linear independence.

Assume that there exists a non-trivial linear dependence

$$\sum_{i=1}^{m} x_i e_i + \sum_{j=m+1}^{n} y_j e'_j + \sum_{k=m+1}^{p} z_k e''_k = 0.$$

Then indices j and k must necessarily exist such that $y_j \neq 0$ and $z_k \neq 0$, for otherwise we would obtain a non-trivial linear dependence between the elements of the bases of L_1 or L_2.

Therefore, the non-zero vector $\sum_{k=m+1}^{p} z_k e''_k \in L_2$ must also lie in L_1, because it equals $-(\sum_{i=1}^{m} x_i e_i + \sum_{j=m+1}^{n} y_j e'_j)$. This means that it lies in $L_1 \cap L_2$ and can therefore be represented as a linear combination of the vectors $\{e_1, \ldots, e_m\}$, comprising the basis of $L_1 \cap L_2$. But this representation gives a non-trivial linear dependence between the vectors $\{e_1, \ldots, e_m, e'_{m+1}, \ldots, e''_p\}$, in contradiction to their definition. The theorem is proved.

5.4. Corollary. *Let $n_1 \leq n_2 \leq n$ be the dimensions of the spaces L_1, L_2, and L respectively. Then the numbers $i = \dim L_1 \cap L_2$ and $s = \dim(L_1 + L_2)$ can assume any values that satisfy the conditions $0 \leq i \leq n_1$, $n_2 \leq s \leq n$ and $i + s = n_1 + n_2$.*

Proof. The necessity follows from the inclusions $L_1 \cap L_2 \subset L_1$, $L_2 \subset L_1 + L_2 \subset L$ and from Theorem 5.3. To prove sufficiency we choose $s = n_1 + n_2 - i$ linearly independent vectors in L: $\{e_1, \ldots, e_i; e'_{i+1}, \ldots, e'_{n_1}; e''_{i+1}, \ldots, \ldots, e''_{n_2}\}$ and denote by L_1 and L_2 the linear spans of $\{e_1, \ldots, e_i; e'_{i+1}, \ldots, e'_{n_1}\}$ and $\{e_1, \ldots, e_i; e''_{i+1}, \ldots, e''_{n_2}\}$ respectively. As in the theorem, it is not difficult to verify that $L_1 \cap L_2$ is the linear span of $\{e_1, \ldots, e_i\}$.

5.5. We can now establish that the invariants $n_1 = \dim L_1$, $n_2 = \dim L_2$, and $i = \dim L_1 \cap L_2$ completely characterize the arrangement of pairs of subspaces (L_1, L_2) in L. For the proof, we take a different pair (L'_1, L'_2) with the same invariants, construct matched pairs of bases for L_1, L_2 and L'_1, L'_2, and then construct their union — the bases of $L_1 + L_2$ and $L'_1 + L'_2$, as in the proof of Theorem 5.3. Finally we extend these unions up to two bases of L. The linear automorphism that transforms the first basis into the second one establishes the fact that L_1, L_2 and L'_1, L'_2 have the same arrangement.

5.6. General position. In the notation of the preceding section, we shall say that the subspaces $L_1, L_2 \subset L$ are in general position if their intersection has the smallest dimension and their sum the greatest dimension permitted by the inequalities of Corollary 5.4.

For example, two planes in three-dimensional space are in the general position if they intersect along a straight line, while two planes in a four-dimensional space are in the general position if they intersect at a point.

The same concept can also be expressed by saying that L_1 and L_2 *intersect transversally*.

The term "general position" originates from the fact that in some sense most pairs of subspaces (L_1, L_2) are arranged in the general position, while other arrangements are degenerate. This assertion can be refined by various methods. One method is to describe the set of pairs of subspaces by some parameters and verify that a pair is not in the general position only if these parameters satisfy additional relations which the general parameters do not satisfy.

Another method, which is suitable for $K = \mathbf{R}$ and $\mathbf{C}$, is to choose a basis of L, define L_1 and L_2 by two systems of linear equations, and show that the coefficients of these equations can be changed infinitesimally ("perturb L_1 and L_2") so that the new pair would be in the general position.

It is also possible to study the invariants characterizing the relative arrangement of triples, quadruples, and higher numbers of subspaces of L. The combinatorial difficulties here grow rapidly, and in order to solve this problem a different technique is required; in addition, beginning with quadruples, the arrangement is no longer characterized just by discrete invariants, such as the dimensions of different sums and intersections.

We note also that, as our "physical" intuition shows, the arrangement, say, of a straight line relative to a plane, is characterized by the angle between them. But as we noted in §1, the concept of angle requires the introduction of an additional structure. In a purely linear situation, there is only the difference between a "zero" and a "non-zero" angle.

We shall now study n-tuples of subspaces.

5.7. Definition. *A space L is a direct sum of its subspaces $L_1, \ldots, L_n$, if every vector $l \in L$ can be uniquely represented in the form $\sum_{i=1}^n l_i$, where $l_i \in L_i$.*

When the conditions of the definition are satisfied, we write $L = L_1 \oplus \ldots \oplus L_n$ or $L = \bigoplus_{i=1}^n L_i$. For example, if $\{e_1, \ldots, e_n\}$ is a basis of L and $L_i = Ke_i$ is the linear span of e_i, then $L = \bigoplus_{i=1}^n L_i$. Evidently, if $L = \bigoplus_{i=1}^n L_i$, then $L = \sum_{i=1}^n L_i$; the last condition is weaker.

5.8. Theorem. *Let $L_1, \ldots, L_n \subset L$ be subspaces of L. Then $L = \bigoplus_{i=1}^n L_i$, if and only if any of the following two conditions holds:*

a) $\sum_{i=1}^n L_i = L$ *and* $L_j \cap \left(\sum_{i \neq j} L_i\right) = \{0\}$ *for all* $1 \leq j \leq n$.

b) $\sum_{i=1}^n L_i = L$ *and* $\sum_{i=1}^n \dim L_i = \dim L$ *(here it is assumed that L is finite-dimensional)*.

Proof. a) The uniqueness of the representation of any vector $l \in L$ in the form $\sum_{i=1}^n l_i \in L$ is equivalent to the uniqueness of such a representation for the zero vector. Indeed, if $\sum_{i=1}^n l_i = \sum_{i=1}^n l'_i$, then $0 = \sum_{i=1}^n (l_i - l'_i)$, and vice versa. If there exists a non-trivial representation $0 = \sum_{i=1}^{n} l_i$ in which, say, $l_j \neq 0$, then $l_j = -\sum_{i \neq j} l_i \in L_j \cap \left(\sum_{i \neq j} L_i\right)$, so that condition a) does not hold. Inverting this argument we find that the violation of the condition a) implies the non-uniqueness of the representation of zero.

b) If $\bigoplus_{i=1}^{n} L_i = L$, then in all cases

$$\sum_{i=1}^{n} L_i = L \quad \text{and} \quad \sum_{i=1}^{n} \dim L_i \geq \dim L,$$

because the union of the bases of L_i generates L and therefore contains a basis of L. According to Theorem 5.3 applied to L_j and $\sum_{i \neq j} L_i$, we have

$$\dim L_j \cap \left(\sum_{i \neq j} L_i\right) + \dim L = \dim L_j + \dim\left(\sum_{i \neq j} L_i\right).$$

But the preceding assertion implies that the dimension of the intersection on the left is zero. In addition, if the sum of all the L_i is a direct sum, then the sum of all the L_i except L_j is also a direct sum, and we may assume by induction that

$$\dim\left(\sum_{i \neq j} L_i\right) = \sum_{i \neq j} \dim L_i.$$

Therefore $\sum_i \dim L_i = \dim L$.

Conversely, if $\dim L_i = \dim L$, then the union of the bases of all the L_i consists of $\dim L$ elements and generates the whole of L and is therefore a basis of L. Indeed a non-trivial representation of zero $0 = \sum_{i=1}^{n} l_i$, $l_i \in L_i$, would furnish a non-trivial linear combination of the elements of this basis which equals zero, which is impossible.

We shall now examine the relationship between decompositions into a direct sum and special linear projection operators.

5.9. Definition. *The linear operator $p : L \to L$ is said to be a projection operator if $p^2 = p \circ p = p$.*

The direct decomposition $L = \bigoplus_{i=1}^{n} L_i$ is naturally associated with n projection operators which are defined as follows: for any $l_j \in L_j$,

$$p_i\left(\sum_{j=1}^{n} l_j\right) = l_i.$$

Since any element $l \in L$ can be uniquely represented in the form $\sum_{j=1}^{n} l_j$, $l_j \in L_j$, then the mappings p_i are well defined. Their linearity and the property $p_i^2 = p_i$ are verified directly from the definition. Evidently, $L_i = \operatorname{im} p_i$.

Moreover, if $i \neq j$, then $p_i p_j = 0$: the vector l_i corresponds to the representation $l_i = \sum_{j=1}^n l'_j$ where $l'_j = 0$ for $i \neq j$, $l'_i = l_i$.

Finally, $\sum_{i=1}^n p_i = \mathrm{id}$, because $(\sum_{i=1}^n p_i)(\sum_{j=1}^n l_j) = \sum_{j=1}^n l_j$ if $l_j \in L_j$. Conversely, such a system of projection operators can be used to define a corresponding direct decomposition.

5.10. Theorem. *Let $p_1, \dots, p_n : L \to L$ be a finite set of projection operators satisfying the conditions*

$$\sum_{i=1}^n p_i = \mathrm{id}, \quad p_i p_j = 0 \quad \text{for} \quad i \neq j.$$

Let $L_i = \mathrm{im}\, p_i$. Then $L = \bigoplus_{i=1}^n L_i$.

Proof. Applying the operator id$= \sum_{i=1}^n p_i$ to any vector $l \in L$, we obtain $l = \sum_{i=1}^n p_i(l)$, where $p_i(l) \in L_i$. Therefore $L = \sum_{i=1}^n L_i$. To prove that this sum is a direct sum we shall apply the criterion a) of Theorem 5.8. Let $l \in L_j \cap \left(\sum_{i \neq j} L_i\right)$. The definition of the spaces $L_i = \mathrm{im}\, p_i$ implies that there exist vectors $l_1, \dots, l_n$ such that

$$l = p_j(l_j) = \sum_{i \neq j} p_i(l_i).$$

We apply the operator p_j to this equality and make use of the fact that $p_j^2 = p_j$, $p_j p_i = 0$ for $i \neq j$. We obtain

$$p_j(l_j) = \sum_{i \neq j} p_j p_i(l_i) = 0.$$

Therefore, $l = 0$, which completes the proof.

5.11. Direct complements. If L is a finite-dimensional space, then for any subspace $L_1 \subset L$ there exists a subspace $L_2 \subset L$ such that $L = L_1 \oplus L_2$. Aside from the trivial cases $L_1 = \{0\}$ or $L_1 = L$, this choice is not unique. In fact, selecting the basis $\{e_1, \dots, e_m\}$ of L_1 and extending it up to the basis $\{e_1, \dots, e_m, e_{m+1}, \dots, e_n\}$ of L, we can take for L_2 the linear span of the vectors $\{e_{m+1}, \dots, e_n\}$.

5.12. External direct sums. Thus far we have started from the set of subspaces $L_1, \dots, L_n$ of the same space L. Now, let $L_1, \dots, L_n$ be spaces which are not imbedded beforehand in a general space. We shall define their *external direct sum* L as follows:

a) L, as a set, is $L_1 \times \dots \times L_n$, i.e., the elements of L are the sets $(l_1, \dots l_n)$, where $l_i \in L_i$.

b) Addition and multiplication by a scalar are performed coordinate-wise:

$$(l_1, \dots, l_n) + (l'_1, \dots, l'_n) = (l_1 + l'_1, \dots, l_n + l'_n),$$

$$a(l_1, \ldots, l_n) = (al_1, \ldots, al_n).$$

It is not difficult to verify that L satisfies the axioms of a linear space. The mapping $f_i : L_i \to L$, $f_i(l) = (0, \ldots, 0, l, 0, \ldots, 0)$ (l is in the ith location) is a linear imbedding of L_i into L, and from the definitions it follows immediately that $L = \bigoplus_{i=1}^n f_i(L_i)$. Identifying L_i with $f_i(L_i)$, we obtain a linear space which contains L_i and decomposes into the direct sum of L_i. This justifies the name external direct sum. It is often convenient to denote the external direct sum by $\bigoplus_{i=1}^n L_i$.

5.13. Direct sums of linear mappings. Let $L = \bigoplus_{i=1}^n L_i$ and $M = \bigoplus_{i=1}^n M_i$, and let $f : L \to M$ be a linear mapping such that $f(L_i) \subset M_i$. We denote by f_i the induced linear mapping $L_i \to M_i$. In this case, it is customary to write $f = \bigoplus_{i=1}^n f_i$. The external direct sum of linear mappings is defined analogously. Choosing bases of L and M that are the union of the bases of L_i and M_i respectively, we find that the matrix of f is the union of blocks, which are matrices representing the mappings f_i lying along the diagonals; the other locations contain zeros.

5.14. Orientation of real linear spaces. Let L be a finite-dimensional linear space over the field of real numbers. Two ordered bases $\{e_i\}$ and $\{e_i'\}$ of it are always identically arranged in the sense that there is a unique linear isomorphism $f : L \to L$ that maps e_i into e_i'. However, we pose a more subtle question: when is it possible to transform the basis $\{e_i\}$ into the basis $\{e_i'\}$ by a *continuous motion* or *deformation*, i.e., to find a family $f_t : L \to L$ of linear isomorphisms, depending continuously on the parameter $t \in [0, 1]$ such that $f_0 =$id and $f_1(e_i) = e_i'$ for all i ? (Only the elements of the matrix of f in some basis must vary continuously as a function of t.) For this there is an obvious necessary condition: since the determinant of f_t is a continuous function of t and does not vanish, the sign of $\det f_t$ must coincide with that of $\det f_0 = 1$, i.e., $\det f_t > 0$.

The converse is also true: *if the determinant of the matrix transforming the basis $\{e_i\}$ into $\{e_i'\}$ is positive, then $\{e_i\}$ can be transformed into $\{e_i'\}$ by a continuous motion.*

This assertion can, obviously, be formulated differently: *any real matrix with a positive determinant can be connected to a unit matrix by a continuous curve, consisting of non-singular matrices (the set of real non-singular matrices with positive determinant is connected).* To transfer from the language of bases to the language of matrices it is sufficient to work not with the pair of bases $\{e_i\}$, $\{f_t(e_i)\}$ but rather with the matrix of the change of basis from the first basis to the second.

We shall prove this assertion by dividing it into a series of steps.

a) *Let $A = B_1 \ldots B_n$, where A and B_i are matrices with positive determinants. If all the B_j can be connected to E by a continuous curve, then so can A.*

Indeed, let $B_j(t)$ be continuous curves in the space of non-singular matrices such that $B_j(0) = B_j$, $B_j(1) = E$. Then the curve $A(t) = B_1(t)\dots B_n(t)$ connects A and E.

b) *If A can be connected by a continuous curve to B and B can be so connected to E, then A can be connected to E.*

Indeed, if $A(t)$ is such that $A(0) = A$ and $A(1) = B$ and $B(t)$ is such that $B(0) = B$ and $B(1) = E$, then the curve

$$t \mapsto \begin{cases} A(2t) & \text{for } 0 \le t \le 1/2, \\ B(2t-1) & \text{for } 1/2 \le t \le 1 \end{cases}$$

connects A to E. The device of changing the scale and the origin of t is used only because we stipulated that the curves of the matrices be parametrized by numbers $t \in [0,1]$. Obviously, any intermediate parametrization intervals can be used, all required deformations can be performed successively, and the scale need be changed only at the end. Therefore, in what follows, we shall ignore the parametrization intervals.

c) *Any non-singular square matrix A can be represented by a product of a finite number of elementary matrices of the types $F_{s,t}, F_{s,t}(\lambda), F_s(\lambda)$, $\lambda \in \mathbf{R}$. We denote by E_{st} the matrix with ones at the location (s,t) and zeros elsewhere. Then by definition,*

$$F_{s,t} = E - E_{ss} - E_{tt} + E_{st} + E_{ts};$$

$$F_{s,t}(\lambda) = E + \lambda E_{st}; \; F_s(\lambda) = E + (\lambda - 1)E_{ss}.$$

This result is proved in §4 of Chapter 2 in "Introduction to Algebra" as a corollary to the theorem of §5.

d) Now, let the matrix A be represented as a product of elementary matrices. Assuming that its determinant is positive, we shall show how to connect it to E with the help of several successive deformations, using the results of the steps a) and b) above.

First of all, $\det F_{s,t}(\lambda) = 1$ for all λ and $F_{s,t}(0) = E$. By varying λ in the starting cofactors from the initial value to zero we can deform all such factors into E, and we can therefore assume that they are absent at the outset.

The matrices $F_s(\lambda)$ are diagonal: λ stands in the location (s,s) and ones elsewhere. We change λ from the initial value to $+1$ or -1, according to the sign of the initial value. The deformation will yield either the unit matrix or the matrix of the linear mapping that changes one of the basis vectors into the opposite vector, leaving the remaining basis vectors unaffected.

At this stage the result of the deformation of A will be the matrix of the composition of two transformations: one reduces to the permutation of the vectors of the basis ($F_{s,t}$ interchanges the sth and tth vectors) and the other changes the signs of some of the vectors (the composition $F_s(+1)$ and $F_t(-1)$).

Any permutation can be decomposed into a product of transpositions. The matrix of a permutation of the basis vectors in the plane $\begin{pmatrix} 0 & 1 \\ 1 & 0 \end{pmatrix}$ can be connected to $\begin{pmatrix} -1 & 0 \\ 0 & 1 \end{pmatrix}$ by the curve $\begin{pmatrix} -\cos t & \sin t \\ \sin t & \cos t \end{pmatrix}$, $\pi/2 \geq t \geq 0$. Clearly, by distributing the elements of the last matrix over the locations (s,s), (s,t), (t,s), and (t,t) we obtain the corresponding deformation in any dimension, annihilating the $F_{s,t}$.

The matrix A has now been transformed into a diagonal matrix with the elements ± 1 standing along the diagonal; in addition, the number of -1's is even, because the determinant of A is positive. The matrix $\begin{pmatrix} -1 & 0 \\ 0 & -1 \end{pmatrix}$ can be connected to $\begin{pmatrix} 1 & 0 \\ 0 & 1 \end{pmatrix}$ by the curve $\begin{pmatrix} \cos t & -\sin t \\ \sin t & \cos t \end{pmatrix}$, $\pi \geq t \geq 0$. The proof is completed by collecting all -1's in pairs and performing such deformations of all pairs.

We now return to the orientation.

We shall say that the bases $\{e_i\}$, $\{e'_i\}$ are *identically oriented* if the determinant of the matrix of the transformation between them is positive. It is clear that the set of ordered bases of L is divided precisely into two classes, consisting of identically oriented bases, while the bases of different classes are oriented differently (or oppositely).

The choice of one of these classes is called *the orientation of the space L*.

The orientation of a one-dimensional space corresponds to indicating the "positive direction in it", or the half-line $\mathbf{R}_+ e = \{ae | a > 0\}$, where e is any vector which determines the orientation.

In a two-dimensional space, fixing the orientation with the help of the basis $\{e_1, e_2\}$ can be regarded as indicating the "positive direction of rotation" of the plane from e_1 to e_2. This agrees intuitively with the fact that the basis $\{e_2, e_1\}$ gives the opposite orientation (the determinant of the transition matrix $\begin{pmatrix} 0 & 1 \\ 1 & 0 \end{pmatrix}$ equals -1) and the opposite direction of rotation.

In the general case, the transformation from the basis $\{e_i\}$ to the basis $\{e'_i\}$, consisting of the same vectors arranged in a different order, preserves the orientation if the permutation is even and changes it if the permutation is odd. Reversing the sign of one of the vectors e_i reverses the orientation.

In the three-dimensional physical space the choice of a specific orientation can be related to human physiology: asymmetry of the right and left sides. In most people the heart is located on the left side. The thumb, the index finger and the middle finger of the left hand, bent towards the palm, together form in a linearly independent position an ordered basis, which determines the orientation ("left-hand rule"). The question of whether or not there exist purely physical processes which would distinguish the orientation of space, i.e., "non-invariant relative to mirror reflection", was answered affirmatively about 20 years ago, to everyone's surprise, by an experiment which established the non-conservation of parity in weak interactions.

EXERCISES

1. Let (L_1, L_2, L_3) be an ordered triple of pairwise distinct planes in K^3. Prove that there exist two possible types of relative arrangement of such triples, characterized by the fact that $\dim L_1 \cap L_2 \cap L_3 = 0$ or 1. Which type should be regarded as the general one ?

2. Prove that the triples of pairwise distinct straight lines in K^3 are all identically arranged and that this assertion is not true for quadruples.

3. Let $L_1 \subset L_2 \subset \ldots \subset L_n$ be a flag in a finite-dimensional space L, $m_i = \dim L_i$. Prove that if $L'_1 \subset \ldots \subset L'_n$ is a different flag, $m'_i = \dim L'_i$, then there exists an automorphism of L transforming the first flag into the second flag, if and only if $m_i = m'_i$ for any i.

4. Do the same problem for direct decompositions.

5. Prove the assertion of the fifth item in §5.6.

6. Let $p : L \to L$ be a projection operator. Prove that $L = \ker p \oplus \operatorname{im} p$. Based on this, show that in an appropriate basis of L any projection operator p can be represented by a matrix of the form

$$\left(\begin{array}{c|c} E_r & 0 \\ \hline 0 & 0 \end{array}\right),$$

where $r = \dim \operatorname{im} p$.

7. Let L be an n-dimensional space over a finite field consisting of q elements.

   a) Calculate the number of k-dimensional subspaces in L, $1 \leq k \leq n$.

   b) Calculate the number of pairs of subspaces $L_1, L_2 \subset L$ with fixed dimensions L_1, L_2 and $L_1 \cap L_2$. Verify that as $q \to \infty$ the relative number of these pairs arranged in the general position, amongst all pairs with given $\dim L_1$ and $\dim L_2$ approaches one.

§6. Quotient Spaces

6.1. Let L be a linear space, $M \subset L$ a linear subspace of L, and $l \in L$ a vector. Different problems lead to the study of sets of the type

$$l + M = \{l + m | m \in M\},$$

"translations" of a linear space M by a vector l. We shall shortly verify that these translations do not necessarily have to be linear subspaces in L; they are called *linear subvarieties*. We shall start by proving the following lemma:

6.2. Lemma . *$l_1 + M_1 = l_2 + M_2$, if and only if $M_1 = M_2 = M$ and $l_1 - l_2 \in$ $\in M$. Thus any linear subvariety uniquely determines a linear subspace M, whose translation it is. The translation vector, however, is determined only to within an element belonging to this subspace.*

Proof. First, let $l_1 - l_2 \in M$. Set $l_1 - l_2 = m_0$. Then

$$l_1 + M = \{l_1 + m | m \in M\}, \quad l_2 + M = \{l_1 + m - m_0 | m \in M\}.$$

But when m runs through all vectors in M, $m - m_0$ also runs through all vectors in M. Therefore, $l_1 + M = l_2 + M$.

Conversely, let $l_1 + M_1 = l_2 + M_2$. Set $m_0 = l_1 - l_2$. It is clear from the definition that then $m_0 + M_1 = M_2$. Since $0 \in M_2$, we must have $m_0 \in M_1$. Hence $m_0 + M_1 = M_1$ according to the argument in the preceding item, so that $M_1 = M_2 = M$. This completes the proof.

6.3. Definition. *The factor (or quotient) space L/M of a linear space L is the set of all linear subvarieties in L that are translations of M, together with the following operations:*

a) $(l_1 + M) + (l_2 + M) = (l_1 + l_2) + M$, and

b) $a(l_1 + M) = al_1 + M$ for any $l_1, l_2 \in L$, $a \in K$.

These operations are well defined and transform L/M into a linear space over the field K.

6.4. Verification of the correctness of the definition. This consists of the following steps:

a) *If $l_1 + M = l'_1 + M$ and $l_2 + M = l'_2 + M$, then $l_1 + l_2 + M = l'_1 + l'_2 + M$.*

Actually, Lemma 6.2 implies that $l_1 - l'_1 = m_1 \in M$ and $l_2 - l'_2 = m_2 \in M$. Therefore, once again according to Lemma 6.2,

$$(l_1 + l_2) + M = (l'_1 + l'_2) + (m_1 + m_2) + M = (l'_1 + l'_2) + M,$$

because $m_1 + m_2 \in M$.

b) *If $l_1 + M = l'_1 + M$, then $al_1 + M = al'_1 + M$.*

Indeed, again setting $l_1 - l'_1 = m \in M$, we have $al_1 - al'_1 = am \in M$, and the application of Lemma 6.2 gives the required result.

Thus addition and multiplication by a scalar are indeed uniquely defined in L/M. It remains to verify the axioms of a linear space. They follow immediately from the corresponding formulas in L. For example, one of the distributivity formulas is verified thus:

$$a[(l_1 + M) + (l_2 + M)] = a((l_1 + l_2) + M) = a(l_1 + l_2) + M =$$

$$= al_1 + al_2 + M = (al_1 + M) + (al_2 + M) = a(l_1 + M) + a(l_2 + M).$$

Here the following are used in succession: definition of addition in L/M, definition of multiplication by a scalar in L/M, distributivity in L, and again the definition of addition and multiplication by a scalar in L/M.

6.5. Remarks. a) It is evident from the definition that the additive group L/M coincides with the quotient group of the additive group L over the additive group M. In particular, the subvariety $M \subset L$ is zero in L/M.

b) There exists a canonical mapping $f : L \to L/M : f(l) = l + M$. It is surjective, and its fibres – the inverse images of the elements – are precisely the subvarieties corresponding to these elements. Indeed, according to Lemma 6.2

$$f^{-1}(l_0 + M) = \{l \in L | l + M = l_0 + M\} = \{l \in L | l - l_0 \in M\} = l_0 + M.$$

We note that in this chain of equalities $l_0 + M$ is regarded for the first time as an *element* of the set L/M, while the others are regarded as *subsets* of L.

From §6.4 it is clear that f is a linear mapping, while Lemma 6.2 shows that $\ker f = M$, because $l_0 + M = M$, if and only if $l_0 \in M$.

6.6. Corollary. *If L is finite-dimensional, then* $\dim L/M = \dim L - \dim M$.

Proof. Apply Theorem 3.12 to the mapping $L \to L/M$ constructed.

Many important problems in mathematics lead to situations when the spaces $M \subset L$ are infinite-dimensional, while the quotient spaces L/M are finite-dimensional. In this case, Corollary 6.6 cannot be used, and the calculation of $\dim L/M$ usually becomes a non-trivial problem. The number $\dim L/M$ is generally called the *codimension* of the subspace M in L and is denoted by $\operatorname{codim} M$ or $\operatorname{codim}_L M$.

6.7. We pose the following problem. Given two mappings $f : L \to M$ and $g : l \to N$, when does a mapping $h : M \to N$ exist such that $g = hf$? In diagramatic language: when can the diagram

be inserted into the commutative triangle.

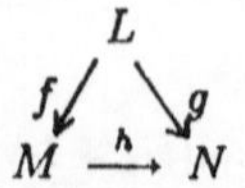

(cf. §13 on commutative diagrams). The answer for linear mappings is given by the following result.

6.8. Proposition. *For* h *to exist it is necessary and sufficient that* $\ker f \subset \ker g$. *If this condition is satisfied and* $\operatorname{im} f = M$, *then* h *is unique.*

Proof. If h exists, then $g = hf$ implies that $g(l) = hf(l) = 0$ if $f(l) = 0$. Therefore, $\ker f \subset \ker g$.

Conversely, let $\ker f \subset \ker g$. We first construct h on the subspace $\operatorname{im} f \subset M$. The only possibility is to set $h(m) = g(l)$, if $m = f(l)$. It is necessary to verify that h is determined uniquely and linearly on $\operatorname{im} f$. The first property follows from the fact that if $m = f(l_1) = f(l_2)$, then $l_1 - l_2 \in \ker f \subset \ker g$, whence $g(l_1) = g(l_2)$. The second property follows automatically from the linearity of f and g.

Now it is sufficient to extend the mapping h from the subspace $\operatorname{im} f \subset M$ into the entire space M, for example, by selecting a basis in $\operatorname{im} f$, extending it up to a basis in M, and setting h equal to zero on the additional vectors.

6.9. Let $g : l \to M$ be a linear mapping. We have already defined the kernel and the image of g. We supplement this definition by setting

$$(\text{coimage of } g) \quad \operatorname{coim} g = L/\ker g,$$

$$(\text{cokernel of } g) \quad \operatorname{coker} g = M/\operatorname{im} g.$$

There exists a chain of linear mappings, which "partition g",

$$\ker g \xrightarrow{i} L \xrightarrow{c} \operatorname{coim} g \xrightarrow{h} \operatorname{im} g \xrightarrow{j} M \xrightarrow{f} \operatorname{coker} g$$

where all mappings, except h, are canonical insertions and factorizations, while h is the only mapping that completes the commutative diagram

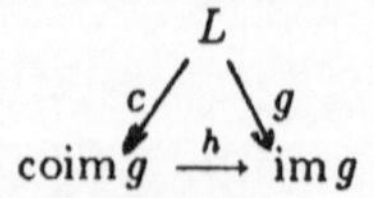

It is unique, because $\ker c = \ker g$, and it is an isomorphism, because the inverse mapping also exists and is defined uniquely.

The point of uniting these spaces into pairs (with and without the prefix "co") is explained in the theory of duality: see the next chapter and Exercise 1 there.

6.10. Fredholm's finite-dimensional alternative. Let $g : L \to M$ be a linear mapping. The number

$$\operatorname{ind} g = \dim \operatorname{coker} g - \dim \ker g$$

is called the *index* of the operator g. It follows from the preceding section that *if L and M are finite-dimensional, then the index g depends only on L and M*:

$$\operatorname{ind} g = (\dim M - \dim \operatorname{im} g) - (\dim L - \dim \operatorname{im} g) = \dim M - \dim L.$$

In particular, if $\dim M = \dim L$, for example, if g is a linear operator on L, then $\operatorname{ind} g = 0$ for any g. This implies the so-called Fredholm alternative:

either the equation $g(x) = y$ is solvable for all y and then the equation $g(x) = 0$ has only zero solutions; or

this equation cannot be solved for all y and then the homogeneous equation $g(x) = 0$ has non-zero solutions.

More precisely, *if $\operatorname{ind} g = 0$, then the dimension of the space of solutions of the homogeneous equation equals the codimension of the spaces on the right hand sides for which the inhomogeneous equation is solvable.*

EXERCISES

1. Let $M, N \subset L$. Prove that the following mapping is a linear isomorphism:

$$(M + N)/N \to M/M \cap N; \quad m + n + N \mapsto m + M \cap N.$$

2. Let $L = M \oplus N$. Then the canonical mapping

$$M \to L/N : m \mapsto m + N$$

is an isomorphism.

§7. Duality

7.1. In §1, we associated every linear space L with its dual space $L^* = \mathcal{L}(L, K)$, while in §3 we showed that if $\dim L < \infty$, then $\dim L^* = \dim L$ and we constructed the canonical isomorphism $\epsilon_L : L \to L^{**}$. Here we shall continue the description of duality, and we shall include in our analysis linear mappings, subspaces, and factor spaces.

The term duality originated from the fact that the theory of duality clarifies a number of properties of the "dual symmetry" of linear spaces, which are very difficult to imagine clearly but which are absolutely fundamental. It is enough to say that the "wave-particle" duality in quantum mechanics is adequately expressed precisely in terms of the linear duality of infinite-dimensional linear spaces (more precisely, the combination of linear and group duality in the Fourier analysis).

It is convenient to trace this symmetry by altering somewhat the notations adopted in §1 and §3.

7.2. Symmetry between L and L^*. Let $l \in L$ and $f \in L^*$. Instead of $f(l)$ we shall write (f, l) (the symbol is analogous to the inner product, but the vectors are from different spaces !) We have thus defined the mapping $L^* \times L \to K$. It is linear with respect to each of the two arguments f, l with the other held fixed:

$$(f_1 + f_2, l) = (f_1, l) + (f_2, l), \ (af_1, l) = a(f_1, l)$$

$$(f, l_1 + l_2) = (f, l_1) + (f, l_2), \ (f, al_1) = a(f, l_1).$$

In general, mappings $L \times M \to K$ with this property are said to be *bilinear*, as well as *pairings* of the spaces L and M. The pairing introduced above between L and L^* is said to be *canonical* (see the discussion of this word in §3.8).

The mapping $\epsilon_L : L \to L^{**}$ in §3.10, as is evident from its definition, can be defined by the condition:

$$(\epsilon_L(l), f) = (f, l),$$

where the symbol denoting the pairing of L^{**} and L^* stands on the left side and the pairing of L^* and L is indicated on the right side. If $\dim L < \infty$, so that ϵ_L is an isomorphism and we identify L^{**} and L by means of ϵ_L, then this formula acquires a symmetric form $(l, f) = (f, l)$. In other words, we can also regard L as the *dual space of L^**.

7.3. Symmetry between dual bases. Let $\{e_1, \ldots, e_n\}$ be a basis of L and let $\{e^1, \ldots, e^n\}$ be its dual basis in L^*. According to §3.9, it is defined by the formulas

$$(e^i, e_k) = \delta_{ik} = \begin{cases} 0 & \text{for } i \neq k, \\ 1 & \text{for } i = k \end{cases}$$

The symmetry $(e^i, e_k) = (e_k, e^i)$, according to the preceding section, indicates that the basis (e_k) is dual to the basis (e^i) if L is regarded as the space of linear functionals on L^*. Thus (e^i) and (e_k) form a *dual pair of bases*, and this relation is symmetric.

Let us represent the vector $l^* \in L^*$ as a linear combination $\sum_{i=1}^n b_i e^i$ and the vector $l \in L$ in the form $\sum_{j=1}^n a_j e_j$. Then

$$(l^*, l) = \sum_{i,j=1}^n a_j b_i (e^i, e_j) = \sum_{i=1}^n a_i b_i = (a_1, \ldots, a_n) \begin{pmatrix} b_1 \\ \vdots \\ b_n \end{pmatrix} =$$

$$= (\vec{a}^t)\vec{b} = (\vec{b}^t)\vec{a} = (b_1, \ldots, b_n) \begin{pmatrix} a_1 \\ \vdots \\ a_n \end{pmatrix} = (l, l^*),$$

where $\vec{a}, \vec{b}$ are column vectors of the corresponding coefficients. This formula is completely analogous to the formula for the inner product of vectors in Euclidean space, but here it relates vectors from different spaces.

7.4. Dual or conjugate mapping. Let $f : L \to M$ be a linear mapping of linear spaces. We shall now show that there exists a unique linear mapping $f^* : M^* \to L^*$ that satisfies the condition

$$(f^*(m^*), l) = (m^*, f(l))$$

for any vectors $m^* \in M^*$, $l \in L$.

a) *Uniqueness of* f^*. Let f_1^* and f_2^* be two such mappings. Then $(f_1^*(m^*), l) = (m^*, f(l)) = (f_2^*(m^*), l)$ for all $m^* \in M^*$, $l \in L$, whence it follows that $((f_1^* - f_2^*)(m^*), l) = 0$. We fix m^* and vary l. Then the element $(f_1^* - f_2^*)(m^*) \in L^*$, as a linear functional on L, assumes only zero values and hence equals zero. Therefore $f_1^* = f_2^*$.

b) *Existence of* f^*. We fix $m^* \in M$ and regard the expression $(m^*, f(l))$ as a function on L. The linearity of f and the bilinearity of $(\, , \,)$ imply that this function is linear. Hence it belongs to L^*. We denote it by $f^*(m^*)$. The equalities

$$f^*(m_1^* + m_2^*) = f^*(m_1^*) + f^*(m_2^*), \; f^*(am^*) = af^*(m^*)$$

follow from the linearity of $(m^*, f(l))$ with respect to m^*. This means that f^* is a linear mapping.

Assume that bases have been selected in L and M and dual bases in L^* and M^*. Let f in these bases be represented by the matrix A. We assert that f^* in the dual bases is represented by the transposed matrix A^t. Indeed, let B be the matrix of f^*. According to the definition and §7.3, we have, denoting the coordinate vectors of m^*, l by $\vec{a}, \vec{b}$,

$$(m^*, f(l)) = \vec{a}^t(A\vec{b}),$$

$$(f^*(m^*), l) = (B\vec{a})^t\vec{b} = (\vec{a}^t B^t)\vec{b}.$$

It follows immediately from the associativity of multiplication of matrices and the uniqueness of f^* that $A = B^t$, i.e., $B = A^t$.

The basic properties of the conjugate mapping are summarized in the following theorem:

7.5. Theorem. a) $(f+g)^* = f^* + g^*$;

b) $(af)^* = af^*$; *here* $f, g : L \to M$ *and* $a \in K$;

c) $(fg)^* = g^* f^*$; *here* $L \xrightarrow{g} M \xrightarrow{f} N$;

d) $\mathrm{id}^* = \mathrm{id}$, $0^* = 0$;

e) L^{**} *and* M *are canonically identified with* L *and* M *respectively, then* $f^{**} : L^{**} \to M^{**}$ *is identified with* $f : L \to M$.

Proof. If it is assumed that L and M are finite-dimensional, then it is simplest to verify all of these asertions by representing f and g by matrices in dual bases and using the simple properties of the transposition operation:

$$(aA + bB)^t = aA^t + bB^t, (AB)^t = B^t A^t, \ E^t = E, \ 0^t = 0, \ (A^t)^t = 0.$$

We leave it as an exercise to the reader to verify invariance.

7.6. Duality between subspaces of L and L^*. Let $M \subset L$ be a linear subspace. We denote by $M^\perp \subset L^*$ the set of functionals which vanish on M and call it the *orthogonal complement of* M. In other words,

$$m^* \in M^\perp \Leftrightarrow (m^*, m) = 0 \text{ for all } m \in M$$

It is easy to see that $M^\perp$ is a linear space. The following assertions summarize the basic properties of this construction (L is assumed to be finite-dimensional).

a) *There exists a canonical isomorphism* $L^*/M^\perp \to M^*$. It is constructed as follows: we associate with the variety $l^* + M^\perp$ the restriction of the functional l^* to M. It is independent of the choice of l^*, because the restrictions of the functionals from $M^\perp$ to M are null restrictions. The linearity of this mapping is obvious. It is surjective, because any linear functional on M extends to a functional on L.

Indeed, let $\{e_1, \ldots, e_m\}$ be a basis of M and let $\{e_1, \ldots, e_m, e_{m+1}, \ldots, e_n\}$ be its extension to a basis of L. The functional f on M given by the values $f(e_1) \ldots, f(e_n)$, is extended onto L, for example, by setting $f(e_{m+1} = \ldots = f(e_n) = 0$.

Finally, the mapping $L^*/M^\perp \to M^*$ constructed above is injective. Indeed, it has a null kernel: if the restriction of l^* to M equals zero, then $l^* \in M^\perp$ and $l^* + M^\perp = M^\perp$ is the zero element in $L^*/M^\perp$.

b) $\dim M + \dim M^\perp = \dim L$. Indeed, this follows from the preceding assertion, Corollary 6.6, and the fact that $\dim L^* = \dim L$ and $\dim M^* = \dim M$.

c) *Under the canonical identification of L^{**} with L, the space $(M^\perp)^\perp$ coincides with M.*

Indeed, because $(m^*, m) = 0$ for all m^* and the given $m \in M$, it is clear that $M \subset (M^\perp)^\perp$. But, in addition, according to the preceding property applied twice,

$$\dim(M^\perp)^\perp = \dim L - \dim M^\perp = \dim M.$$

Hence, $M = (M^\perp)^\perp$.

d) $(M_1 + M_2)^\perp = M_1^\perp \cap M_2^\perp$; $(M_1 \cap M_2)^\perp = M_1^\perp + M_2^\perp$.

The proof is left as an exercise.

EXERCISES

1. Let the linear mapping $g : L \to M$ of finite-dimensional spaces be associated with the chain of mappings constructed in §6.8. Construct the canonical isomorphisms

$$\ker g^* \to \operatorname{coker} g,\ \operatorname{coim} g^* \to \operatorname{im} g,\ \operatorname{im} g^* \to \operatorname{coim} g, \operatorname{coker} g^* \to \ker g.$$

2. Hence derive "Fredholm's third theorem": in order for the equation $g(x) = y$ to be solvable (with respect to x with y held fixed), it is necessary and sufficient that y be orthogonal to the kernel of the conjugate mapping $g^* : M^* \to L^*$.

3. The sequence of linear spaces and mappings $L \xrightarrow{f} M \xrightarrow{g} N$ is said to be exact in the term M, if $\operatorname{im} f = \ker g$. Check the following assertions:

a) the sequence $0 \to L \xrightarrow{f} M$ is exact in the term L, if and only if f is an injection.

b) The sequence $M \xrightarrow{g} N \to 0$ is exact in the term N, if and only if g is a surjection.

c) The sequence of finite-dimensional spaces $0 \to L \xrightarrow{f} M \xrightarrow{g} N \to 0$ is exact (in all terms) if and only if the dual sequence $0 \to N^* \xrightarrow{g^*} M^* \xrightarrow{f^*} L^* \to 0$ is exact.

4. We know that if the mapping $f : L \to M$ in some bases can be represented by the matrix A, then the mapping f^* in the dual bases can be represented by the matrix A^t. Deduce that the rank of a matrix equals the rank of the transposed matrix, i.e., that the maximal numbers of linearly independent rows and columns of the matrix are equal.

§8. The Structure of a Linear Mapping

8.1. In this section we shall begin the study of the following problem: is it possible to construct a better geometric picture of the structure of a linear mapping $f : L \to M$? When L and M are entirely unrelated to one another, the answer is very simple: it is given by Theorem 8.2. A much more interesting and varied picture is obtained when $M = L$ (this is the case considered in the next section) and $M = L^*$ (Chapter 2). In matrix language, we are talking about putting the matrix of f into its simplest form with the help of an appropriate basis, specially adapted to the structure of f. In the first case, the bases in L and M can be selected independently; in the second case, we are talking about one basis in L or one basis in L and its dual basis in L^*: the lower degree of freedom of choice leads to a greater variety of answers.

Our problem can be reformulated as follows in the language of §5. Let us construct the exterior direct sum of spaces $L \oplus M$ and associate with the mapping f its graph Γ_f: the set of all vectors of the form $(l, f(l)) \in L \oplus M$. It is easy to verify that Γ_f is a subspace of $L \oplus M$. We are interested in the invariants of the arrangement of Γ_f in $L \oplus M$. For the case when the bases in L and M can be selected independently, the answer is given by the following theorem.

8.2. Theorem. *Let $f : L \to M$ be a linear mapping of finite-dimensional spaces. Then*

a) there exist direct decompositions $L = L_0 \oplus L_1$, $M = M_1 \oplus M_2$ such that $\ker f = L_0$ and f induces an isomorphism of L_1 to M_1.

b) There exist bases in L and M such that the matrix of f in these bases has the form (a_{ij}), where $a_{ii} = 1$ for $1 \le i \le r$ and $a_{ij} = 0$ for the remaining values of i, j.

c) Let A be an $m \times n$ matrix. Then there exist non-singular square matrices B and C with dimensions $m \times m$ and $n \times n$ and a number $r \le \min(m, n)$ such that the matrix BAC has the form described in the preceding item. The number r is unique and equals the rank of A.

Proof. a) Set $L_0 = \ker f$ and let L_1 be the direct complement of L_0: this is possible by virtue of §5.10. Next, set $M_1 = \operatorname{im} f$ and let M_2 be the direct complement of M_1. We need only verify that f determines an isomorphism of L_1 to M_1. The mapping $f : L_1 \to M_1$ is injective, because the kernel of f, i.e., L_0, intersects L_1 only at the origin. It is surjective because if $l = l_0 + l_1 \in L$, $l_0 \in L_0$, $l_1 \in L_1$, then $f(l) = f(l_1)$.

b) We set $r = \dim L_1 = \dim M_1$ and choose a basis $\{e_1, \ldots, e_r, e_{r+1}, \ldots, e_n\}$ of L, where the first r vectors form a basis of L_1 and the remaining vectors form a basis of L_0. Furthermore, the vectors $e_i' = f(e_i)$, $1 \le i \le r$ form a basis of

$M_1 = \operatorname{im} f$. We extend it to a basis of M with the vectors $\{e'_{r+1}, \ldots, e'_m\}$. Obviously,

$$f(e_1, \ldots, e_r; e_{r+1}, \ldots, e_n) = (e'_1, \ldots, e'_r;\ 0, \ldots, 0) =$$

$$= (e'_1, \ldots, e'_r;\ e'_{r+1}, \ldots, e'_m) \left(\begin{array}{c|c} E_r & 0 \\ \hline 0 & 0 \end{array} \right),$$

so that the matrix of f in these bases has the required form.

c) Based on the matrix A, we construct a linear mapping f of the coordinate spaces $K^n \to K^m$ with this matrix and then apply to it the assertion b) above. In the new bases the matrix of f will have the required form and will be expressed in terms of A in the form BAC, where B and C are the transition (change of basis) matrices (see §4.8). Finally, $\operatorname{rank} A = \operatorname{rank} BAC = \operatorname{rank} f = \dim \operatorname{im} f$. This completes the proof.

We now proceed to the study of linear operators. We begin by introducing the simplest class: *diagonalizable* operators.

We shall say that the subspace $L_0 \subset L$ is invariant with respect to the operator f, if $f(L_0) \subset L_0$.

8.3. Definition. *The linear operator $f : L \to L$ is diagonalizable if either one of the following two equivalent conditions holds:*

a) L decomposes into a direct sum of one-dimensional invariant subspaces.

b) There exists a basis of L in which the matrix of f is diagonal.

The equivalence of these conditions is easily verified. If in the basis (e_i) the matrix of f is diagonal, then $f(e_i) = \lambda_i e_i$, so that the one-dimensional subspaces spanned by e_i are invariant and L decomposes into their direct sum. Conversely, if $L = \bigoplus L_i$ is such a decomposition and e_i is any non-zero vector in L_i, then the $\{e_i\}$ form a basis of L.

Diagonalizable operators form the simplest and, in many respects, the most important class of operators. For example, over the field of complex numbers, as we shall show below, any operator can be diagonalized by changing infinitesimally its matrix so that the operator "in the general position" is diagonalizable.

To understand what can prevent an operator from being diagonalizable, we shall introduce two definitions and prove one theorem.

8.4. Definition. a) *A one-dimensional subspace $L_1 \subset L$ is said to be a proper subspace for the operator f if it is invariant, i.e., $f(L_1) \subset L_1$. If L_1 is such a subspace, then the effect of f on it is equivalent to multiplication by a scalar $\lambda \in K$. This scalar is called the eigenvalue of f (on L_1).*

b) *The vector $l \in L$ is said to be an eigenvector of f if $l \neq 0$ and the linear span Kl is a proper subspace. In other words, $f(l) = \lambda l$ for an appropriate $\lambda \in K$.*

According to Definition 8.3, diagonalizable operators f admit a decomposition of L into a direct sum of its proper subspaces. We shall determine when f has at least one proper subspace.

8.5. Definition. *Let L be a finite-dimensional linear subspace. Let $f : L \to L$ be a linear operator and A its matrix in some basis. We denote by $P(t)$ the polynomial* $\det(tE - A)$ *with coefficients in the field K* (det *denotes the determinant) and call it the characteristic polynomial of the operator f and of the matrix A.*

8.6. Theorem. a) *The characteristic polynomial of f does not depend on the choice of basis in which its matrix is represented.*

b) *Any eigenvalue of f is a root of $P(t)$ and any root of $P(t)$ lying in K is an eigenvalue of f, corresponding to some (not necessarily the only) proper subspace of L.*

Proof. a) According to §4.8, the matrix of f in a different basis has the form $B^{-1}AB$. Therefore, using the multiplicativity of the determinant, we find

$$\det(tE - B^{-1}AB) = \det(B^{-1}(tE - A)B) =$$

$$= (\det B)^{-1} \det(tE - A) \det B = \det(tE - A).$$

We note that $P(t) = t^n - \operatorname{Tr} f \cdot t^{n-1} + \ldots + (-1)^n \det f$ (the notation of §4.9).

b) Let $\lambda \in K$ be a root of $P(t)$. Then the mapping $\lambda \cdot \mathrm{id} - f$ is represented by a singular matrix and its kernel is therefore non-trivial. Let $l \neq 0$ be an element from the kernel. Then $f(l) = \lambda l$ so that λ is an eigenvalue of f and l is the corresponding eigenvector. Conversely, if $f(l) = \lambda l$, then l lies in the kernel of $\lambda \cdot \mathrm{id} - f$ so that $\det(\lambda \cdot \mathrm{id} - f) = P(\lambda) = 0$.

8.7. We now see that the operator f, in general, does not have eigenvalues and is therefore not diagonalizable if its characteristic polynomial $P(t)$ does not have roots in the field K. This is entirely possible over fields that are not algebraically closed, such as $\mathbf{R}$ and finite fields. For example, let the elements of the matrix $A = \begin{pmatrix} a & b \\ c & d \end{pmatrix}$ be real. Then

$$\det(tE - A) = t^2 - (a + d)t + (ad - bc),$$

and if $(a + d)^2 - 4(ad - bc) = (a - d)^2 + 4bc < 0$, A is not diagonalizable.

We have thus encountered here for the first time a case when the properties of linear mappings depend significantly on the properties of the field.

In order to be able to ignore the last point as long as possible, in the next section §9 we shall assume that the field K is algebraically closed. The reader who is not familiar with other algebraically closed fields (with the exception of **C**) may assume everywhere that $K = \mathbf{C}$. The algebraic closure of K is equivalent to either of the following two conditions: a) any polynomial of a single variable (of degree ≥ 1) $P(t)$ with coefficients in K has a root $\lambda \in K$; b) any such polynomial $P(t)$ with leading coefficient equal to unity, can be represented in the form $\prod_{i=1}^{n}(t-\lambda_i)^{r_i}$, where $a, \lambda_i \in K$; $\lambda_i \neq \lambda_j$ for $i \neq j$; this representation is unique if $P(t) \not\equiv 0$. In this case, the number r_i is called the multiplicity of the root λ_i of the polynomial $P(t)$. *The set of all roots of the characteristic polynomial is called the spectrum of the operator f.* If all multiplicities are equal to one, the spectrum of f is said to be *simple.*

If the field K is algebraically closed, then according to Theorem 8.6 any linear operator $f : L \to L$ has a proper subspace. However, it may nevertheless happen that it is non-diagonalizable, because the sum of all proper subspaces may happen to be less than L, whereas for a diagonalizable operator it always equals L. Before considering the general case, we shall examine complex 2×2 matrices.

8.8. Example. Let L be a two-dimensional complex space with a basis. In this basis the operator $f : L \to L$ is represented by the matrix $A = \begin{pmatrix} a & b \\ c & d \end{pmatrix}$. The characteristic polynomial of f is $t^2 - (a+d)t + (ad - bc)$, and its roots are $\lambda_{1,2} = -\frac{a+d}{2} \pm \sqrt{\frac{(a-d)^2}{4} + bc}$. We shall examine separately the following two cases:

a) $\lambda_1 \neq \lambda_2$. Let e_1 and e_2 be the characteristic vectors corresponding to λ_1 and λ_2 respectively. They are linearly independent, because if $ae_1 + be_2 = 0$, then

$$f(ae_1 + be_2) = a\lambda_1 e_1 + b\lambda_2 e_2 = 0,$$

whence $\lambda_1(ae_1+be_2)-(a\lambda_1 e_1+b\lambda_2 e_2) = b(\lambda_1-\lambda_2)e_2 = 0$, i.e., $b = 0$ and analogously $a = 0$. Therefore, in the basis $\{e_1, e_2\}$ the matrix of f is diagonal.

b) $\lambda_1 = \lambda_2 = \lambda$. Here the operator f is diagonalizable, only if it multiplies by λ all vectors from L. Hence $\begin{pmatrix} a & b \\ c & d \end{pmatrix} = \begin{pmatrix} \lambda & 0 \\ 0 & \lambda \end{pmatrix}$, i.e. $a = d = \lambda$, $b = c = 0$. If on the other hand, these conditions are not satisfied and only the weaker condition $(a-d)^2 + 4bc = 0$ holds, guaranteeing that $\lambda_1 = \lambda_2$, then f can have only one eigenvector and f is obviously not diagonalizable.

An example of such a matrix is $\begin{pmatrix} \lambda & 1 \\ 0 & \lambda \end{pmatrix}$. This matrix is called a *Jordan block with the dimension* 2×2 *(or rank 2)* .

In §9 we shall show that these matrices are the "building bloc..." for the normal form of a general linear operator over an algebraically closed field. We give the following general definition:

8.9. Definition. a) *A matrix of the form*

$$J_r(\lambda) = \begin{pmatrix} \lambda & 1 & 0 & \dots & 0 \\ 0 & \lambda & 1 & \dots & 0 \\ \dots & \dots & \dots & \dots & \dots \\ 0 & 0 & 0 & \dots & \lambda \end{pmatrix}$$

is called an $r \times r$ Jordan block $J_r(\lambda)$ with the eigenvalue λ.

b) *A Jordan matrix is a matrix consisting of diagonal blocks $J_{r_i}(\lambda_i)$ with zeros outside these blocks:*

$$J = \left(\begin{array}{c|c|c} J_{r_1}(\lambda_1) & 0 & \dots \\ \hline 0 & J_{r_2}(\lambda_2) & \dots \\ \hline \dots & \dots & \dots \end{array}\right).$$

c) *A Jordan basis for the operator $f : L \to L$ is a basis of the space L in which the matrix of f is a Jordan matrix or, as it is customarily said, has the Jordan normal form.*

d) *The solution of a matrix equation of the form $X^{-1}AX = J$, where A is a square matrix, X is an unknown non-singular matrix, and J is an unknown Jordan matrix, is called the reduction of A to Jordan normal form.*

8.10. Example. Let $L_n(\lambda)$ be a linear space of complex functions of the form $e^{\lambda x} f(x)$, where $\lambda \in C$ and $f(x)$ runs through the polynomials of degree $\leq n-1$. Since $\frac{d}{dx}(e^{\lambda x} f(x)) = e^{\lambda x}(\lambda f(x) + f'(x))$, the derivative $\frac{d}{dx}$ is a linear operator in this space. We set $e_{i+1} = \frac{x^i}{i!} e^{\lambda x}$ (recall that $0! = 1$), $i = 0, \dots, n-1$. Obviously,

$$\frac{d}{dx}(e_{i+1}) = \frac{x^{i-1}}{(i-1)!} e^{\lambda x} + \lambda \frac{x^i}{i!} e^{\lambda x} = e_i + \lambda e_{i+1}$$

(the first term is absent for $i = 0$). Therefore,

$$\frac{d}{dx}(e_1, \dots, e_n) = (e_1, \dots, e_n) \begin{pmatrix} \lambda & 1 & 0 & \dots & 0 \\ 0 & \lambda & 1 & \dots & 0 \\ \dots & \dots & \dots & \dots & \dots \\ \dots & \dots & \dots & \dots & \lambda \end{pmatrix}.$$

Thus the functions $\left(\frac{x^i}{i!} e^{\lambda x}\right)$ form a Jordan basis for the operator $\frac{d}{dx}$ in our space.

This example demonstrates the special role of Jordan matrices in the theory of linear differential equations (see Exercises 1–3 in §9).

8.11. Aside from the geometric considerations examined above, in the next chapter we shall need algebraic information about polynomial functions of operators. Let $f : L \to L$ be a fixed operator.

a) For any polynomial $\sum_{i=0}^{n} a_i t^i = Q(t)$ with coefficients from the field K the expression $\sum_{i=0}^{n} a_i f^i$ makes sense in the ring $\mathcal{L}(L,L)$ of endomorphisms of the space L; we shall denote it by $Q(f)$.

b) We shall say that the polynomial $Q(t)$ *annihilates* the operator f, if $Q(f) = 0$. Non-zero polynomials that annihilate f always exist if L is finite-dimensional. Indeed, if $\dim L = n$, then $\dim \mathcal{L}(L,L) = n^2$ and the operators id, $f, \ldots, f^{n^2}$ are linearly dependent over K. This discussion shows that there exists a polynomial of degree $\leq n^2$ that annihilates f. In reality the Cayley-Hamilton theorem, which we shall prove below, establishes the existence of an annihilating polynomial of degree n.

c) Consider a polynomial $M(t)$ whose leading coefficient equals 1, which approximates f, and which has the lowest possible degree. It is called the *minimal polynomial* of f. Obviously, it is uniquely defined: if $M_1(t)$ and $M_2(t)$ are two such polynomials, then $M_1(t) - M_2(t)$ annihilates f and has a strictly lower degree, so that $M_1(t) - M_2(t) = 0$.

d) We shall show that any polynomial that annihilates f can be decomposed into the minimal polynomials of f. Indeed, let $Q(f) = 0$. We decompose Q with a remainder on M: $Q(t) = X(t)M(t) + R(t)$, $\deg R(t) < \deg M(t)$. Then $R(f) = Q(f) - X(f)M(f) = 0$, so that $R = 0$.

8.12. Cayley-Hamilton theorem. *The characteristic polynomial $P(t)$ of an operator f annihilates f.*

Proof. We shall make use of this theorem and we shall prove it only for the case of an algebraically closed field K though it is also true without this restriction.

We perform induction on $\dim L$. If L is one-dimensional, then f is a multiplication by a scalar λ, $P(t) = t - \lambda$ and $P(f) = 0$.

Let $\dim L = n \geq 2$ and suppose that the theorem is proved for spaces with dimension $n-1$. We select an eigenvalue λ of the operator f and a one-dimensional proper subspace $L_1 \subset L$ corresponding to λ. Let $\{e_1\}$ be a basis of L_1; we extend it to the basis $\{e_1, \ldots, e_n\}$ in the space L. The matrix f in this base has the form

$$\left(\begin{array}{c|c} \lambda & * \ldots * \\ \hline 0 & A \end{array}\right).$$

Therefore $P(t) = (t-\lambda)\det(tE - A)$. The operator f determines the linear mapping $\bar{f} : L/L_1 \to L/L_1$, $\bar{f}(l + L_1) = f(l) + L_1$. The vectors $\bar{e}_i = e_i + L_1 \in L/L_1$, $i \geq 2$ form a basis of L/L_1, and the matrix of $\bar{f}$ in this basis equals A. Therefore, $\bar{P}(t) =$ $= \det(tE - A)$ is the characteristic polynomial of $\bar{f}$ and, according to the induction hypothesis, $\bar{P}(\bar{f}) = 0$. Hence, $\bar{P}(f)l \in L_1$ for any vector $l \in L$. Therefore

$$P(f)l = (f - \lambda)\bar{P}(f)l = 0,$$

because $f - \lambda$ reduces to zero any vector in L_1. This completes the proof.

8.13. Examples. a) Let $f = \mathrm{id}_L$ and $\dim L = n$. Then the characteristic polynomial of f is $(t-1)^n$ and the minimal polynomial is $(t-1)$, so that they are not equal for $n > 1$.

b) Let f be represented by the Jordan block $J_r(\lambda)$. The characteristic polynomial of f is $(t-\lambda)^r$. To calculate the minimal polynomial we note that $J_r(\lambda) = \lambda E_r + J_r(0)$. Furthermore,

$$J_k(0)^k = \begin{pmatrix} 0 & 0 & \dots & 1 & 0 & \dots & 0 \\ 0 & 0 & \dots & 0 & 1 & \dots & 0 \\ \dots & \dots & \dots & \dots & \dots & \dots & \dots \\ 0 & 0 & \dots & 0 & 0 & \dots & 0 \end{pmatrix}$$

where ones stand along the kth diagonal above the principal diagonal; $J_r(0)^k = 0$ for $k \geq r$. On the other hand, λE_r and $J_r(0)$ commute, so that

$$(J_r(\lambda) - \lambda E_r)^k = J_r(0)^k$$

for $0 \leq k \leq r-1$, and since the minimal polynomial is a divisor of the characteristic polynomial, this proves that they are equal.

EXERCISES

1. Let $f : L \to L$ be a diagonalizable operator with a simple spectrum.

   a) Prove that any operator $g : L \to L$ such that $gf = fg$ can be represented in the form of a polynomial of f.

   b) Prove that the dimension of the space of such operators g equals $\dim L$.

   Are these assertions true if the spectrum of f is not simple ?

2. Let $f, g : L \to L$ be linear operators in an n-dimensional space over a field with characteristic zero. Assume that $f^n = 0$, $\dim\ker f = 1$, and $gf - fg = f$. Prove that the eigenvalues of g have the form $a, a-1, a-2, \dots, a-(n-1)$ for some $a \in K$.

§9. The Jordan Normal Form

The main goal of this section is to prove the following theorem on the existence and uniqueness of the Jordan normal form for matrices and linear operators.

9.1. Theorem. *Let K be an algebraically closed field, L a finite-dimensional linear space over K, and $f : L \to L$ a linear operator. Then:*

a) *A Jordan basis exists for the operator f, i.e., the matrix of the operator A in the original basis can be reduced by a change of basis X to the Jordan form $X^{-1}AX = J$.*

b) *The matrix J is unique, apart from a permutation of its constituent Jordan blocks.*

9.2. The proof of the theorem is divided into a series of intermediate steps. We begin by constructing the direct composition $L = \bigoplus_{i=1}^{n} L_i$, where the L_i are invariant subspaces for f, which will later correspond to the *set of Jordan blocks for f with the same number λ along the diagonal.* In order to characterize these subspaces in an invariant manner, we recall that

$$(J_r(\lambda) - \lambda E_r)^n = 0.$$

An operator which when raised to some power is equal to zero is said to be *nilpotent*. Thus the operator $f - \lambda$ is nilpotent on the subspace corresponding to the block $J_r(\lambda)$. The same is true for its restriction to the sum of subspaces for fixed λ. This motivates the following definition.

9.3. Definition. *The vector $l \in L$ is called a root vector of the operator f, corresponding to $\lambda \in K$, if there exists an r such that $(f - \lambda)^r l = 0$)* (here $f - \lambda$ denotes the operator $f - \lambda$ id).

All eigenvectors are evidently root vectors.

9.4. Proposition. *We denote by $L(\lambda)$ the set of root vectors of the operator f in L corresponding to λ. Then $L(\lambda)$ is a linear subspace in L and $L(\lambda) \neq \{0\}$ if and only if λ is an eigenvalue of f.*

Proof. Let $(f - \lambda)^{r_1} l_1 = (f - \lambda)^{r_2} l_2 = 0$. Setting $r = \max(r_1, r_2)$, we find that $(f - \lambda)^r (l_1 + l_2) = 0$ and $(f - \lambda)^{r_1}(al_1) = 0$. Therefore, $L(\lambda)$ is a linear subspace.

If λ is an eigenvalue of f, then there exists an eigenvector corresponding to λ such that $L\lambda \neq \{0\}$. Conversely, let $l \in L(\lambda)$, $l \neq 0$. We select the *smallest* value of r for which $(f - \lambda)^r l = 0$. Obviously, $r \geq 1$. The vector $l' = (f - \lambda)^{r-1} l$ is an eigenvector of f with eigenvalue $\lambda : l' \neq 0$ according to the choice of r and $(f - \lambda)l' = 0$, whence $f(l') = \lambda l'$.

9.5. Proposition. *$L = \bigoplus L(\lambda_i)$, where λ_i runs through all the eigenvalues of the operator f, i.e., the different roots of the characteristic polynomial of f.*

Proof. Let $P(t) = \prod_{i=1}^{s} (t - \lambda_i)^{r_i}$ be the characteristic polynomial of $f, \lambda_i \neq \lambda_j$ for $i \neq j$. Set $F_i(t) = P(t)(t - \lambda_i)^{-r_i}$, $f_i = F_i(f)$, $L_i = \operatorname{im} f_i$. We check the fol' ,wing series of assertions.

a) $f - \lambda_i^{r_i} L_i = \{0\}$, that is, $L_i \subset L(\lambda_i)$. Indeed,

$$(f - \lambda_i)^{r_i} f_i = (f - \lambda_i)^{r_i} F_i(f) = P(f) = 0$$

according to the Cayley-Hamilton theorem.

b) $L = L_1 + \ldots + L_s$. Indeed, since the polynomials $F_i(t)$ in aggregate are relatively prime, there exist polynomials $X_i(t)$ such that $\sum_{i=1}^{s} F_i(t)X_i(t) = 1$. Therefore, substituting f for t, we have

$$\sum_{i=1}^{s} F_i(f)X_i(f) = \mathrm{id}.$$

Applying this identity to any vector $l \in L$, we find

$$l = \sum_{i=1}^{s} f_i(X_i)(f)l) \in \sum_{i=1}^{s} L_i.$$

c) $L = L_1 \oplus \ldots \oplus L_s$. Indeed, we choose $1 \leq i \leq s$ and verify that $L_i \cap \cap \left(\sum_{j \neq i} L_j\right) = \{0\}$. Let l be a vector from this intersection. Then

$$(f - \lambda_i)^{r_i} l = 0, \text{ since } l \in L_i;$$

$$F_i(f)l = \prod_{j \neq i} (f - \lambda_j)^{r_j} l = 0, \text{ since } l \in \sum_{j \neq i} L_j.$$

Since $(t - \lambda_i)^{r_i}$ and $F_i(t)$ are relatively prime polynomials, there exist polynomials $X(t)$ and $Y(t)$ such that $X(t)(t - \lambda_i)^{r_i} + Y(t) \times F_i(t) = 1$. Substituting here f for t and applying the operator identity obtained to l, we obtain $X(f)(0) + Y(f)(0) = = l = 0$.

d) $L_i = L(\lambda_i)$. Indeed we have already verified that $L_i \subset L(\lambda_i)$. To prove the converse we choose a vector $l \in L(\lambda_i)$ and represent it in the form $l = l' + l''$, $l' \in L_i$, $l'' \in \bigoplus_{j \neq i} L_i$. There exists a number r' such that $(f - \lambda_i)^{r'} l'' = 0$, because $l'' = l - l' \in L(\lambda_i)$. In addition, $F_i(f)l'' = 0$. Writing the identity $X(t)(t - \lambda_i)^{r'} + +Y(t)F_i(t) = 1$ and replacing t by f, we find that $l'' = 0$, so that $l = l' \in L_i$.

9.6. Corollary. *If the spectrum of an operator f is simple, then f is diagonalizable.*

Proof. Indeed, the number of different eigenvalues of f then equals $n = \deg P(t) = = \dim L$. Hence, in the decomposition $L = \oplus_{i=1}^{n} L(\lambda_i)$ all spaces $L(\lambda_i)$ are one-dimensional and since each of them contains an eigenvector, the operator f becomes diagonal in a basis consisting of these vectors.

We now *fix* one of the eigenvalues λ and prove that the restriction of f to $L(\lambda)$ has a Jordan basis, corresponding to this value of λ. To avoid introducing a new notation *we shall assume up to the end of* §9.7 *that f has only one eigenvalue λ and*

$L = L(\lambda)$. Moreover, since any Jordan basis for the operator f is simultaneously a Jordan basis for the operator $f + \mu$, where μ is any constant, we can even assume that $\lambda = 0$. Then, according to the Cayley-Hamilton theorem, the operator f is nilpotent: $P(t) = t^n, f^n = 0$. We shall now prove the following proposition.

9.7. Proposition. *A nilpotent operator f on a finite-dimensional space L has a Jordan basis; the matrix of f in this basis is a combination of blocks of the form $J_r(0)$.*

Proof. If we already have a Jordan basis in the space L, it is convenient to represent it by a diagram D, similar to the one shown here.

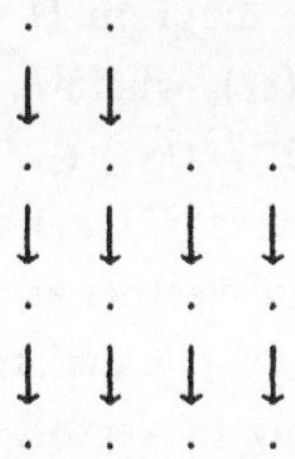

In this diagram, the dots denote elements of the basis and the arrows describe the action of f (in the general case, the action of $f - \lambda$). The operator f transforms to zero the elements in the lowest row, that is, the eigenvectors of f entering into the basis occur in this row. Each column thus stands for a basis of the invariant subspace, corresponding to one Jordan block, whose dimension equals the height of this column (the number of points in it): if

$$f(e_h) + e_{h-1},\ f(e_{h-1}) = e_{h-2}, \dots,\ f(e_1) = 0,$$

then

$$f(e_1 \dots, e_h) = (e_1, \dots, e_h) \begin{pmatrix} 0 & 1 & 0 & \dots & 0 \\ 0 & 0 & 1 & \dots & 0 \\ \dots & \dots & \dots & \dots & \dots \\ 0 & 0 & 0 & \dots & 0 \end{pmatrix}.$$

Conversely, if we find a basis of L whose elements are transformed by f into other elements of the basis or into zero, so that the elements of this basis together with the action of f can be described by such a diagram, then it will be the Jordan basis for L.

We shall prove existence by induction on the dimension of L. If $\dim L = 1$, then the nilpotent operator f is a zero operator and any non-zero vector in L forms its Jordan basis. Now, let $\dim L = n > 1$ and assume that the existence of a Jordan basis has already been proved for dimensions less than n. We denote by

$L_0 \subset L$ the subspace of eigenvectors for f, that is, $\ker f$. Since $\dim L_0 > 0$, we have $\dim L/L_0 < n$, while the operator $f : L \to L$ induces the operator

$$\bar{f} : L/L_0 \to L/L_0 : \bar{f}(l + L_0) = f(l) + L_0.$$

(The correctness of the definition of f and its linearity are obvious.)

According to the induction hypothesis, $\bar{f}$ has a Jordan basis. We can assume that it is non-empty. Otherwise $L = L_0$ and any basis of L_0 will be a Jordan basis for $\bar{f}$. Let us construct the diagram $\bar{D}$ for elements of the Jordan basis of $\bar{f}$. We take the uppermost vector $\bar{e}_i$, $i = 1, \ldots, m$ in each column, and set $\bar{e}_i = e_i + L_0$, $e_i \in L$. We shall now construct the diagram D of vectors of the space L as follows. For $i = 1, \ldots, m$ the ith column in the diagram D will consist (top to bottom) of the vectors $e_i, f(e_i), \ldots, f^{h_i-1}(e_i), f^{h_i}(e_i)$, where h_i is the height of the ith column in the diagram $\bar{D}$. Since $\bar{f}^{h_i}(\bar{e}_i) = 0$, $f^{h_i}(e_i) \in L_0$ and $f^{h_i+1}(e_i) = 0$. We select a basis of the linear span of the vectors $f^{h_1}(e_1), \ldots, f^{h_m}(e_m)$ in L_0, extend it to a basis of L_0, and insert the additional vectors as additional columns (of unit height) in the bottom row of the diagram D; f transforms them into zero.

Thus the diagram D consisting of vectors of the space L together with the action of f on its elements has exactly the form required for a Jordan basis. We have only to check that the elements of D actually form a basis of L.

We shall first show that the linear span of D equals L. Let $l \in L, \bar{l} = l + L_0$. By assumption, $\bar{l} = \sum_{i=1}^{m} \sum_{j=0}^{h_i-1} a_{ij} \bar{f}^j(\bar{e}_i)$. Since L_0 is invariant under f, it follows that

$$l - \sum_{i=1}^{m} \sum_{j=0}^{h_i-1} a_{ij} f^j(e_i) \in L_0.$$

But all the vectors $f^j(e_i)$, $j \le h_i - 1$ lie in the rows of the diagram D, beginning with the second from the bottom, and the subspace L_0 is generated by the elements of the first row of D by construction. Therefore l can be represented as a linear combination of the elements of D.

It remains to verify that the elements of D are linearly independent. First of all, the elements in the bottom row of D are linearly independent. Indeed, if some non-trivial linear combination of them equals zero, then it must have the form $\sum_{i=1}^{m} a_i f^{h_i}(e_i) = 0$, because the remaining elements of the bottom row extend the basis of the linear span of $\{f^{h_1}(e_1), \ldots, f^{h_m}(e_m)\}$ up to a basis of L_0. But all the $h_i \ge 1$, therefore

$$f\left(\sum_{i=1}^{m} a_i f^{h_i-1}(e_i)\right) = 0,$$

so that

$$\sum_{i=1}^{m} a_i f^{h_i-1}(e_i) \in L_0 \quad \text{and} \quad \sum_{i=1}^{m} a_i \bar{f}^{h_i-1}(\bar{e}_i) = 0.$$

It follows from the last relation that all the $a_i = 0$, because the vectors $\bar{f}^{h_i-1}(\bar{e}_i)$ comprise the bottom row of the diagram $\bar{D}$ and are part of a basis of L/L_0.

Finally, we shall show that if there exists a non-trivial linear combination of the vectors of D equal to zero, then it is possible to obtain from it a non-trivial linear dependence between the vectors in the bottom row of D. Indeed, consider the top row of D, which contains the non-zero coefficients of this imagined linear combination. Let the number of this row (counting from the bottom) be h. We apply to this combination the operator f^{h-1}. Evidently, the part of this combination corresponding to the hth row will transform into a non-trivial linear combination of elements of the bottom row, while the remaining terms will vanish. This completes the proof of the proposition.

Now we have only to verify the part of Theorem 9.1 that refers to uniqueness.

9.8. Let an arbitrary Jordan basis of the operator f be fixed. Any diagonal element of the matrix f in this basis is obviously one of the eigenvalues λ of this operator. Examine the part of the basis corresponding to all of the blocks of matrices with this value of λ and denote by L_λ its linear span. Since $(J_r(\lambda) - \lambda)^r = 0$, we have $L_\lambda \subset L(\lambda)$, where $L(\lambda)$ is the root space of L. In addition, $L = \bigoplus L_{\lambda_i}$ by definition of the Jordan basis and $L = \bigoplus L(\lambda_i)$ by Proposition 9.5, where in both cases λ_i runs through all eigenvalues of f once. Therefore, $\dim L_{\lambda_i} = \dim L(\lambda_i)$ and $L_{\lambda_i} = L(\lambda_i)$. Hence the sum of the dimensions of the Jordan blocks, corresponding to each λ_i, is independent of the choice of Jordan basis and, moreover, the linear spans of the corresponding subsets of the basis L_{λ_i} are basis-independent. It is thus sufficient to check the uniqueness theorem for the case $L = L(\lambda)$ or even for $L = L(0)$.

We construct the diagram D corresponding to a given Jordan basis of $L = L(0)$. The dimensions of the Jordan blocks are the heights of its columns; if the columns in the diagram are arranged in decreasing order, these heights are uniquely determined if the lengths of the rows in the diagram are known, beginning with the bottom row, in decreasing order. We shall show that the length of the bottom row equals the dimension of $L_0 = \ker f$. Indeed, we take any eigenvector l for f and represent it as a linear combination of the elements of D. All vectors lying above the bottom row will appear in this linear combination with zero coefficients. Indeed, if the highest vectors with non-zero coefficients were to lie in a row with number $h \geq 2$, then the vector $f^{h-1}(l) = 0$ would be a non-trivial linear combination of the elements of the bottom row of D (cf. the end of the proof of Proposition 9.7), and this contradicts the linear independence of the elements of D. This means that the bottom row of D forms a basis of L_0, so that its length equals $\dim L_0$; hence, this length is the same for all Jordan bases. In exactly the same way, the length of the second row does not depend on the choice of basis, so that, in the notation used in this section, it equals the dimension of $\ker \bar{f}$ in L/L_0. This completes the proof of uniqueness and of Theorem 9.1.

9.9. Remarks. a) Let the operator f be represented by the matrix A in some basis. Then the problem of reducing A to Jordan form can be solved as follows.

Calculate the characteristic polynomial of A and its roots.

Calculate the dimensions of the Jordan blocks, corresponding to the roots λ. For this, it is sufficient to calculate the lengths of the rows of the corresponding diagrams, that is, $\dim\ker(A-\lambda)$, $\dim\ker(A-\lambda)^2 - \dim\ker(A-\lambda)$, $\dim\ker(A-\lambda)^3 - \dim\ker(A-\lambda)^2, \ldots$.

Construct the Jordan form J of the matrix A and solve the matrix equation $AX - XJ = 0$. The space of solutions of this linear system of equations will, generally speaking, be multidimensional, and the solutions will also include singular matrices. But according to the existence theorem, non-singular solutions necessarily exist; any one can be chosen.

b) One of the most important applications of the Jordan form is for the calculation of functions of a matrix (thus far we have considered only polynomial functions). Assume, for example, that we must find a large power A^N of the matrix A. Since the degree of the Jordan matrix is easy to calculate (see §8.13), an efficient method is to use the formula $A^N = XJ^NX^{-1}$, where $A = XJX^{-1}$. The point is that the matrix X is calculated once and for all and does not depend on N. The same formula can be used to estimate the growth of the elements of the matrix A^N.

c) It is easy to calculate the minimal polynomial of a matrix A in terms of the Jordan form. Indeed, we shall restrict ourselves for simplicity to the case of a field with zero characteristic. Then the minimal polynomial of $J_r(\lambda)$ equals $(t-\lambda)^r$ (see §8.13), the minimal polynomial of the block matrix $(J_{r_i}(\lambda))$ equals $(t-\lambda)^{\max(r_i)}$, and finally the minimal polynomial of the general Jordan matrix with diagonal elements $\lambda_1, \ldots, \lambda_s$, $(\lambda_i \neq \lambda_j \quad \text{for} \quad i \neq j)$ equals $\prod_{j=1}^{s}(t-\lambda_j)^{r_j}$, where r_j is the smallest dimension of the Jordan block corresponding to λ_j.

9.10. Other normal forms. In this section, we shall briefly describe other normal forms of matrices which are useful, in particular, for algebraically open fields.

a) *Cyclic spaces and cyclic blocks.* The space L is said to be *cyclic* with respect to an operator f, if L contains a vector l, also called a cyclic vector, such that $l, f(l), \ldots, f^{n-1}(l)$ form a basis of L. Setting $e_i = f^{n-i}(l)$, $i = 1, \ldots, n = \dim L$, we have

$$f(e_1, \ldots, e_n) = (e_1, \ldots, e_n)\begin{pmatrix} a_{n-1} & 1 & 0 & \ldots & 0 \\ a_{n-2} & 0 & 1 & \ldots & 0 \\ \ldots & \ldots & \ldots & \ldots & \ldots \\ a_0 & 0 & 0 & \ldots & 0 \end{pmatrix},$$

where $a_i \in K$ are uniquely determined from the relation $f^n(l) = \sum_{i=0}^{n-1} a_i f^i(l)$. The matrix of f in this basis is called a cyclic block. Conversely, if the matrix of f in the basis $(e_1, \ldots, e_n)$ is a cyclic block, then the vector $l = e_n$ is cyclic and $e_i = f^{n-i}(e_n)$ (induction downwards on i).

We shall show that the form of the cyclic block corresponding to f is independent of the choice of the starting cyclic vector. For this, we shall verify that the first column of the block consists of the coefficients of the minimal polynomial of the operator $f : M(t) = t^n - \sum_{i=0}^{n-1} a_i t^i$.

Indeed, $M(f) = 0$ because $M(f)[f^i(l)] = f^i[M(f)l] = 0$, and the vectors $f^i(l)$ generate L. On the other hand, if $N(t)$ is a polynomial of degree $< n$, then $N(f) \neq 0$, because otherwise, applying the operator $N(f) = 0$ to the cyclic vector l, we would obtain a non-trivial linear relation between the vectors of the basis $l, f(l), \ldots, f^{n-1}(l)$.

b) *Criterion for a space to be cyclic.* According to the preceding analysis, if the space L is cyclic with respect to f, then its dimension n equals the degree of the minimal polynomial of f and, consequently, the minimal polynomial coincides with the characteristic polynomial. The converse is also true: if the operators $\mathrm{id}, f, \ldots, f^{n-1}$ are linearly independent, then there exists a vector l such that the vectors $l, f(l), \ldots, f^{n-1}(l)$ are linearly independent, so that L is cyclic. We shall not prove this assertion.

c) *Any operator in an appropriate basis can be reduced to a direct sum of cyclic blocks.* The proof is analogous to the proof of the Jordan form theorem. Instead of the factors $(t - \lambda_i)^{r_i}$ of the characteristic polynomial, we study the factors $p_i(t)^{r_i}$, where the $p_i(t)$ are irreducible (over the field K) divisors of the characteristic polynomial. The uniqueness theorem is also valid here, if we restrict our attention to the case when the minimal polynomials of all cyclic blocks are irreducible. Without this restriction it is not true: a cyclic space can be the direct sum of two cyclic subspaces whose minimal polynomials are relatively prime.

EXERCISES

1. Let L be the finite-dimensional space of differentiable functions of a complex variable x with the property that if $f \in L$, then $\frac{df}{dx} \in L$. Prove that there exist combinations of numbers $\lambda_1, \ldots, \lambda_s$ and integers $r_1, \ldots, r_s \geq 1$ such that $L = \bigoplus L_i$, where L_i is the space of functions of the form $e^{\lambda_i x} P_i(x)$, $P_i(x)$ being an arbitrary polynomial of degree $\leq r_i - 1$. (Hint: examine the Jordan basis for the operator $\frac{d}{dx}$ on L and calculate successively the form of the functions entering into it, beginning with the bottom row of its diagram.)

2. Let $y(x)$ be a function of a complex variable x, satisfying a differential equation of the form

$$\frac{d^n y}{dx^n} + \sum_{i=0}^{n-1} a_i \frac{d^i y}{dx^i} = 0, \ a_i \in \mathbf{C}.$$

We denote by L the linear space of functions spanned by $d^i y/dx^i$ for all $i \geq 0$. Prove that it is finite-dimensional and that the operator d/dx transforms it into itself.

3. Using the results of Exercises 1 and 2, prove that $y(x)$ can be represented in the form $\sum e^{\lambda_i x} P_i(x)$, where P_i are polynomials. How are the numbers λ_i related to the form of the differential equation ?

4. Let $J_r(\lambda)$ be a Jordan block on $\mathbf{C}$. Prove that the matrix obtained by introducing appropriate infinitesimal displacements of its elements will be diagonalizable. (Hint: change the elements along the diagonal, making them pairwise unequal).

5. Extend the results of Exercise 4 to arbitrary matrices on $\mathbf{C}$, using the facts that the coefficients of the characteristic polynomial are continuous functions of the elements of the matrix and that the condition for the polynomial not to have degenerate roots is equivalent to the condition that its discriminant does not vanish.

6. Give a precise meaning to the following assertions and prove them:

a) A general 2×2 matrix over $\mathbf{C}$ is diagonalizable.

b) A general 2×2 matrix with identical characteristic roots is not diagonalizable.

§10. Normed Linear Spaces

In this section we shall study the special properties of linear spaces over real and complex numbers that are associated with the possibility of defining in them the concept of a limit and constructing the foundation of analysis. These properties play a special role in the infinite-dimensional case, so that the material presented is essentially an elementary introduction to functional analysis.

10.1. Definition. *The pair (E, d), where E is a set and $d : E \times E \to \mathbf{R}$ is a real-valued function, is called a metric space, if the following conditions are satisfied for all $x, y, z \in E$:*

a) $d(x,y) = d(y,x)$ *(symmetry).*

b) $d(x,x) = 0$; $d(x,y) > 0$, *if* $x \neq y$ *(positivity);*

c) $d(x,z) \leq d(x,y) + d(y,z)$ *(triangle inequality).*

A function d with these properties is called a metric, and $d(x,y)$ is the distance between the points x and y.

10.2. Examples. a) $E = \mathbf{R}$ or $\mathbf{C}$, $d(x,y) = |x-y|$.

b) $E = \mathbf{R}^n$ or $\mathbf{C}^n$, $d(\vec{x},\vec{y}) = \left(\sum_{i=1}^n |x_i - y_i|^2\right)^{1/2}$. This is the so-called natural metric. In Chapter 2 we shall study it systematically and we shall study its extensions to arbitrary basic fields in the theory of quadratic forms. Examples of other metrics are

$$d_1(\vec{x},\vec{y}) = \max(|x_i - y_i|), \quad d_2(\vec{x},\vec{y}) = \sum_{i=1}^n |x_i - y_i|.$$

c) $E = C(a,b)$, the space of continuous functions on the interval [a,b]. Here are three of the most important metrics:

$$d_1(f,g) = \max_{a \le t \le b} |f(t) - g(t)|,$$

$$d_2(f,g) = \int_a^b |f(t) - g(t)| dt,$$

$$d_3(f,g) = \left(\int_a^b |f(t) - g(t)|^2 dt \right)^{1/2}.$$

(Verify the axioms. The triangle inequality for d_2 in example b) and d_3 in example c) will be proved in Chapter 2.)

d) E is any set, $d(x,y) = 1$ for $x \neq y$. This is one of the *discrete* metrics on E.

(Each metric is associated with some topology on E, and the last metric described above induces the discrete topology.)

10.3. Balls, boundedness and completeness. In a metric space E with metric d the sets

$$B(x_0,r) = \{x \in E | d(x_0,x) < r\},$$
$$\overline{B}(x_0,r) = \{x \in E | d(x_0,x) \le r\},$$
$$S(x_0,r) = \{x \in E | d(x_0,x) = r\}$$

are called, correspondingly, *an open ball, a closed ball*, and *a sphere* with radius r centered at the point x_0. One should not associate with them intuitive representations which are too close to three-dimensional space. For example, in Example §10.2d, all spheres of radius $r \neq 1$ are empty.

A subset $F \subset E$ is said to be *bounded* if it is contained in a ball (of finite radius).

The sequence of points $x_1, x_2, \ldots, x_n, \ldots$ in E *converges to the point* $a \in E$ if $\lim_{n\to\infty} d(x_n,a) = 0$. The sequence is said to be *fundamental* (or a *Cauchy sequence*), if for all $\epsilon > 0$ there exists an $N = N(\epsilon)$ such that $d(x_m, x_n) < \epsilon$ for $m, n > N(\epsilon)$.

A metric space E is said to be *complete* if any Cauchy sequence in it converges. From the completeness of $\mathbf{R}$ and $\mathbf{C}$, proved in analysis, it follows that the spaces $\mathbf{R}^n$ and $\mathbf{C}^n$ with any of the metrics d, d_1, and d_2 in §10.2b are complete.

10.4. Normed linear spaces. Now let L be a linear space over $\mathbf{R}$ or $\mathbf{C}$. Metrics on L which satisfy the following conditions play an especially important role:

a) $d(l_1, l_2) = d(l_1 + l, l_2 + l)$ for any $l, l_1, l_2 \in L$ (invariance with respect to translation);

b) $d(al_1, al_2) = |a| d(l_1, l_2)$ (multiplication by a scalar a increases the distance by a factor of $|a|$).

Let d be such a metric. We shall call the number $d(l, 0)$ the *norm* of the vector l (with respect to d) and we shall denote it by $\|l\|$. The following properties of the norm follow from the axioms of the metric (§10.2) and the conditions a) and b):

$$\|0\| = 0, \|l\| > 0, \quad if \quad l \neq 0;$$

$$\|al\| = |a| \, \|l\| \quad for\ all \quad a \in K,\ l \in L;$$

$$\|l_1 + l_2\| \leq \|l_1\| + \|l_2\| \quad for\ all \quad l_1, l_2 \in L.$$

The first two properties are obvious and the third is verified as follows:

$$\|l_1 + l_2\| = d(l_1 + l_2, 0) = d(l_1, -l_2) \leq d(l_1, 0) + d(0, -l_2) = \|l_1\| + \|l_2\|.$$

A linear space L equipped with a norm function $\| \ \| : L \to \mathbf{R}$, satisfying the three requirements enumerated above, is called a *normed* space.

Conversely, the metric can be reconstructed from the norm: setting $d(l_1, l_2) = \|l_1 - l_2\|$, it is easy to verify the axioms of the metric. For it, $d(l, 0) = \|l\|$.

A complete normed linear space is called a Banach space. The spaces $\mathbf{R}^n$ and $\mathbf{C}^n$ with any norms corresponding to the metrics in §10.2, are Banach spaces.

The general concept of convergence of a sequence in a metric space given in §10.3 can be specialized to the case of normed linear spaces and is called *convergence in the norm.* The linear structure makes it possible to give a stronger definition of the concept of convergence of a series than the convergence of its partial sums in the norm. Namely, the series $\sum_{i=0}^{\infty} l_i$ is said *to converge absolutely* if the series $\sum_{i=1}^{\infty} \|l_i\|$ converges.

10.5. The norm and convexity. It is not difficult to describe all norms on a one-dimensional space L: *any two of them differ from one another by a positive constant factor.* Indeed, let $l \in L$ be a non-zero vector and $\| \ \|_1, \| \ \|_2$ two norms. If $\|l\|_1 = c\|l\|_2$, then $\|al\|_1 = |a| \, \|l\|_1 = c|a| \, \|l\|_2 = c\|al\|_2$ for all $a \in K$

We shall call balls (spheres) with non-zero radius centred at zero with respect to any of the norms in a one-dimensional space *disks* (*circles*). As follows from the preceding discussion, *the set of all disks and circles in L does not depend on the choice of the starting norm.* Instead of giving a norm, one can indicate its unit disk B or unit circle S: S is reconstructed from B as the boundary of B, while B is reconstructed from S as the set of points of the form $\{al \mid l \in S, |a| \le 1\}$. We note that when $K = \mathbf{R}$ disks are segments centred at zero, and circles are pairs of points that are symmetric relative to zero.

To extend this description to spaces of any number of dimensions we shall require the concept of convexity. A subset $E \subset L$ is said to be *convex*, if for any two vectors $l_1, l_2 \in E$ and for any number $0 \le a \le 1$, the vector $al_1 + (1-a)l_2$ lies in E. This agrees with the usual definition of convexity in $\mathbf{R}^2$ and $\mathbf{R}^3$: together with any two points ("tips of the vectors l_1 and l_2"), the set E must contain the entire segment connecting them ("tips of the vectors $al_1 + (1-a)l_2$").

Let $\| \ \|$ be some norm on L. Set $B = \{l \in L \mid \|l\| \le 1\}$, $S = \{l \in L \mid \|l\| = 1\}$. The restriction of $\| \ \|$ to any linear subspace $L_0 \subset L$ induces a norm on L_0. From here it follows that for any *one-dimensional* subspace $L_0 \subset L$ the set $L_0 \cap B$ is a disk in L_0, while the set $L_0 \cap S$ is a circle in the sense of the definition given above. In addition, from the triangle inequality it follows that if $l_1, l_2 \in B$ and $0 \le a \le 1$, then

$$\|al_1 + (1-a)l_2\| \le a\|l_1\| + (1-a)\|l_2\| \le 1,$$

that is, $al_1 + (1-a)l_2 \in B$, so that B is a convex set.

The converse theorem is also valid:

10.6. Theorem. *Let $S \subset L$ be a set satisfying the following two conditions:* a) *The intersection $S \cap L_0$ with any one-dimensional subspace L_0 is a circle.*

b) *The set $B = \{al \mid |a| \le 1, \ l \in S\}$ is convex. Then there exists on L a unique norm $\| \ \|$, for which B is the unit ball, while S is the unit sphere.*

Proof. We denote by $\| \ \| : L \to \mathbf{R}$ a function which in each one-dimensional subspace L_0 is a norm with a unit sphere $S \cap L_0$. It is clear that such a function exists and is unique, and it is only necessary to verify the triangle inequality for it. Let $l_1, l_2 \in L, \|l_1\| = N_1, \ \|l_2\| = N_2, \ N_i \ne 0$. We apply the condition of convexity of B to the vectors $N_1^{-1}l_1$ and $N_2^{-1}l_2 \in S$. We obtain

$$\left\| \frac{N_1}{N_1 + N_2} N_1^{-1} l_1 + \frac{N}{N_1 + N_2} N_2^{-1} l_2 \right\| \le 1,$$

whence

$$\|l_1 + l_2\| \le N_1 + N_2 = \|l_1\| + \|l_2\|.$$

10.7. Theorem. *Any two norms $\| \ \|_1$ and $\| \ \|_2$ on a finite-dimensional space L are equivalent in the sense that there exist positive constants $0 < c \le c'$ with the* *condition*

$$c\|l\|_2 \le \|l\|_1 \le c'\|l\|_2$$

for all $l \in L$. In particular, the topologies, that is concepts of convergence corresponding to any two norms, coincide and all finite-dimensioal normed spaces are Banach spaces.

Proof. Choose a basis in L and examine the natural norm $\|\vec{x}\| = \left(\sum_{i=1}^{n} |x_i|^2\right)^{1/2}$ with respect to the coordinates in this basis. It is sufficient to verify that any norm $\| \ \|_1$ is equivalent to this norm. Its restriction to the unit sphere of the norm $\| \ \|$ is a continuous function of the coordinates $\vec{x}$ which assumes only positive values (continuity follows from the triangle inequality). Therefore, this function is separated from zero by the constant $c > 0$ and is bounded by the constant $c' > 0$ by the Bolzano–Weierstrass theorem (the unit sphere S for $\| \ \|$ is closed and bounded). The inequalities $c \le \|l\|_1 \le c'$ for all $l \in S$, imply the inequality $c\|l\| \le \|l\|_1 \le c'\|l\|$ for all $l \in L$. Since L is complete in the topology corresponding to the norm $\| \ \|$ and the concepts of convergence for equivalent norms coincide, L is complete in any norm.

10.8. The norm of a linear operator. Let L and M be normed linear spaces over one and the same field **R** or **C**.

We shall study the linear mapping $f : L \to M$. It is said to be *bounded*, if there exists a real number $N \ge 0$ such that the inequality $\|f(l)\| \le N\|l\|$ holds for all $l \in L$ (the norm on the left – in M and the norm on the right – in L). We denote by $\mathcal{L}^1(L, M)$ the set of bounded linear operators. For all $f \in \mathcal{L}^1(L, M)$ we denote by $\|f\|$ the lower bound of all N, for which the inequalities $\|f(l)\| \le N\|l\|$, $l \in L$ hold.

10.9. Theorem. a) *$\mathcal{L}^1(L, M)$ is a normed linear space with respect to the function $\|f\|$, which is called the induced norm.*

b) *If L is finite-dimensional, then $\mathcal{L}^1(L, M) = \mathcal{L}(L, M)$, that is, any linear mapping is bounded.*

Proof. a) Let $f, g \in \mathcal{L}^1(L, M)$. If $\|f(l)\| \le N_1\|l\|$ and $\|g(l)\| \le N_2\|l\|$ for all l, then

$$\|(f+g)(l)\| \le (N_1 + N_2)\|l\|, \ \|af(l)\| \le |a|N_1\|l\|.$$

Therefore $f + g$ and af are bounded and, moreover, inserting the lower bounds, we have

$$\|f + g\| \le \|f\| + \|g\|, \ \|af\| = |a| \, \|f\|.$$

If $\|f\| = 0$, then for any $\epsilon > 0$, $\|f(l)\| \le \epsilon\|l\|$. Hence, $\|f(l)\| = 0$, so that $f = 0$.

c) On the unit sphere in L the mapping $l \mapsto \|f(l)\|$ is a continuous function. Since this sphere is bounded and closed, this function is bounded and, moreover, its upper bound is attained. Therefore, $\|f(l)\| \leq N$ on the sphere, so that $\|f(l)\| \leq$ $\leq N\|l\|$ for all $l \in L$.

We have discovered at the same time that

$$\|f\| = \max\{\|f(l)\|,\ l \text{ belongs to the unit sphere} \in L\}.$$

10.10. Examples. a) In a finite-dimensional space L, the sequence of vectors $l_1, \ldots, l_n, \ldots$ converges to the vector l, if and only if in some (and, therefore, in any) basis the sequence of the ith coordinates of the vectors l_j converges to the ith coordinate of the vector l, that is, if $f(l_1), \ldots, f(l_n) \ldots$ converges for any linear functional $f \in L^*$. The last condition can be transferred to an infinite-dimensional space by requiring that $f(l_i)$ converge only for bounded functionals f. This leads, generally speaking, to a *new topology* on L, called the *weak topology*.

b) Let L be the space of real differentiable functions on $[0,1]$ with the norm $\|f\| = \left(\int_0^1 f(t)^2 dt\right)^{1/2}$. Then the operator of multiplication by t is bounded, because $\int_0^1 t^2 f(t)^2 dt \leq \int_0^1 f(t)^2 dt$ but the operator $\frac{d}{dt}$ is not bounded. Indeed, for any integer $n \geq 0$ the function $\sqrt{2n+1}\,t^n$ lies on the unit sphere, while the norm of its derivative equals $n\sqrt{\frac{2n+1}{2n-1}} \to \infty$ as $n \to \infty$.

10.11. Theorem. *Let $L \xrightarrow{f} M \xrightarrow{g} N$ be bounded linear mappings of normed spaces. Then, their composition is bounded and*

$$\|g \circ f\| \leq \|g\|\,\|f\|.$$

Proof. If $\|f(l)\| \leq N_1\|l\|$ and $\|g(m)\| \leq N_2\|m\|$ for all $l \in L$ and $m \in M$, then

$$\|g \circ f(l)\| \leq N_2\|f(l)\| \leq N_2 N_1 \|l\|,$$

whence, inserting the lower bounds, we obtain the assertion.

EXERCISES

1. Calculate the norms in $\mathbf{R}^2$ for which the unit balls are the following sets:
   a) $x^2 + y^2 \leq 1$.
   b) $x^2 + y^2 \leq r^2$.
   c) The square with the vertices $(\pm 1,\ \pm 1)$.
   d) The square with the vertices $(0,\ \pm 1), (\pm 1,\ 0)$.

2. Let $f(x) \geq 0$ be a twice differentiable real function on $[a, b] \subset \mathbf{R}$ and let $f''(x) \leq 0$. Show that the set $\{(x, y) | a \leq x \leq b,\ 0 \leq y \leq f(x)\} \subset \mathbf{R}^2$ is convex.

3. Using the result of Exercise 2, prove that the set $|x|^p + |y|^p \leq 1$ for $p > 1$ in $\mathbf{R}^2$ is a unit ball for some norm. Calculating this norm, prove the Minkowski inequality

$$(|x_1 + y_1|^p + |x_2 + y_2|^p)^{1/p} \leq (|x_1|^p + |x_2|^p)^{1/p} + (|y_1|^p + |y_2|^p)^{1/p}.$$

4. Generalize the results of Exercise 3 to the case $\mathbf{R}^n$.

5. Let B be a unit ball with some norm in L and let B^* be the unit ball with the induced norm in $L^* = \mathcal{L}(L, K)$, $K = \mathbf{R}$ or $\mathbf{C}$. Give an explicit description of B^* and calculate B^* for the norms in Exercises 1 and 3.

§11. Functions of Linear Operators.

11.1. In §§8 and 9 we defined the operators $Q(f)$, where $f : L \to L$ is a linear operator and Q is any polynomial with coefficients from the basic field K. If $K = \mathbf{R}$ or $\mathbf{C}$, the space L is normed, and the operator f is bounded, then $Q(f)$ can be defined for a more general class of functions Q with the help of a limiting process.

We shall restrict our analysis to holomorphic functions Q, defined by power series with a non-zero radius of convergence: $Q(t) = \sum_{i=0}^{\infty} a_i t^i$. We set $Q(f) = \sum_{i=0}^{\infty} a_i f^i$, if this series of operators converges absolutely, that is, if the series $\sum_{i=0}^{\infty} a_i \|f^i\|$ converges. (In the case $\dim L < \infty$, with which we shall primarily be concerned here, $\mathcal{L}^1(L, L) = \mathcal{L}(L, L)$, and the space of all operators is finite-dimensional and a Banach space; see Theorem 10.9b).

11.2. Examples. a) Let F be a nilpotent operator. Then $\|f^i\| = 0$ for sufficiently large i, and the series $Q(f)$ always converges absolutely. Indeed, it equals one of its partial sums.

b) Let $\|f\| < 1$. The series $\sum_{i=0}^{\infty} f^i$ converges absolutely and

$$(\mathrm{id} - f) \sum_{i=0}^{\infty} f^i = \left(\sum_{i=0}^{\infty} f^i \right) (\mathrm{id} - f) = \mathrm{id}.$$

Indeed,

$$(\mathrm{id} - f) \sum_{i=0}^{N} f^i = \mathrm{id} - f^{N+1} = \left(\sum_{i=0}^{N} f^i \right) (\mathrm{id} - f)$$

and passage to the limit $N \to \infty$ gives the required result. In particular, if $\|f\| < 1$, then the operator id$-f$ is invertible.

c) We define the exponential of a bounded operator f by the operator

$$e^f = \exp(f) = \sum_{n=0}^{\infty} \frac{1}{n!} f^n.$$

Since $\|f^n\| \le \|f\|^n$ and the numerical series for the exponential converges uniformly on any bounded set, the function $\exp(f)$ is defined for any bounded operator f and is continuous with respect to f.

For example, the Taylor series $\sum_{i=0}^{\infty} \frac{(\Delta t)^i}{i!} \phi^{(i)}(t)$ for $\phi(t+\Delta t)$ can be formally written in the form $\exp\left(\Delta t \frac{d}{dt}\right)\phi$. In order for this notation to be precisely defined, it is, of course, necessary to choose a space of infinitely differentiable functions ϕ with a norm and check the convergence in the induced norm.

A particular case: $\exp(a\,\mathrm{id}) = e^a \mathrm{id}$ (a is a scalar); $\exp(\mathrm{diag}(a_1, \ldots, a_n)) = \mathrm{diag}(\exp a_1, \ldots, \exp a_n)$.

The basic property of the numerical exponential $e^a e^b = e^{a+b}$, generally speaking, *no longer holds* for exponentials of operators. There is, however, an important particular case, when it does hold:

11.3. Theorem. *If the operators $f, g : L \to L$ commute, that is, $fg = gf$, then* $(\exp f)(\exp g) = \exp(f+g)$.

Proof. Applying the binomial expansion and making use of the possibility of rearranging the terms of an absolutely converging series, we obtain

$$(\exp f)(\exp g) = \left(\sum_{i\ge 0} \frac{1}{i!} f^i\right)\left(\sum_{k\ge 0} \frac{1}{k!} g^k\right) = \sum_{i,k\ge 0} \frac{1}{i!k!} f^i g^k =$$

$$= \sum_{m\ge 0} \sum_{i=0}^{m} \frac{1}{i!(m-i)!} f^i g^{m-i} = \sum_{m\ge 0} \frac{1}{m!} \sum_{i=0}^{m} \frac{m!}{i!(m-i)!} f^i g^{m-i} =$$

$$= \sum_{m\ge 0} \frac{1}{m!} (f+g)^m = \exp(f+g).$$

The commutativity of f and g is used when $(f+g)^m$ is expanded in the binomial expansion.

11.4. Corollary. *Let $f : L \to L$ be a bounded operator. Then the mapping $\mathbf{R} \to \mathcal{L}^1(L, L) : t \mapsto \exp(tf)$ is a homomorphism from the group $\mathbf{R}$ to the subgroup of invertible operators $\mathcal{L}^1(L, L)$ with respect to multiplication.*

The set of operators $\{\exp tf | t \in R\}$ is called a *one-parameter subgroup of operators.*

11.5. Spectrum. Let f be an operator in a finite-dimensional space and let $Q(t)$ be a power series such that $Q(f)$ converges absolutely. It is easy to see that if $Q(t)$ is a polynomial, then in the Jordan basis for f the matrix $Q(f)$ is upper triangular and the numbers $Q(\lambda_i)$, where λ_i are the eigenvalues of f, lie along its diagonal. Applying this argument to the partial sums of Q and passing to the limit, we find that this is also true for any series $Q(t)$. In particular, if $S(f)$ is the spectrum of f, then $S(Q(f)) = Q(S(f)) = \{Q(\lambda)|\lambda \in S(f)\}$. Furthermore, if the multiplicities of the characteristic roots λ_i are taken into account, then $Q(S(f))$ will be the spectrum of $Q(f)$ with the correct multiplicities. In particular,

$$\det(\exp f) = \prod_{i=1}^{n} \exp \lambda_i = \exp\left(\sum_{i=1}^{n} \lambda_i\right) = \exp \operatorname{Tr} f.$$

Changing over to the language of matrices, we note two other simple properties, which can be proved in a similar manner:

a) $Q(A^t) = Q(A)^t$;

b) $Q(\bar{A}) = \overline{Q(A)}$, where the overbar denotes complex conjugation; it is assumed here that the coefficients in the series for Q are real.

Using these properties and the notation of §4, we prove the following theorem, which concerns the theory of classical Lie groups (here $K = \mathbf{R}$ or $\mathbf{C}$).

11.6. Theorem. *The mapping* exp *maps* $gl(n,K)$, $sl(n,K)$, $o(n,K)$, $u(n)$, *and* $su(n)$ *into* $Gl(n,K)$, $Sl(n,K)$, $SO(n,K)$, $U(n)$, *and* $SU(n)$ *respectively.*

Proof. The space $gl(n,K)$ is mapped into $GL(n,K)$ because according to Corollary 11.4 the matrices $\exp A$ are invertible. If $\operatorname{Tr} A = 0$, then $\det \exp A = 1$, as was proved in the preceding item. The condition $A + A^t = 0$ implies that $(\exp A)(\exp A)^t = 1$, and the condition $A + \bar{A}^t = 0$ implies that $\exp A \overline{(\exp A)}^t = 1$. This completes the proof.

11.7. Remark. In all cases, the image of exp covers some neighbourhood of unity in the corresponding group. For the proof, we can define the logarithm of the operators f with the condition $\|f - \mathrm{id}\| < 1$ by the usual formula $\log f = \sum_{n\geq 0}(-1)^n \frac{(f-\mathrm{id})^n}{n}$ and show that $f = \exp(\log f)$.

On the whole, however, the mappings exp are not, generally speaking, surjective. For example, there does not exist a matrix $A \in sl(2,\mathbf{C})$ for which $\exp A = \begin{pmatrix} -1 & 1 \\ 0 & -1 \end{pmatrix} \in SL(2,\mathbf{C})$. Indeed, A cannot be diagonalizable, because otherwise $\exp A$ would be diagonalizable. Hence, all eigenvalues of A are equal, and since the trace of A equals zero, these eigenvalues must be zero eigenvalues. But, then the eigenvalues of $\exp A$ equal 1, whereas the eigenvalues of $\begin{pmatrix} -1 & 1 \\ 0 & -1 \end{pmatrix}$ equal -1.

§12. Complexification and Decomplexification

12.1. In §§8 and 9, we showed that the study of an algebraically closed field elucidates the geometric structure of linear operators and provides a convenient canonical form for matrices. Therefore, even when working with a real field, it is sometimes convenient to make use of complex numbers. In this section we shall study two basic operations: extension and restriction of the field of scalars in application to linear spaces and linear mappings. We shall work mostly with the transformation from **R** to **C** (*complexification*) and from **C** to **R** (*decomplexification*), and we shall briefly discuss a more general case.

12.2. Decomplexification. Let L be a linear space over **C**. Let us ignore the possibility of multiplying vectors in L by all complex numbers, and retain only multiplication over **R**. Then we obviously obtain a linear space over **R**, which we denote by $L_{\mathbf{R}}$; we shall call this space the decomplexification of L.

Let L and M be two linear spaces over **C**, and let $f : L \to M$ be a linear mapping. Regarded as a mapping of $L_{\mathbf{R}} \to M_{\mathbf{R}}$ it obviously remains linear. We shall denote it by $f_{\mathbf{R}}$ and call it the decomplexification of f. It is clear that

$$\mathrm{id}_{\mathbf{R}} = \mathrm{id},\ (fg)_{\mathbf{R}} = f_{\mathbf{R}} g_{\mathbf{R}};\ (af + bg)_{\mathbf{R}} = af_{\mathbf{R}} + bg_{\mathbf{R}},\ \text{ if } a, b \in \mathbf{R}.$$

12.3. Theorem. a) *Let $\{e_1, \ldots, e_m\}$ be a basis of a space L over* **C**. *Then $\{e_1, \ldots, e_m, ie_1, \ldots, ie_m\}$ is a basis of the space $L_{\mathbf{R}}$ over* **R**. *In particular, $\dim_{\mathbf{R}} L_{\mathbf{R}} = 2 \dim_{\mathbf{C}} L$.*

b) Let $A = B + iC$ be the matrix of the linear mapping $f : L \to M$ in the bases $\{e_1, \ldots, e_m\}$ and $\{e'_1, \ldots, e'_n\}$ over **C**, where B and C are real matrices. Then the matrix of the linear mapping $f_{\mathbf{R}} : L_{\mathbf{R}} \to M_{\mathbf{R}}$ in the bases $\{e_1, \ldots, e_m, ie_1, \ldots, ie_m\}$, $\{e'_1, \ldots, e'_n, ie'_1, \ldots, ie'_n\}$ will be given by

$$\begin{pmatrix} B & -C \\ C & B \end{pmatrix}.$$

Proof. a) For any element $l \in L$ we have

$$l = \sum_{k=1}^{m} a_k e_k = \sum_{k=1}^{m} (b_k + ic_k) e_k = \sum_{k=1}^{m} b_k e_k + \sum_{k=1}^{m} c_k (ie_k),$$

where b_k, c_k are the real and imaginary parts of a_k. Therefore, $\{e_k, ie_k\}$ generate $L_{\mathbf{R}}$. If $\sum_{k=1}^{m} b_k e_k + \sum_{k=1}^{m} c_k(ie_k) = 0$, where $b_k, c_k \in \mathbf{R}$, then $b_k + ic_k = 0$ by virtue of the linear independence of $\{e_1, \ldots, e_k\}$ over **C**, whence it follows that $b_k = c_k = 0$ for all k.

b) The definition of A implies that

$$f(e_1, \ldots, e_m) = (e'_1, \ldots, e'_n)(B + iC),$$

whence, because of the linearity of f over $\mathbf{C}$,

$$f(ie_1, \ldots, ie_m) = (e'_1, \ldots, e'_n)(-C + iB).$$

Therefore,

$$(f(e_1), \ldots, f(e_m), f(ie_1), \ldots, f(ie_m)) =$$
$$= (e'_1, \ldots, e'_n, ie'_1, \ldots, ie'_m) \begin{pmatrix} B & -C \\ C & B \end{pmatrix},$$

which completes the proof.

Corollary. *Let $f : L \to L$ be a linear operator over the finite-dimensional complex space L. Then $\det f_{\mathbf{R}} = |\det f|^2$.*

Proof. Let f be represented by the matrix $B + iC$ (B and C are real) in the basis $\{e_1, \ldots, e_m\}$. Then, applying elementary transformations (directly in block form) first to the rows and then to the columns, we obtain

$$\det f_{\mathbf{R}} = \det \begin{pmatrix} B & -C \\ C & B \end{pmatrix} = \det \begin{pmatrix} B + iC & -C + iB \\ C & B \end{pmatrix} =$$

$$= \det \begin{pmatrix} B + iC & 0 \\ C & B - iC \end{pmatrix} = \det(B + iC)\det(B - iC) = \det f\, \overline{\det f} = |\det f|^2.$$

12.4. Restriction of the field of scalars: general situation. It is quite obvious how to extend the definitions of §12.2. Let K be a field, $\mathcal{K}$ a subfield of it, and L a linear space over K. Ignoring multiplication of vectors by all elements of the field K and retaining only multiplication by the elements of $\mathcal{K}$ we obtain the linear space $L_{\mathcal{K}}$ over $\mathcal{K}$. Analogously, the linear mapping $f : L \to M$ over K transforms into the linear mapping $f_{\mathcal{K}} : L_{\mathcal{K}} \to M_{\mathcal{K}}$. One name for these operations is *restriction of the field of scalars* (from K to $\mathcal{K}$). Obviously $\mathrm{id}_{\mathcal{K}} = \mathrm{id}$, $(fg)_{\mathcal{K}} = f_{\mathcal{K}} g_{\mathcal{K}}$, $(af + bg)_{\mathcal{K}} = af_{\mathcal{K}} + bg_{\mathcal{K}}$, if $a, b \in \mathcal{K}$. The field K itself can also be viewed as a linear space over $\mathcal{K}$. If it is finite-dimensional, then the dimensions $\dim_K L$ and $\dim_{\mathcal{K}} L_{\mathcal{K}}$ are related by the formula

$$\dim_{\mathcal{K}} L_{\mathcal{K}} = \dim_{\mathcal{K}} K \dim_K L.$$

For the proof, it is sufficient to verify that if $\{e_1, \ldots, e_n\}$ is a basis in L over K and $\{b_1, \ldots, b_m\}$ is a basis of K over $\mathcal{K}$, then $\{b_1 e_1, \ldots, b_1 e_n; \ldots; b_m e_1, \ldots, b_m e_n\}$ form a basis of $L_{\mathcal{K}}$ over $\mathcal{K}$.

12.5. Complex structure on a real linear space. Let L be a complex linear space and $L_{\mathbf{R}}$ its decomplexification. To restore completely multiplication by complex numbers in $L_{\mathbf{R}}$ it is sufficient to know the operator $J : L_{\mathbf{R}} \to L_{\mathbf{R}}$ for multiplication by $i : J(l) = il$. This operator is obviously linear over $\mathbf{R}$ and satisfies the condition $J^2 = -\mathrm{id}$; if it is known, then for any complex number $a+bi, a, b \in \mathbf{R}$, we have

$$(a+bi)l = al + bJ(l).$$

This argument leads to the following important concept.

12.6. Definition. *Let L be a real space. The assignment of a linear operator $J : L \to L$ satisfying the condition $J^2 = -\mathrm{id}$, is called a complex structure on L.*

The complex structure on $L_{\mathbf{R}}$ described above is called canonical. This definition is justified by the following theorem.

12.7. Theorem. *Let (L, J) be a real linear space with a complex structure. We introduce on L the operation of multiplication by complex numbers from $\mathbf{C}$ according to the formula*

$$(a+bi)l = al + bJ(l).$$

Then L will be transformed into a complex linear space $\tilde{L}$, for which $\tilde{L}_{\mathbf{R}} = L$.

Proof. Both axioms of distributivity are easily verified starting from the linearity of J and the formulas for adding complex numbers. We verify the axiom of associativity for multiplication:

$$(a+bi)[(c+di)l] = (a+bi)[cl + dJ(l)] = a[cl + dJ(l)] +$$

$$+bJ[cl + dJ(l)] = acl + adJ(l) + bcJ(l) - bdl =$$

$$= (ac - bd)l + (ad + bc)J(l) = [ac - bd + (ad + bc)i]l = [(a+bi)(c+di)]l.$$

All the remaining axioms are satisfied because L and $\tilde{L}$ coincide as additive groups.

12.8. Corollary. If (L, J) is a finite-dimensional real space with a complex structure, then $\dim_{\mathbf{R}} L = 2n$ is even and the matrix of J in an appropriate basis has the form

$$\begin{pmatrix} 0 & -E_n \\ E_n & 0 \end{pmatrix}.$$

Proof. Indeed, Theorems 12.7 and 12.3a imply that $\dim_{\mathbf{R}} L = 2 \dim_{\mathbf{C}} \tilde{L}$ (the finiteness of $\tilde{L}$ follows from the fact that any basis of L over $\mathbf{R}$ generates $\tilde{L}$ over $\mathbf{C}$). Next, we select a basis $\{e_1, \ldots, e_n\}$ of the space $\tilde{L}$ over $\mathbf{C}$. The matrix of multiplication by i in this basis equals iE_n. Therefore, Theorem 12.3b implies that the matrix of the operator J in the basis $\{e_1, \ldots, e_n; ie_1, \ldots, ie_n\}$ of the space L has the required form.

12.9. Remarks. a) Let L be a complex space and let $g : L_{\mathbf{R}} \to L_{\mathbf{R}}$ be a real linear mapping. We pose the following question: when does a complex linear mapping $f : L \to L$ such that $g = f_{\mathbf{R}}$ exist ? Obviously for this, g must commute with the operator J of the natural complex structure on $L_{\mathbf{R}}$, because $g(il) = g(Jl) = ig(l) =$ $= Jg(l)$ for all $l \in L$ This condition is also sufficient, because it automatically implies that g is linear over $\mathbf{C}$:

$$g((a+bi)l) = ag(l) + bg(il) = ag(l) + bgJ(l) =$$

$$= ag(l) + bJg(l) = (a+bJ)g(l) = (a+bi)g(l).$$

b) Now let L be an even-dimensional real space, and let $f : L \to L$ be a real linear operator. We pose the question: when does there exist a complex structure J such that f is the decomplexification of a complex linear mapping $g : \tilde{L} \to \tilde{L}$, where $\tilde{L}$ is the complex space constructed with the help of J ? A partial answer for the case $\dim_{\mathbf{R}} L = 2$ is as follows: such a structure exists, if f does not have eigenvectors in L.

Indeed, in this case f has two complex conjugate eigenvalues $\lambda \pm i\mu$, $\lambda, \mu \in R, \mu \neq 0$. Let $J = \mu^{-1}(f - \lambda \mathrm{id})$. According to the Cayley-Hamilton theorem, $f^2 - 2\lambda f + (\lambda^2 + \mu^2)\mathrm{id} = 0$, whence

$$J^2 = \mu^{-2}(f^2 - 2\lambda f + \lambda^2 \mathrm{id}) = -\mathrm{id}.$$

In addition, J commutes with f. This completes the proof.

12.10. Complexification. Now we fix the real linear space L and introduce the complex structure J, defined by the formula

$$J(l_1, l_2) = (-l_2, l_1).$$

on the external direct sum $L \oplus L$. Obviously, $J^2 = -1$. By the *complexification of the space* L, we mean the complex space $\widetilde{L \oplus L}$ associated with this structure. We shall denote it by $L^{\mathbf{C}}$. Other standard notations are: $\mathbf{C} \otimes_{\mathbf{R}} L$ or $\mathbf{C} \oplus L$; their origin will become clear after we become familiar with tensor products of linear spaces. Identifying L with the subset of vectors of the form $(l, 0)$ in $L \oplus L$ and using the fact that $i(l,0) = J(l,0) = (0,l)$, we can write any vector from $L^{\mathbf{C}}$ in the form

$$(l_1, l_2) = (l_1, 0) + (0, l_2) = (l_1, 0) + i(l_2, 0) = l_1 + il_2.$$

In other words, $L^{\mathbf{C}} = L \oplus iL$, and the last sum is a direct sum over $\mathbf{R}$, but not over $\mathbf{C}$!

Any basis of L over $\mathbf{R}$ will be a basis of $L^{\mathbf{C}}$ over $\mathbf{C}$, so that $\dim_{\mathbf{R}} L = \dim_{\mathbf{C}} L^{\mathbf{C}}$.

Now let $f : L \to M$ be a linear mapping of linear spaces over $\mathbf{R}$. Then the mapping $f^{\mathbf{C}}$ (or $f \otimes \mathbf{C}$): $L^{\mathbf{C}} \to M^{\mathbf{C}}$, defined by the formula

$$f(l_1, l_2) = (f(l_1), f(l_2)),$$

is linear over $\mathbf{R}$ and commutes with J, because

$$fJ(l_1, l_2) = f(-l_2, l_1) = (-f(l_2), f(l_1)) = Jf(l_1, l_2).$$

Therefore it is complex-linear. It is called the complexification of the mapping f. Obviously, $\mathrm{id}^{\mathbf{C}} = \mathrm{id}$, $(af + bg)^{\mathbf{C}} = af^{\mathbf{C}} + bg^{\mathbf{C}}$; $a, b \in \mathbf{R}$; and $(fg)^{\mathbf{C}} = f^{\mathbf{C}} g^{\mathbf{C}}$. Regarding the pair of bases of L and M as bases of $L^{\mathbf{C}}$ and $M^{\mathbf{C}}$, respectively, we verify that the matrix of f in the starting pair of bases equals the matrix of $f^{\mathbf{C}}$ in this "new" pair. In particular, the (complex) eigenvalues of f and $f^{\mathbf{C}}$ and the Jordan forms are the same.

We shall now see what happens when the operations of decomplexification and complexification are combined in the two possible orders.

12.11. First complexification, then decomplexification. Let L be a real space. We assert that there exists a natural isomorphism

$$(L^{\mathbf{C}})_{\mathbf{R}} \to L \oplus L.$$

Indeed, by construction $L^{\mathbf{C}}$ coincides with $L \oplus L$ as a real space. Analogously, $(f^{\mathbf{C}})_{\mathbf{R}} \to f \oplus f$ (in the sense of this identification) for any real linear mapping $f : L \to M$.

Composition in the reverse order leads to a somewhat less obvious answer. We introduce the following definition.

12.12. Definition. *Let L be a complex space. The complex conjugate space $\bar{L}$ is the set L with the same additive group structure, but with a new multiplication by a scalar from $\mathbf{C}$, which we temporarily denote by $a * l$:*

$$a * l = \bar{a} l \text{ for any } a \in \mathbf{C},\ l \in L.$$

The axioms are easily verified, using the fact that $\overline{ab} = \bar{a}\bar{b}$, and $\overline{a+b} = \bar{a} + \bar{b}$.

Similarly, if (L, J) is a real space with a complex structure, then the operator J also defines a complex structure, which is said to be *conjugate* with respect to the initial structure. In the notation of Theorem 12.7, if $\tilde{L}$ is the complex space corresponding to (L, J), then $\bar{\tilde{L}}$ is the complex space corresponding to $(L, -J)$.

12.13. First decomplexification, then complexification. We can now construct for any complex linear space L the canonical complex-linear isomorphism

$$f : (L_{\mathbf{R}})^{\mathbf{C}} \to L \oplus \bar{L}.$$

To this end, we note that there are two real linear operators on $(L_{\mathbf{R}})^{\mathbf{C}}$: the operator of the canonical complex structure $J(l_1, l_2) = (-l_2, l_1)$ and the operator of multiplication by i, corresponding to the starting complex structure of $L : i(l_1, l_2) = (il_1, il_2)$. Since J commutes with i, it is complex-linear in this structure. Since $J^2 = -\mathrm{id}$, its eigenvalues equal $\pm i$. We introduce the standard notation for the two subspaces corresponding to these eigenvalues:

$$L^{1,0} = \{(l_1, l_2) \in (L_{\mathbf{R}})^{\mathbf{C}} | J(l_1, l_2) = i(l_1, l_2)\},$$

$$L^{0,1} = \{(l_1, l_2) \in (L_{\mathbf{R}})^{\mathbf{C}} | J(l_1, l_2) = -i(l_1, l_2)\}.$$

Both of the sets $L^{1,0}$ and $L^{0,1}$ are complex subspaces of $(L_{\mathbf{R}})^{\mathbf{C}}$: they are obviously closed under addition and multiplication by real numbers, while closure under multiplication by J follows from the fact that J and i commute. We shall show that $L = L^{1,0} \oplus L^{0,1}$ and also that $L^{1,0}$ is naturally isomorphic to L, while $L^{0,1}$ is naturally isomorphic to $\bar{L}$.

It follows immediately from the definitions that $L^{1,0}$ consists of vectors of the form $(l, -il)$, while $L^{0,1}$ consists of vectors of the form (m, im). For given $l_1, l_2 \in L$ the equation $(l_1, l_2) = (l, -il) + (m, im)$ for l, m has a unique solution $l = \frac{l_1 + il_2}{2}$, $m = \frac{l_1 - il_2}{2}$. Therefore, $L = L^{1,0} \oplus L^{0,1}$. The mappings $L \to L^{1,0} : l \mapsto (l, -il)$ and $\bar{L} \to L^{0,1} : l \mapsto (l, il)$ are real linear isomorphisms. In addition, they are commutative with respect to the action of i on $L, \bar{L}$ and the action of J on $L^{1,0}, L^{0,1}$, by definition of the operations. This completes our construction.

12.14. Semilinear mappings of complex spaces. Let L and M be complex linear spaces. A linear mapping $f : L \to \overline{M}$ is called *semilinear* (or *antilinear*) as a mapping $f : L \to M$. In other words, f is a homomorphism of additive groups, and

$$f(al) = \bar{a} f(l)$$

for all $a \in \mathbf{C}$, $l \in L$. The special role of semilinear mappings will become clear in Chapter 2, when we study Hermitian complex spaces.

12.15. Extension of the field of scalars: general situation. As in §12.4, let K be a field and $\mathcal{K}$ a subfield of K. Then, for any linear space L over $\mathcal{K}$ it is possible to define a linear space $K \otimes_{\mathcal{K}} L$ or L^K over K with the same dimension. It is impossible to give a general definition of L^K without introducing the language of tensor products, but the following temporary definition is adequate for practical purposes: if $\{e_1, \ldots, e_n\}$ is a basis of L over $\mathcal{K}$, then L^K consists of all formal linear combinations $\{\sum_{i=1}^n a_i e_i | a_i \in K\}$, that is, it has the same basis over K. In particular, $(\mathcal{K}^n)^K = K^n$. The K-linear mapping $f^K : L^K \to M^K$ is defined from the $\mathcal{K}$-linear mapping $f : L \to M$: if f is defined by a matrix in some bases of L and M, then f^K is defined by the same matrix.

In conclusion, we give an application of complexification.

12.16. Proposition. *Let $f : L \to L$ be a linear operator in a real space with dimension ≥ 1. Then f has an invariant subspace of dimension 1 or 2.*

Proof. If f has a real eigenvalue, then the subspace spanned by the corresponding eigenvector is invariant. Otherwise, all eigenvalues are complex. Choose one of them $\lambda + i\mu$. It will also be an eigenvalue of $f^{\mathbf{C}}$ in $L^{\mathbf{C}}$. Choose the corresponding eigenvector $l_1 + il_2$ in $L^{\mathbf{C}}$, $l_1, l_2 \in L$. By definition

$$f^{\mathbf{C}}(l_1 + il_2) = f(l_1) + if(l_2) = (\lambda + i\mu)(l_1 + il_2) = (\lambda l_1 - \mu l_2) + i(\mu l_1 + \lambda l_2).$$

Therefore, $f(l_1) = \lambda l_1 - \mu l_2$, $f(l_2) = \mu l_1 + \lambda l_2$, and the linear span of $\{l_1, l_2\}$ in L is invariant under f.

§13. The Language of Categories

13.1. Definition of a category. A category C consists of the following objects:

a) a set (or collection) $\mathrm{Ob}\, C$, whose elements are called *objects* of the category;

b) a set (or collection) $\mathrm{Mor}\, C$, whose elements are called *morphisms* of the category, or *arrows*;

c) for every ordered pair of objects $X, Y \in \mathrm{Ob}\, C$, a set $\mathrm{Hom}_C(X,Y) \subset \mathrm{Mor}\, C$, whose elements are called *morphisms from X into Y* and are denoted by $X \to Y$ or $f : X \to Y$ or $X \xrightarrow{f} Y$; and,

d) for every ordered triplet of objects $X, Y, Z \in \mathrm{Ob}\, C$ a mapping

$$\mathrm{Hom}_C(X,Y) \times \mathrm{Hom}_C(Y,Z) \to \mathrm{Hom}_C(X,Z),$$

which associates to the pair of morphisms (f, g) the morphism gf or $g \circ f$, called their *composition* or *product*.

These data must satisfy the following conditions:

e) $\mathrm{Mor}\, C$ is a disjoint union $\bigcup \mathrm{Hom}_C(X,Y)$ for all ordered pairs $X, Y \in \mathrm{Ob}\, C$. In other words, for every morphism f there exist uniquely defined objects X, Y such that $f \in \mathrm{Hom}\, C(X,Y)$: X is the *starting* point and Y is the *terminal* point of the arrow f.

f) The composition of morphisms is associative.

g) For every object X there exists an identity morphism $\mathrm{id}_X \in \mathrm{Hom}_C(X,X)$ such that $\mathrm{id}_X \circ f = f \circ \mathrm{id}_X = f$ whenever these compositions are defined. It is not difficult to see that such a morphism is unique: if id'_X is another morphism with the same property, then $\mathrm{id}'_X = \mathrm{id}'_X \circ \mathrm{id}_X = \mathrm{id}_X$.

The morphism $f : X \to Y$ is called an *isomorphism* if there exists a morphism $g : Y \to X$ *such that* $gf = \mathrm{id}_X, fg = \mathrm{id}_Y$.

13.2. Examples. a) *The category of sets* Set. Its objects are sets and the morphisms are mappings of these sets.

b) *The category* $\mathcal{L}in_{\mathcal{K}}$ *of linear spaces over the field* $\mathcal{K}$. Its objects are linear spaces and the morphisms are linear mappings.

c) *The category of groups.*

d) *The category of abelian groups.*

The differences between sets and collections are discussed in axiomatic set theory and are linked to the necessity of avoiding Russell's famous paradox. Not every collection of objects forms in aggregate a set, because the concept "the set of all sets not containing themselves as an element" is contradictory. In the axiomatics of Gödel–Bernays, such collections of sets are called classes. The theory of categories requires collections of objects which lie dangerously close to such paradoxical situations. We shall, however, ignore these subtleties.

13.3. Diagrams. Since in the axiomatics of categories nothing is said about the set-theoretical structure of the objects, we cannot in the general case work with the "elements" of these objects. All basic general-categorical constructions and their applications to specific categories are formalized predominantly in terms of morphisms and their compositions. A convenient language for such formulations is the language of diagrams. For example, instead of saying that the four objects X, Y, U and V, the four morphisms $f \in \mathrm{Hom}_C(X,Y)$, $g \in \mathrm{Hom}_C(Y,V)$, $h \in \mathrm{Hom}_C(X,U)$, and $d \in \mathrm{Hom}_C(U,V)$ and in addition $gf = dh$ are given, it is said that the *commutative square*

$$\begin{array}{ccc} X & \xrightarrow{f} & Y \\ {\scriptstyle h}\downarrow & & \downarrow{\scriptstyle g} \\ U & \xrightarrow{d} & V \end{array}$$

is given. Here, "commutativity" means the equality $gf = dh$, which indicates that the "two paths along the arrows" from X to V lead to the same result. More generally, the diagram is an oriented graph, whose vertices are the objects of C and whose edges are morphisms, for example

$$\begin{array}{ccccc} X & \longrightarrow & Y & \longrightarrow & Z \\ & & \downarrow & & \downarrow \\ U & \longrightarrow & V & \longrightarrow & W \end{array}$$

The diagram is said to be commutative if any paths along the arrows in it with common starting and terminal points correspond to identical morphisms.

In the category of linear spaces, as well as in the category of abelian groups, a class of diagrams called complexes is especially important. A complex is a finite or infinite sequence of objects and arrows

$$\ldots X \longrightarrow Y \longrightarrow U \longrightarrow V \longrightarrow \ldots$$

satisfying the following condition: *the composition of any two neighbouring arrows is a zero morphism.* We note that the concept of a zero morphism is not a general-categorical concept: it is specific to linear spaces and abelian groups and to a special class of categories, the so-called additive categories. Often, the objects comprising a complex and the morphisms are enumerated by some interval of integers:

$$\ldots \longrightarrow X_{-1} \xrightarrow{f_{-1}} X_0 \xrightarrow{f_0} X_1 \xrightarrow{f_1} X_2 \xrightarrow{f_2} \ldots$$

Such a complex of linear spaces (or abelian groups) is said to be *exact in the term* X_i, if $\operatorname{im} f_{i-1} = \ker f_i$ (we note that in the definition of a complex the condition $f_i \circ f_{i-1} = 0$ merely means that $\operatorname{im} f_{i-1} \subset \ker f_i$). A complex that is exact in all terms is said to be *exact* or *acyclic* or *an exact sequence.*

Here are three very simple examples:

a) The sequence $0 \to L \xrightarrow{i} M$ is always a complex; it is exact in the term L if and only if $\ker i$ is the image of the null spce 0. In other words, here exactness means that i is an injection.

b) The sequence $M \xrightarrow{i} N \to 0$ is always a complex; the fact that it is exact in the term N means that $\operatorname{im} j = N$, that is, that j is a surjection.

c) The complex $0 \to L \xrightarrow{i} M \xrightarrow{j} N \to 0$ is exact if i is an injection, j is a surjection, and $\operatorname{im} i = \ker j$. Identifying L with the image of i which is a subspace in M, we can therefore identify N with the factor space M/L, so that such "*exact triples*" or *short exact sequences* are categorical representatives of the triples $(L \subset \subset M,\ M/L)$.

13.4. Natural constructions and functors. Constructions which can be applied to objects of a category so that in so doing, objects of a category (a different one or the same one) are obtained again are very important in mathematics. If these constructions are unique (they do not depend on arbitrary choices) and universally applicable, then it often turns out that they can also be transferred to morphisms. The axiomatization of a number of examples led to the important concept of a functor, which, however, is also natural from the purely categorical viewpoint.

13.5. Definition of a functor. Let C and D be categories. Two mappings (usually denoted by F): $\operatorname{Ob} C \to \operatorname{Ob} D$ and $\operatorname{Mor} C \to \operatorname{Mor} D$ are said to form a *functor* from C into D if they satisfy the following conditions:

a) if $f \in \mathrm{Hom}_C(X, Y)$, then $F(f) \in \mathrm{Hom}_D(F(X), F(Y))$;

b) $F(gf) = F(g)F(f)$, whenever the composition gf is defined, and $F(\mathrm{id}_X) = \mathrm{id}_{F(X)}$ for all $X \in \mathrm{Ob}\, C$.

The functors which we have defined are often called *covariant* functors. *Contravariant* functors which "invert the arrows" are also defined. For them the conditions a) and b) above are replaced by the following ones:

a') if $f \in \mathrm{Hom}_C(X, Y)$, then $F(f) \in \mathrm{Hom}_D(F(Y), F(X))$;

b') $F(gf) = F(f)F(g)$ and $F(\mathrm{id}_X) = \mathrm{id}_{F(X)}$.

This distinction can be avoided by introducing a construction which to each category C associates a *dual category* C^o according to the following rule: $\mathrm{Ob}\, C = \mathrm{Ob}\, C^o$, $\mathrm{Mor}\, C = \mathrm{Mor}\, C^o$, and $\mathrm{Hom}_C(X, Y) = \mathrm{Hom}_{C^o}(Y, X)$; in addition, the composition gf of morphisms in C corresponds to the composition fg of the same morphisms in C^o, taken in reverse order. It is convenient to denote by X^o and f^o objects and morphisms in C^o which correspond to the objects and morphisms X and f in C. Then, the commutative diagram in C

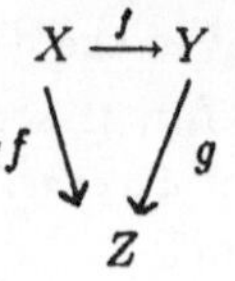

corresponds to the commutative diagram

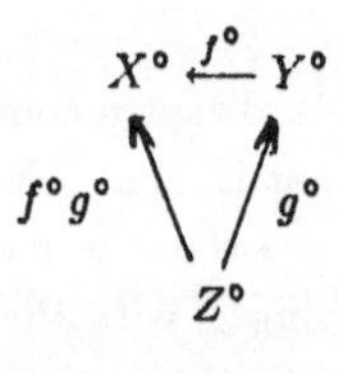

in C^o.

A (covariant) functor $F : C \to D^o$ can be identified with the contravariant functor $F : C \to D$ in the sense of the definition given above.

13.6. Examples. a) Let $\mathcal{K}$ be a field, $\mathcal{L}in_{\mathcal{K}}$ the category of linear spaces over $\mathcal{K}$, and let Set be the category of sets. In §1 we explained how to form a correspondence between any set $S \in \mathrm{Ob}\,\mathrm{Set}$ and the linear space $F(S) \in \mathrm{Ob}\, \mathcal{L}in_{\mathcal{K}}$ of functions on S with values in $\mathcal{K}$. Since this is a natural construction, it should be expected that it can be extended to a functor. Such is the case. The functor turns out to be contravariant: it establishes a correspondence between the morphism $f : S \to T$ and the linear mapping $F(f) : F(T) \to F(S)$, most often denoted by f^* and called the *pull back* or *reciprocal image*, on the functions

$$f^*(\phi) = \phi \circ f, \quad \text{where} \quad f : S \to T,\ \phi : T \to \mathcal{K}.$$

In other words, $f^*(\phi)$ is a function on S, whose values are constant along the "fibres" $f^{-1}(t)$ of the mapping f and are equal to $\phi(t)$ on such a fibre. A good exercise for the reader is to verify that we have indeed constructed a functor.

b) The duality mapping $\mathcal{L}in_{\mathcal{K}} \to \mathcal{L}in_{\mathcal{K}}$, on objects defined by the formula $L \mapsto L^* = \mathcal{L}(L, \mathcal{K})$ and on morphisms by the formula $f \mapsto f^*$, is a *contravariant functor* from the category $\mathcal{L}in_{\mathcal{K}}$ into itself. This was essentially proved in §7.

c) The operations of complexification and decomplexification, studied in §12, define the functors $\mathcal{L}in_{\mathbf{R}} \to \mathcal{L}in_{\mathbf{C}}$ and $\mathcal{L}in_{\mathbf{C}} \to \mathcal{L}in_{\mathbf{R}}$ respectively. This is also true for the general constructions of extending and restricting the field of scalars, briefly described in §12.

d) For any category C and any object $X \in \mathrm{Ob}\, C$, two functors from C into the category of sets are defined: a covariant functor $h_X : C \to \mathrm{Set}$ and a contravariant functor $h^X : C \to \mathrm{Set}^o$.

They are defined as follows: $h_X(Y) = \mathrm{Hom}_C(X, Y)$, $h_Y(f : Y \to Z)$ is the mapping $h_X(Y) = \mathrm{Hom}_C(X, Y) \to h_X(Z) = \mathrm{Hom}_C(X, Z)$, which associates with the morphism $X \to Y$ its composition with the morphism $f : Y \to Z$.

Analogously, $h^X(Y) = \mathrm{Hom}_C(Y, X)$ and $h^X(f : Y \to Z)$ is the mapping $h^X(Z) = \mathrm{Hom}_C(Z, X) \to h^X(Y) = \mathrm{Hom}_C(Y, X)$, which associates with the morphism $Z \to X$ its composition with the morphism $f : Y \to Z$.

Verify that h_X and h^X are indeed functors. They are called functors *representing* the object X of the category.

We note that if $C + \mathcal{L}in_{\mathcal{K}}$, then h^X and h_X may be regarded as functors whose values also lie in $\mathcal{L}in_{\mathcal{K}}$ and not in Set.

13.7. Composition of functors. If $C_1 \xrightarrow{F} C_2 \xrightarrow{G} C_3$ are three categories and two functors between them, then the composition $GF : C_1 \to C_3$ is defined as the set-theoretic composition of mappings on objects and morphisms. It is trivial to verify that it is a functor.

It is possible to introduce a "category of categories", whose objects are categories while the morphisms are functors !

The next step of this high ladder of abstractions is, however, more important: *the category of functors.* We shall restrict ourselves to explaining what morphisms of functors are.

13.8. Natural transformations of natural constructions or functorial morphisms. Let $F, G : C \to D$ be two functors with a common starting point and a common terminal point. A *functorial morphism* $\phi : F \to G$ is a collection of morphisms of objects $\phi(X) : F(X) \to G(X)$ in the category D, one for each object X of the category C, with the property that for every morphism $f : X \to Y$ in the category C the square

$$\begin{array}{ccc} F(X) & \xrightarrow{\phi(X)} & G(X) \\ {\scriptstyle F(f)}\downarrow & & \downarrow{\scriptstyle G(f)} \\ F(Y) & \xrightarrow{\phi(Y)} & G(Y) \end{array}$$

is commutative. A functorial morphism is an *isomorphism* if all $\phi(X)$ are isomorphisms.

13.9. Example. Let $** : \mathcal{L}in_{\mathcal{K}} \to \mathcal{L}in_{\mathcal{K}}$ be the functor of "double conjugation": $L \to L^{**}, f \to f^{**}$. In §7 we constructed for every linear space L a canonical linear mapping $\epsilon_L : L \to L^{**}$. It defines the functor morphism $\epsilon_M : Id \to **$, where Id is the identity functor on $\mathcal{L}in_{\mathcal{K}}$, which to every linear space associates the space itself and to every linear mapping associates the mapping itself. Indeed, by definition, we should verify the commutativity of all-possible squares of the form

$$\begin{array}{ccc} L & \xrightarrow{\epsilon_L} & L^{**} \\ {\scriptstyle f}\downarrow & & \downarrow{\scriptstyle f^{**}} \\ M & \xrightarrow{\epsilon_M} & M^{**} \end{array}$$

For finite-dimensional spaces L and M, this is asserted by Theorem 7.5. We leave to the reader the verification of the general case.

EXERCISES

1. Let Set_0 be a category whose objects are sets while the morphisms are mappings of sets $f : S \to T$, such that for any point $t \in T$ the fibre $f^{-1}(t)$ is finite. Show that the following conditions define the *covariant* functor $F_0 : \mathrm{Set}_0 \to \mathcal{L}in_{\mathcal{K}}$:

   a) $F_0(S) = F(S)$: functions on S with values in $\mathcal{K}$.

   b) For any morphism $f : S \to T$ and function $\phi : S \to K$ the function $F_0(f)(\phi) = f_*(\phi) \in F(T)$ is defined as

$$f_*(\phi)(t) = \sum_{s \in f^{-1}(t)} \phi(s)$$

("integration along fibres").

2. Prove that the lowering of the field of scalars from K to $\mathcal{K}$ (see §12) defines the functor $\mathcal{L}in_K \to \mathcal{L}in_{\mathcal{K}}$.

§14. The Categorical Properties of Linear Spaces

14.1. In this section we collect some assertions about categories of all linear spaces $\mathcal{L}in_{\mathcal{K}}$ or finite-dimensional spaces $\mathcal{L}inf_{\mathcal{K}}$ over a given field $\mathcal{K}$. Most of them are a reformulation of assertions which we have already proved in the language of categories. These assertions are chosen based on the following peculiar criterion: these are precisely the properties of the category $\mathcal{L}in_{\mathcal{K}}$ that are *violated* for the closest categories, such as the category of modules over general rings (for example, over **Z**, that is, the category of abelian groups), or even the category of infinite-dimensional topological spaces. The detailed study of these violations for the category of modules is the basic subject of *homological algebra*, while in functional analysis it often leads to a search for new definitions which would permit reconstructing the "good" properties of $\mathcal{L}inf_{\mathcal{K}}$ (this is the concept of nuclear topological spaces).

14.2. Theorem on the extension of mappings. a) *Let P, M and N be linear spaces. Let P be finite-dimensional and let $j : M \to N$ be a surjective linear mapping. Then, any mapping $g : P \to N$ can be lifted to the mapping $h : P \to M$ such that $g = jh$. In other words, the diagram with the exact row*

$$\begin{array}{ccccc} & & P & & \\ & & \downarrow{\scriptstyle g} & & \\ M & \xrightarrow{j} & N & \longrightarrow & 0 \end{array}$$

can be inserted into the commutative diagram

$$\begin{array}{ccccc} & & P & & \\ & \swarrow{\scriptstyle h} & \downarrow{\scriptstyle g} & & \\ M & \xrightarrow{j} & N & \longrightarrow & 0 \end{array}$$

b) *Let P, L, and M be linear spaces. Let M be finite-dimensional and let $i : L \to M$ be an injective mapping. Then, any mapping $g : L \to P$ can be extended to a linear mapping $h : M \to P$ so that $g = hi$. In other words, the diagram with the exact bottom row*

$$\begin{array}{ccccc} & & P & & \\ & & \uparrow{\scriptstyle g} & & \\ M & \xleftarrow{i} & L & \longleftarrow & 0 \end{array}$$

can be inserted into the commutative diagram

$$\begin{array}{ccccc} & & P & & \\ & \nearrow{\scriptstyle h} & \uparrow{\scriptstyle g} & & \\ M & \xleftarrow{i} & L & \longleftarrow & 0 \end{array}$$

Proof. a) We select a basis $\{e_1, \ldots, e_n\}$ in P and set $e_i' = g(e_i) \in N$. The fact that j is surjective implies that there exist vectors $e_i'' \in M$ such that $j(e_i'') = e_i'$, $i = 1, \ldots, n$. Definition 3.3 implies that there exists a unique linear mapping $h : P \to M$ such that $h(e_i) = e_i''$, $i = 1, \ldots, n$. By construction, $jh(e_i) = j(e_i'') = e_i' = g(e_i)$. Since $\{e_i\}$ form a basis of P, we have $jh = g$.

b) We choose the basis $\{e_1', \ldots, e_m'\}$ of the space L and extend $e_k = i(e_k')$, $1 \le k \le m$ to the basis $\{e_1, \ldots, e_m; e_{m+1}, \ldots, e_n\}$ of the space M. We set $h(e_i) = g(e_i')$ for $1 \le i \le m$ and $h(e_j) = 0$ for $m + 1 \le j \le n$. Such a mapping exists according to the same Proposition 3.3. It is also possible to apply directly Proposition 6.8. The theorem is proved.

In the category of modules the objects P, satisfying the condition a) of the theorem (for all M, N), are said to be *projective* and those satisfying the condition b) are said to be *injective*. We have proved that in the category of finite-dimensional linear spaces all objects are projective and injective.

14.3. Theorem on the exactness of the functor $\mathcal{L}$. *Let $0 \to L \xrightarrow{i} M \xrightarrow{j} N \to 0$ be an exact triple of finite-dimensional linear spaces and let P be any finite-dimensional space. Then $\mathcal{L}$ as a functor induces, separately with respect to the first and second argument, the exact triples of linear spaces*

$$\text{a)}\quad 0 \longrightarrow \mathcal{L}(P, L) \xrightarrow{i_1} \mathcal{L}(P, M) \xrightarrow{j_1} \mathcal{L}(P, N) \longrightarrow 0,$$

$$\text{b)}\quad 0 \longleftarrow \mathcal{L}(L, P) \xleftarrow{i_2} \mathcal{L}(M, P) \xleftarrow{j_2} \mathcal{L}(N, P) \longleftarrow 0\,.$$

Proof. a) We recall that i_1 is a composition of the variable morphism $P \to L$ with $i : L \to M$, j_i is the composition of $P \to M$ with $j : M \to N$. The mapping i_1 is injective, because i is an injection, so that if the composition $P \to L \to M$ is a zero composition, then $P \to L$ is a zero morphism. The first part of Theorem 14.2 implies that the mapping j_1 is surjective: any morphism $g : P \to N$ can be extended to a morphism $P \to M$, whose composition with j gives the starting morphism. The composition $j_1 i_1$ is a zero composition: it transforms the arrow $P \to L$ into the arrow $P \to N$, which is the composition $P \to L \xrightarrow{i} M \xrightarrow{i} N$, but $ji = 0$.

We have thus verified that the sequence a) is a complex, and it remains to establish its exactness with respect to the central term, that is, $\ker j_1 = \operatorname{im} i_1$. We already know that $\ker j_1 \supset \operatorname{im} i_1$. To prove the reverse inclusion we note that if the arrow $P \to M$ lies in the kernel of j_1, then the composition of this arrow with $j : M \to N$ equals zero, and therefore the image of P in M lies in the kernel of j. But the kernel of j coincides with the image of $i(L) \subset M$ by virtue of the exactness at the starting arrow. Hence P is mapped into the subspace $i(L)$ and therefore the arrow $P \to M$ can be extended up to the arrow $P \to L$, whose composition with i gives the starting arrow. Hence the latter lies in the image of i_1.

b) Here the arguments are entirely analogous or, more precisely, reciprocal. The mapping i_2 is surjective according to the second part of Theorem 14.2. The mapping j_2 is injective because if the composition $M \xrightarrow{j} N \to P$ equals zero, then the arrow $N \to P$ also equals zero, since j is surjective. The composition $i_2 j_2$ equals zero, because the composition $L \to M \xrightarrow{j} N \to P$ equals zero for any final arrow. Therefore it remains to be proved that $\ker i_2 \subset \operatorname{im} j_2$ (the reverse inclusion has just been proved). But if the composition $l \xrightarrow{i} M \xrightarrow{f} P$ equals zero, the arrow $M \xrightarrow{f} P$ lies in the kernel of i_2. Hence, $L = \ker j$ lies in the kernel of f. We shall define the mapping $\bar{f} : N \to P$ by the formula $\bar{f}(n) = f(j^{-1}(n))$, where $j^{-1}(n) \in M$ is any inverse image of n. Nothing depends on the choice of this inverse image, because $\ker j \subset \ker f$. It is easy to verify that $\bar{f}$ is linear and that $j_2(\bar{f}) = f$; indeed, $j_2(\bar{f})$ is the composition $M \xrightarrow{j} N \xrightarrow{\bar{f}} P$, which transforms $m \in M$ into $\bar{f}j(m) = f(j^{-1}(j(m))) = f(m)$. The theorem is proved.

14.4. The categorical characterization of dimension. Let G be some algebra of groups, written additively, and let $\chi : \operatorname{Ob} \mathcal{L}in f_{\mathcal{K}} \to G$ be an arbitrary function, defined on finite-dimensional linear spaces and satisfying the following two conditions:

a) if L and M are isomorphic, then $\chi(L) = \chi(M)$ and

b) for any exact triple of spaces $0 \to L \to M \to N \to 0$, $\chi(M) = \chi(L) + \chi(N)$ (such functions are called additive). The following theorem holds.

14.5. Theorem. *For any additive function χ we have*

$$\chi(L) = \dim_{\mathcal{K}} L \cdot \chi(\mathcal{K}^1),$$

where L is an arbitrary finite-dimensional space.

Proof. We perform induction on the dimension of L. If L is one-dimensional, then L is isomorphic to $\mathcal{K}^1$, so that

$$\chi(L) = \chi(\mathcal{K}^1) = \dim_{\mathcal{K}} L \cdot \chi(\mathcal{K}^1).$$

Assume that the theorem has been proved for all L of dimension n. If the dimension of L equals $n+1$, we select a one-dimensional subspace $L_0 \subset L$ and study the exact triple

$$0 \to L_0 \xrightarrow{i} L \xrightarrow{j} L/L_0 \to 0,$$

where i is the embedding of L_0, while $j(l) = l + L_0 \in L/L_0$. By virtue of the additivity of χ and the induction hypothesis,

$$\chi(L) = \chi(L_0) + \chi(L/L_0) = \chi(\mathcal{K}^1) + \dim_{\mathcal{K}}(L/L_0)\chi(\mathcal{K}^1) =$$

$$= \chi(\mathcal{K}^1) + n\chi(\mathcal{K}^1) = (n+1)\chi(\mathcal{K}^1) = \dim_{\mathcal{K}} L \cdot \chi(\mathcal{K}^1).$$

The theorem is proved.

This result is the beginning of an extensive algebraic theory, which is now being actively developed: the so-called K theory, which lies on the frontier between topology and algebra.

EXERCISES

1. Let $K : 0 \xrightarrow{d_0} L_1 \xrightarrow{d_1} L_2 \xrightarrow{d_2} \dots \xrightarrow{d_{n-1}} L_n \xrightarrow{d_n} 0$ be complex finite-dimensional linear spaces. The factor space $H^i(K) = \ker d_i / \operatorname{im} d_{i-1}$ is called the ith space of homologies of this complex. The number $\chi(K) = \sum_{i=1}^n (-1)^i \dim L_i$ is called the Euler characteristic of the complex. Prove that

$$\chi(K) = \sum_{i=1}^n (-1)^i \dim H^i(K).$$

2. "Snake lemma". Given the following commutative diagram of linear spaces

$$\begin{array}{ccccccc} & & L & \xrightarrow{d_1} & M & \xrightarrow{d_2} & N & \longrightarrow 0 \\ & & f\downarrow & & g\downarrow & & \downarrow h & \\ 0 & \longrightarrow & L' & \xrightarrow{d_1'} & M' & \xrightarrow{d_2'} & N' & \end{array}$$

with exact rows, show that there exists an exact sequence of spaces

$$\ker f \to \ker g \to \ker h \xrightarrow{\delta} \operatorname{coker} f \to \operatorname{coker} g \to \operatorname{coker} h,$$

in which all arrows except δ are induced by d_1, d_2, d_1', d_2' respectively, while the connecting homomorphism δ (also called the coboundary operator) is defined as follows: to define $\delta(n)$ for $n \in \ker h$, it is necessary to find $m \in M$ with $n = d_2(m)$, construct $g(m) \in \mathrm{M}'$, find $l' \in L'$ with $d_1'(l') = g(m)$, and set $\delta(n) = l' + \operatorname{im} f \in \operatorname{coker} f$. In particular, it is necessary to check the existence of $\delta(n)$ and its independence from any arbitrariness in the intermediate choices.

3. Let $K : \dots \to L_i \xrightarrow{d_i} L_{i+1} \to \dots$ and $K' : \dots \to L_i' \xrightarrow{d_i'} L_{i+1}' \to \dots$ be two complexes. A morphism $f : K \to K'$ is a set of linear mappings $f_i : L_i \to L_i'$ such that all squares

$$\begin{array}{ccc} L_i & \xrightarrow{d_i} & L_{i+1} \\ f_i\downarrow & & \downarrow f_{i+1} \\ L_i' & \xrightarrow{d_i'} & L_{i+1}' \end{array}$$

are commutative. Show that the complexes and their morphisms form a category.

4. Show that the mapping $K \to H^i(K)$ can be extended to a functor from categories of complexes into the category of linear spaces.

5. Let $0 \to K \xrightarrow{f} K' \xrightarrow{g} K'' \to 0$ be an exact triple of complexes and their morphisms. By definition, this means that the triples of linear spaces

$$0 \to L_i \xrightarrow{f_i} L_i' \xrightarrow{g_i} L_i'' \to 0$$

are exact for all i. Let H^i be the corresponding space of cohomologies. Using the snake lemma, construct the sequence of spaces of cohomologies

$$\dots \to H^i(K) \to H^i(K') \to H^i(K'') \xrightarrow{\delta} H^{i+1}(K) \to \dots$$

and show that it is exact.

CHAPTER 2

Geometry of Spaces with an Inner Product

§1. On Geometry

1.1. This part of our course and the next one are devoted to a subject which can be called "linear geometries". It is appropriate to introduce it with a brief discussion of the modern meaning of the words "geometry" and "geometric". For many hundreds of years geometry was understood to mean Euclid's geometry in a plane and in space. It still forms the foundation of the standard course in schools, and it is convenient to follow the evolution of geometric concepts for the example of characteristic features of this, now very specialized, geometric discipline.

1.2. "Figures". High-school geometry begins with the study of figures in a plane, such as straight lines, angles, triangles, circles, discs, etc. A natural generalization of this situation is to choose a space M which "envelopes the space" of our geometry and a collection of subsets in M – the "figures" studied in this space.

1.3. "Motion". The second important component of high-school geometry is the measurement of lengths and angles and the clarification of the relations between the linear and angular elements of different figures. A long historical development was required before it was recognized that these measurements are based on the existence of a separate mathematical object – the group of motions in the Euclidean plane or Euclidean space as a whole – and that all metric concepts can be defined in terms of this group. For example, the distance between points is the only function of a pair of points that is invariant with respect to the group of Euclidean motions (if it is required to be continuous and the distance between a selected pair of points is chosen to be the "unit of length"). F. Klein's "Erlangen program" (1872) settled the concept of this remarkable principle, and "geometry" for a long time was considered to be the study of spaces M equipped with a quite large symmetry group and the properties of figures that are invariant with respect to the action of this group, including angles, distances and volumes.

1.4. "Numbers". A discovery of equal fundamental significance (and a much earlier one) was the Cartesian "method of coordinates" and the analytic geometry of planes and surfaces based on it. From the modern viewpoint coordinates are functions over a space M (or over its subsets) with real, complex, or even more general values. The specification of values of these functions fixes a point in space, while the specification of relations between these values determines a set of points. In this geometry the description of the set of figures in M can be replaced by a description of the class of relations between coordinates which describe the figures of interest. The amazing flexibility and power of Descartes' method stems from the fact that the functions over the space can be added, multiplied, integrated, and differentiated, and other limiting processes can be applied to them; ultimately, all of the power of mathematical analysis can be employed. All general modern geometric disciplines – topology, differential and complex-analytic geometry, and algebraic geometry – start from the concept of a geometric object as a collection of spaces M and a collection F of (local) functions given on it as a starting definition.

1.5. "Mappings". If (M_1, F_1) and (M_2, F_2) are two geometric objects of the type described above, then one can study the mappings $M_1 \to M_2$ which have the property that the inverse mapping on functions maps elements from F_2 into elements from F_1. In the most logically complete schemes such mappings include both the symmetry groups of F. Klein and the coordinate functions themselves (as mappings of M into $\mathbf{R}$ or $\mathbf{C}$). Geometric objects form a category, and its morphisms serve as a quite subtle substitute for symmetry even in those cases when there are not many symmetries (like in general Riemannian spaces, where lengths, angles, and volumes can be measured, but motions, generally speaking, are not enough).

1.6. Linear geometries. We can now characterize the place of linear geometries in this general scheme. In a well-known sense, the words linear geometries refer to the direct descendants of Euclidean geometry. The spaces M studied in them are either *linear spaces* (this time over general fields, though $\mathbf{R}$ or $\mathbf{C}$ remain, as before, at the centre of attention, especially in view of the numerous applications) or spaces derived from linear spaces: *affine spaces* ("linear spaces without an origin of coordinates') and *projective spaces* ("affine spaces, supplemented with infinitely separated points"). Symmetry groups are subgroups of the linear group which preserve a fixed "inner product", and also their enlargement with translations (affine groups) or factor groups over homotheties (projective groups). The functions studied are linear or nearly linear, and sometimes quadratic. Figures are linear subspaces and manifolds (generalization of neighbourhoods). One can imagine these generalizations of Euclidean geometry as following from purely logical analysis, and the established formalism of linear geometries does indeed exhibit a surprising orderliness and compactness. However, the viability of this branch of mathematics is to a certain extent linked to its diverse applications in the natural sciences. The concept of an inner product, which forms the basis for the entire second chapter of this course, can serve as a means for measuring angles in abstract Euclidean spaces. But a mathematician who is unaware that it also measures probabilities (in quantum mechanics), velocities (in Minkowski's space in the special theory of relativity), and the correlation coefficients for random quantities (in the theory of probability) not only loses a broad range of vision but also the flexibility of purely mathematical intuition. For this reason we considered it necessary to include in this course information about these interpretations.

2. Inner Products

2.1. Multilinear mappings. Let $L_1, \ldots, L_n$ and M be linear spaces over a general field $\mathcal{K}$. A mapping

$$f : L_1 \times \ldots \times L_n \to M, (l_1, \ldots, l_n) \mapsto f(l_1, \ldots, l_n) \in M,$$

which is linear as a function of any of its arguments $l_i \in L_i$ with the remaining arguments $l_j \in L_j$, $j = 1, \ldots, n, j \neq i$, held fixed is called a *multilinear mapping* (*bilinear* for $n = 2$). In other words,

$$f(l_1, \ldots, l_i + l_i', l_{i+1}, \ldots, l_n) =$$

$$= f(l_1, \ldots, l_i, \ldots, l_n) + f(l_1, \ldots, l_i', \ldots, l_n),$$

$$f(l_1, \ldots, al_i, l_{i+1}, \ldots, l_n) = af(l_1, \ldots, l_i, \ldots, l_n)$$

for $i = 1, \ldots, n$; $a \in \mathcal{K}$. In the case that $M = \mathcal{K}$ multilinear mappings are also called *multilinear functionals* or *forms*.

In Chapter 1 we already encountered bilinear mappings

$$L^* \times L \to \mathcal{K} : (f, l) \mapsto f(l),\ f \in L^*, l \in L;$$

$$\mathcal{L}(L, M) \times L \to M : (f, l) \mapsto f(l),\ f \in \mathcal{L}(L, M), l \in L.$$

The determinant of a square matrix is multilinear as a function of its rows and columns. Another example:

$$\mathcal{K}^n \times \mathcal{K}^m \to \mathcal{K} : (\vec{x}, \vec{y}) \mapsto \sum_{i,j} g_{ij} x_i y_j = \vec{x}^t G \vec{y},$$

where G is any $n \times m$ matrix over $\mathcal{K}$, and the vectors from $\mathcal{K}^n$ and $\mathcal{K}^m$ are represented as columns of their coordinates.

We shall study general multilinear mappings later, in the part of the book devoted to tensor algebra. Here we shall be concerned with bilinear functions, which are most important for applications, $L \times L \to \mathcal{K}$ and also for $\mathcal{K} = \mathbf{C}$ the functions $L \times \bar{L} \to \mathbf{C}$, where $\bar{L}$ is the space that is the complex conjugate of L (see Chapter I,§12). Every such function is also called an *inner product* (dot product) or *metric* on the space L and the pair (L, inner product) is regarded as one geometric object. The metrics studied in this part are metrics in the sense of the definition of §10.1 of Chapter I in special cases only, and the reader should not confuse these homonyms.

The inner product $L \times \bar{L} \to \mathbf{C}$ is most often viewed as a *sesquilinear mapping* $g : L \times L \to \mathbf{C}$: linear with respect to the first argument and semilinear with respect to the second argument: $g(al_1, bl_2) = a\bar{b}g(l_1, l_2)$.

2.2. Methods for specifying the inner product. a) Let $g : L \times L \to \mathcal{K}$ (or $L \times \bar{L} \to \mathbf{C}$) be an inner product over a finite-dimensional space L. We choose a basis $\{e_1, \dots, e_n\}$ in L and define the matrix

$$G = (g(e_i, e_j)); \; i, j = 1, \dots, n.$$

It is called the *Gram matrix* of the basis $\{e_1, \dots, e_n\}$ with respect to g, as well as the *matrix of* g *in the basis* $\{e_1, \dots, e_n\}$. The specification of $\{e_i\}$ and G completely defines g, because by virtue of the properties of bilinearity,

$$g(\vec{x}, \vec{y}) = g\left(\sum_{i=1}^{n} x_i e_i, \sum_{j=1}^{n} y_j e_j\right) = \sum_{i,j=1}^{n} x_i y_j g(e_i, e_j) = \vec{x}^t G \vec{y}.$$

In the case of a sesquilinear form, the analogous formula assumes the form

$$g(\vec{x}, \vec{y}) = g\left(\sum_{i=1}^{n} x_i e_i, \sum_{j=1}^{n} y_j e_j\right) = \sum_{i,j=1}^{n} x_i \bar{y}_j g(e_i, e_j) = \vec{x}^t G \bar{\vec{y}}.$$

Conversely, if the basis $\{e_1, \dots, e_n\}$ is fixed and G is an arbitrary $n \times n$ matrix over $\mathcal{K}$, then the mapping $(\vec{x}, \vec{y}) \to \vec{x}^t G \vec{y}$ (or $\vec{x}^t G \bar{\vec{y}}$ in the sesquilinear case) defines an inner product on L with the matrix G in this basis, as obvious checks show. Thus our construction establishes a bijection between the inner prooducts (bilinear or sesquilinear) on an n-dimensional space with a basis and $n \times n$ matrices.

We shall clarify how G changes under a change of basis. Let A be the matrix of the transformation to the primed basis: $\vec{x} = A\vec{x}'$, where $\vec{x}$ are the coordinates of the vector in the old basis and $\vec{x}'$ are its coordinates in the new one. Then in the bilinear case

$$g(\vec{x}, \vec{y}) = \vec{x}^t G \vec{y} = (A\vec{x}')^t G (A\vec{y}') = (\vec{x}')^t A^t G A \vec{y}',$$

so that the Gram matrix in the primed basis equals $A^t G A$. Analogously, in the sesquilinear case it equals $A^t G \bar{A}$.

In Chapter I we used matrices primarily to express linear mappings and it is interesting to determine whether or not there exists a natural linear mapping associated with g and corresponding to the Gram matrix G. Such a mapping does indeed exist, and its construction provides an equivalent method for defining the inner product.

b) Let $g : L \times L \to \mathcal{K}$ be an inner product. We associate with each vector $l \in L$ a function $g_l : L \to \mathcal{K}$, for which

$$g_l(m) = g(l, m), \ m \in L.$$

This function is linear with respect to m in the bilinear case and antilinear in the sesquilinear case, that is, $g_l \in L^*$ or, correspondingly, $g_l \in \bar{L}^*$ for all l. In addition, the mapping

$$\tilde{g} : L \to L^* \text{ or } \quad L \to \bar{L}^* : l \mapsto g_l = \tilde{g}(l)$$

is linear, it is canonical with respect to g and uniquely defines g according to the formula

$$g(l, m) = (\tilde{g}(l), m),$$

where the outer parentheses on the right indicate a canonical bilinear mapping $L^* \times L \to \mathcal{K}$ or $\bar{L}^* \times L \to \mathbf{C}$.

Conversely, any linear mapping $\tilde{g} : L \to L^*$ (or $L \to \bar{L}^*$) uniquely reconstructs the bilinear mapping $g : L \times L \to \mathcal{K}$ (or $g : L \times \bar{L} \to \mathbf{C}$) according to the same formula

$$g(l, m) = (\tilde{g}(l), m).$$

The tie-up with the preceding construction is as follows: *if a basis* $\{e_1, \dots, e_n\}$ *is selected in* L *and* g *is specified by the matrix* G *in this basis, then* $\tilde{g}$ *is specified by the matrix* G^t *in the bases* $\{e_1, \dots, e_n\}$ *and* $\{e^1, \dots, e^n\}$ *which are mutually dual.*

Indeed, if $\tilde{g}$ is specified by the matrix G^t, then the corresponding inner product g has the following form in the dual bases:

$$g(\vec{x}, \vec{y}) = (\tilde{g}(\vec{x}), \vec{y}) = (\tilde{g}(\vec{x}))^t \vec{y} \text{ (or } (\tilde{g}(\vec{x}))^t \bar{\vec{y}}) =$$

$$= (G^t \vec{x})^t \vec{y} \text{ (or } (G^t \vec{x})^t \bar{\vec{y}}) = \vec{x}^t G \vec{y} \text{ (or } \vec{x}^t G \bar{\vec{y}}),$$

which proves the required result. Here, we made use of the remark made in §7 of Chapter I that the canonical mapping $L^* \times L \to \mathcal{K}$ in the dual basis is defined by the formula $(\vec{x}, \vec{y}) = \vec{x}^t \vec{y}$.

2.3. Symmetry properties of inner products. A permutation of the arguments in a bilinear inner product g defines a new inner product g^t:

$$g^t(l, m) = g(m, l).$$

In the sesquilinear case this operation also changes the position of the "linear" and "sesquilinear" arguments; if we do not want this to happen, then it is more convenient to study $\bar{g}^t$:

$$\bar{g}^t(l,m) = \overline{g(m,l)}.$$

In $\bar{g}^t$ the linear argument will occupy the first position, if it was in the first position in g, while the semilinear argument will correspondingly occupy the second position. The operation $g \to g^t$ or $\bar{g}^t$ is easily described in the language of Gram matrices: it corresponds to the operation $G \to G^t$ or $G \to \bar{G}^t$, respectively (it is assumed that g, g^t and $\bar{g}^t$ are written in the same basis of L). Indeed:

$$g^t(\vec{x},\vec{y}) = g(\vec{y},\vec{x}) = \vec{y}^t G\vec{x} = (\vec{y}^t G\vec{x})^t = \vec{x}^t G^t \vec{y},$$

$$\bar{g}^t(\vec{x},\vec{y}) = \overline{g(\vec{y},\vec{x})} = \overline{\vec{y}^t G\bar{\vec{x}}} = (\bar{\vec{y}}^t \bar{G}\vec{x})^t = \vec{x}^t \bar{G}^t \bar{\vec{y}}.$$

We shall be concerned almost exclusively with inner products that satisfy one of the special conditions of symmetry relative to this operation:

a) $g^t = g$. Such inner products are called ***symmetric***, and the geometry of spaces with a symmetric inner product is called an ***orthogonal geometry***. Symmetric inner products are defined by symmetric Gram matrices G.

b) $g^t = -g$. Such inner products are called ***antisymmetric*** or ***symplectic***, and the corresponding geometries are called ***symplectic geometries***. They correspond to antisymmetric Gram matrices.

Sesquilinear case:

c) $\bar{g}^t = g$. Such inner products are called ***Hermitian symmetric*** or simply ***Hermitian***, and the corresponding geometries are called Hermitian geometries. They correspond to Hermitian Gram matrices. It follows from the condition $\bar{g}^t = g$ that $\bar{g}(l,l) = g(l,l)$ for all $l \in L$, that is, all values of $g(l,l)$ are real.

Hermitian antisymmetric inner products are usually not specially studied, because the mapping $g \mapsto ig$ establishes a bijection between them and Hermitian symmetric inner products:

$$\bar{g}^t = g \Leftrightarrow \overline{(ig)}^t = -ig.$$

The geometric properties of inner products, differing from one another by only a factor, are practically identical. On the contrary, an orthogonal geometry differs in many ways from a symplectic geometry: it is impossible to reduce the relations $g^t = g$ and $g^t = -g$ into one another in this simple manner.

2.4. Orthogonality. Let (L, g) be a vector space with an inner product. The vectors $l_1, l_2 \in L$ are said to be ***orthogonal*** (relative to g), if $g(l_1, l_2) = 0$. The subspaces $L_1, L_2 \subset L$ are said to be orthogonal if $g(l_1, l_2) = 0$ for all $l_1 \in L_1$ and

$l_2 \in L_2$. The main reason that only inner products with one of the symmetry properties mentioned in the preceding section are important is that for them *the property of orthogonality of vectors or subspaces is symmetric relative to these vectors or subspaces.* Indeed: if $g^t = \pm g$ or $g^t = \bar{g}$, then

$$g(l,m) = 0 \Leftrightarrow \pm g^t(l,m) = 0 \Leftrightarrow g(m,l) = 0$$

and analogously in the Hermitian case (with regard to the converse assertion, see Exercise 3).

Unless otherwise stated, in what follows we shall be concerned only with orthogonal, symplectic, or Hermitian inner products. The first application of the concept of orthogonality is contained in the following definition.

2.5. Definition. a) *The kernel of an inner product g in a space L is the set of all vectors $l \in L$ orthogonal to all vectors in L.*

b) *g is non-degenerate if the kernel of g is trivial, that is, it consists only of zero.*

Obviously, the kernel of g coincides with the kernel of the linear mapping $\tilde{g} : L \to L^*$ (or $L \to \bar{L}^*$) and is therefore a linear subspace of L. Therefore instead of the *non-degenerate* form g one can specify the isomorphism $L \to L^*$ (or $\bar{L}^*$). Since the matrix of $\tilde{g}$ is the transposed Gram matrix G^t of a basis of L, *the non-degeneracy of g is equivalent to non-singularity of the Gram matrix* (in any basis). The fact that a non-degenerate orthogonal form g defines an isomorphism $L \to L^*$ is very widely used in tensor algebra and its applications to differential geometry and physics: it provides the basic technique for raising and lowering indices.

The *rank* of g is defined as the dimension of the image of $\tilde{g}$, or as the rank of the Gram matrix G.

2.6. Problem of classification. Let (L_1, g_1) and (L_2, g_2) be two linear spaces with inner products over a field $\mathcal{K}$. By an *isometry* between them, we mean any linear ismorphism $f : L_1 \simeq L_2$ that preserves the values of all inner products, that is,

$$g_1(l,l') = g_2(f(l), f(l')) \quad \text{for all} \quad l, l' \in L_1.$$

We shall call such spaces *isometric* spaces, if an isometry exists between them. Obviously, the identity mapping is an isometry, a composition of isometries is an isometry, and the linear mapping that is the inverse of an isometry is an isometry. In the next section, we shall solve the problem of classification of spaces up to isometry; we shall then study groups of isometries of a space with itself, and we shall show that they include the classical groups described in §4 of Chapter 1.

The classical solution of the problem of classification consists in the fact that any space with an inner product can be partitioned into a direct sum of pairwise orthogonal subspaces of low dimension (one in the orthogonal and Hermitian case, one or two in the symplectic case). We shall therefore complete this section with a direct description of such low-dimensional spaces with a metric.

2.7. One-dimensional orthogonal spaces. Let $\dim L = 1$ and let g be an orthogonal inner product in L. Choose any non-zero vector $l \in L$. If $g(l,l) = 0$ then $g \equiv 0$ so that g is degenerate and equal to zero. If $g(l,l) = a \neq 0$, then for any $x \in \mathcal{K}$, $g(xl, xl) = ax^2$, so that all values of $g(l,l)$ with non-zero vectors in L form in the multiplicative group $\mathcal{K}^* = \mathcal{K}\backslash\{0\}$ of the field $\mathcal{K}$ a coset, with respect to the subgroup, consisting of squares: $\{ax^2 | x \in \mathcal{K}^*\} \in \mathcal{K}^*/(\mathcal{K}^*)^2$. *This coset completely characterizes the non-degenerate symmetric inner product in the one-dimensional space* L: for (L_1, g_1) and (L_2, g_2) two such cosets coincide if and only if these spaces are isometric. Indeed, if $g_1(l_1, l_1) = ax^2$, $g_2(l_2, l_2) = ay^2$, where $l_i \in L_i$, then the mapping $f : l_1 \to y^{-1}xl_2$ defines an isometry of L_1 with L_2, which proves sufficiency. The necessity is obvious.

Since $\mathbf{R}^*/(\mathbf{R}^*)^2 = \{\pm 1\}$ and $\mathbf{C}^* = (\mathbf{C}^*)^2$, we obtain the following important particular cases of classification.

Any 1-dimensional orthogonal space over $\mathbf{R}$ *is isometric to the 1-dimensional coordinate space with one of three scalar products:* xy, $-xy$, 0.

Any 1-dimensional orthogonal space over $\mathbf{C}$ *is isometric to the 1-dimensional coordinate space with one of two scalar products:* xy, 0.

2.8. One-dimensional Hermitian spaces. Here the arguments are analogous. The main field is $\mathbf{C}$; degeneracy of the form is equivalent to its vanishing. If, on the other hand, the form is non-degenerate, then the set of values of $g(l,l)$ for non-zero vectors $l \in L$ is the coset of the subgroup $\mathbf{R}_+^* = \{x \in R^* | x > 0\}$ in the group $\mathbf{C}^*$, because $g(al, al) = a\bar{a}g(l,l) = |a|^2 g(l,l)$ and $|a|^2$ runs through all values in $\mathbf{R}_+^*$, when $a \in \mathbf{C}^*$. But every non-zero complex number z is uniquely represented in the form $re^{i\phi}$, where $r \in R_+^*$, while $e^{i\phi}$ lies on the unit complex circle, which we shall denote by

$$\mathbf{C}_1^* = \{z \in \mathbf{C}^* | \, |z| = 1\}.$$

In the language of groups this defines a direct decomposition $\mathbf{C}^* = \mathbf{R}_+^* \times \mathbf{C}_1^*$ and an isomorphism $\mathbf{C}^*/\mathbf{R}_+^* \to \mathbf{C}_1^*$. Thus non-degenerate sesquilinear forms are classified by complex numbers with unit modulus. However, we have not yet completely taken into account the Hermitian property, which implies that $g(l,l) = \overline{g(l,l)}$, that is, the values of $g(l,l)$ are all real. For this reason, Hermitian forms correspond only to the numbers ± 1 in $\mathbf{C}_1^*$, as in the orthogonal case over $\mathbf{R}$. Final answer:

Any one-dimensional Hermitian space over $\mathbf{C}$ *is isometric to a one-dimensional coordinate space with one of three inner products:* $x\bar{y}, -x\bar{y}, 0$.

We shall call a one-dimensional orthogonal space over $\mathbf{R}$ (or Hermitian space over $\mathbf{C}$) with inner products $xy, -xy, 0$ (or $x\bar{y}, -x\bar{y}, 0$) in an appropriate basis *positive*, *negative*, and *null*. Inner products of non-zero vectors with themselves in these spaces assume only positive, only negative, or only zero values, respectively.

2.9. One-dimensional symplectic spaces. Here we encounter a new situation: any antisymmetric form in a one-dimensional space over a field with characteristic $\neq 2$ is identically equal to zero, in particular, it is singular ! Indeed,

$$g(l,l) = -g(l,l) \Rightarrow 2g(l,l) = 0,$$

$$g(al,bl) = abg(l,l) = 0.$$

As far as the characteristic 2 is concerned, the condition of antisymmetry $g(l,m) = -g(m,l)$ in this case is equivalent to the symmetry condition $g(l,m) = g(m,l)$, so that over such fields a symplectic geometry is identical to an orthogonal geometry. Besides, an orthogonal geometry also has its own peculiarities, and we shall usually ignore this case.

It is therefore clear that one-dimensional symplectic spaces cannot be the building blocks for the construction of general symplectic spaces, and it is necessary to go at least one step further.

2.10. Two-dimensional symplectic spaces. Let (L,g) be a two-dimensional space with a skew-symmetric form g over the field $\mathcal{K}$ with characteristic $\neq 2$. If it is degenerate, then it is automatically a null space. Indeed, let $l \neq 0$ be a vector such that $g(l,m) = 0$ for all $m \in L$. We extend l to the basis $\{l, l'\}$ of L and take into account the fact that $g(l',l') = g(l,l) = 0$ by the preceding subsection. Then, for any $a, b, a', b' \in \mathcal{K}$ we have

$$g(al + a'l', bl + b'l') = abg(l,l) + ab'g(l,l') - a'bg(l,l') + a'b'g(l',l') = 0.$$

Now let g be non-zero and, hence non-degenerate. Then there exists a pair of vectors e_1, e_2 with $g(e_1,e_2) = a \neq 0$ and even with $a = 1$: $g(a^{-1}e_1, e_2) = a^{-1}a = 1$.

Let $g(e_1,e_2) = 1$. Then the vectors e_1 and e_2 are linearly independent and hence form a basis of L: if, say, $e_1 = ae_2$, then $g(ae_2,e_2) = ag(e_2,e_2) = 0$. In the coordinates relative to this basis the inner product g is written in the form

$$g(x_1e_1 + x_2e_2,\ y_1e_1 + y_2e_2) = x_1y_2 - x_2y_1$$

and its Gram matrix is

$$G = \begin{pmatrix} 0 & 1 \\ -1 & 0 \end{pmatrix}.$$

Finally we obtain the following result:

Any two-dimensional symplectic space over a field $\mathcal{K}$ with characteristic $\neq 2$ is isometric to the coordinate space $\mathcal{K}^2$ with the inner product $x_1y_2 - x_2y_1$ or zero.

EXERCISES

1. Let L and M be finite dimensional linear spaces over the field $\mathcal{K}$ and let $g : L \times M \to \mathcal{K}$ be a bilinear mapping. We shall call the set $L_0 = \{l \in L | g(l,m) = 0$ for all $m \in M\}$, the left kernel of g and the set $M_0 = \{m \in M | g(l,m) = 0$ for all $l \in L\}$ the right kernel of g. Prove the following assertions:

a) $\dim L/L_0 = \dim M/M_0$.

b) g induces the bilinear mapping $g' : L/L_0 \times M/M_0 \to \mathcal{K}, g'(l+L_0, m+M_0) = g(l,m)$, for which the left and right kernels are zero.

2. Prove that any bilinear inner product $g : L \times L \to \mathcal{K}$ (over the field $\mathcal{K}$ with characteristic $\neq 2$) can be uniquely decomposed into a sum of symmetric and antisymmetric inner products.

3. Let $g : L \times L \to \mathcal{K}$ be a bilinear inner product such that the property of orthogonality of a pair of vectors is symmetric: from $g(l_1, l_2) = 0$ it follows that $g(l_2, l_1) = 0$. Prove that then g is either symmetric or antisymmetric. (Hint: a) let $l, m, n \in L$. Prove that $g(l, g(l,n)m - g(l,m)n) = 0$. Using the symmetry of orthogonality, deduce that $g(l,n)g(m,l) = g(n,l)g(l,m)$. b) Set $n = l$ and deduce that if $g(l,m) \neq g(m,l)$, then $g(l,l) = 0$. c) Show that $g(n,n) = 0$ for any vector $n \in L$ if g is non-symmetric. To this end, choose l, m with $g(l,m) \neq g(m,l)$ and study separately the cases $g(l,n) \neq g(n,l)$, $g(l,n) = g(n,l)$. d) Show that if $g(n,n) = 0$ for all $n \in L$, then g is antisymmetric.)

4. Give the classification of one-dimensional orthogonal spaces over a finite field $\mathcal{K}$ with characteristic $\neq 2$, by showing that $\mathcal{K}^*/(\mathcal{K}^*)^2$ is the cyclical group of order 2. (Hint: show that the kernel of the homomorphism $\mathcal{K}^* \to \mathcal{K}^* : x \mapsto x^2$ is of order 2, using the fact that the number of roots of any polynomial over a field does not exceed the degree of the polynomial).

5. Let (L, g) be an n-dimensional linear space with a non-degenerate inner product. Prove that the set of vectors $\{e_1, \ldots, e_n\}$ in L is linearly independent if and only if the matrix $(g(e_i, e_j))$ is non-singular.

§3. Classification Theorems.

3.1. The main goal of this section is to classify the finite-dimensional orthogonal, Hermitian, and symplectic spaces up to isometry. Let (L, g) be such a space and let $L_0 \subset L$ be a subspace of it. The restriction of g to L_0 is an inner product in L_0. We shall call L_0 *non-degenerate* if the restriction of g to L_0 is non-degenerate and *isotropic* if the restriction of g to L_0 equals zero. It is significant that even if L is non-degenerate, the restrictions of g to non-trivial subspaces can be degenerate or zero. For example in the symplectic case, all one-dimensional subspaces are degenerate and in an orthogonal space $\mathbf{R}^2$ with the product $x_1y_1 - x_2y_2$ the subspace spanned by the vector (1,1) is degenerate.

The *orthogonal complement* of the subspace $L_0 \subset L$ is the set

$$L_0^\perp = \{l \in L \mid g(l_0, l) = 0 \quad \text{for} \quad \text{all} \quad l_0 \in L_0\}$$

(not to be confused with the orthogonal complement to L_0, lying in L^*, introduced in Chapter I ! Here we shall not make use of it). It is easy to see that $L_0^\perp$ is a linear subspace of L.

3.2. Proposition. *Let (L, g) be finite-dimensional.*

a) *If the subspace $L_0 \subset L$ is non-degenerate, then $L = L_0 \bigoplus L_0^\perp$.*

b) *If both subspaces L_0 and $L_0^\perp$ are non-degenerate, then $(L_0^\perp)^\perp = L_0$.*

Proof. a) Let $\tilde{g} : L \to L^*$ (or $\bar{L}^*$) be the mapping associated with g, as in the preceding section. We denote by $\tilde{g}_0$ its restriction to L_0, $\tilde{g}_0 : L_0 \to L^*$ (or $\bar{L}^*$). If L_0 is non-degenerate, then $\ker \tilde{g}_0 = 0$; otherwise L_0 contains a vector that is orthogonal to all of L and, in particular, to L_0. Therefore, $\dim \operatorname{im} \tilde{g}_0 = \dim L_0$. Hence when l_0 runs through L_0, the linear forms $g(l_0, \cdot)$ as a function of the second argument from L or $\bar{L}$ run through a $\dim L_0$-dimensional space of linear forms on L or $\bar{L}$. Since $L_0^\perp$ is the intersection of the kernels of these forms, $\dim L_0^\perp = \dim L - \dim L_0$, that is,

$$\dim L_0 + \dim L_0^\perp = \dim L.$$

On the other hand, it follows from the non-degeneracy of L_0 that $L_0 \cap L_0^\perp = \{0\}$, because $L_0 \cap L_0^\perp$ is the kernel of the restriction of g to L_0. For this reason, the sum $L + L_0^\perp$ is a direct sum; but its dimension equals $\dim L$ so that $L_0 \bigoplus L_0^\perp = L$.

b) It is clear from the definitions that $L_0 \subset (L_0^\perp)^\perp$. On the other hand, if $L_0, L_0^\perp$ are non-degenerate, then from the preceding result

$$\dim(L_0^\perp)^\perp = \dim L - \dim L_0^\perp = \dim L_0.$$

This completes the proof.

3.3. Theorem. *Let (L, g) be a finite-dimensional orthogonal (over a field with characteristic $\neq 2$) Hermitian or symplectic space. Then there exists a decomposition of L into a direct sum of pairwise orthogonal subspaces*

$$L = L_1 \oplus \ldots \oplus L_m,$$

that are 1-dimensional in the orthogonal and Hermitian case and 1-dimensional degenerate or 2-dimensional non-degenerate in the symplectic case.

Proof. We perform induction on the dimension of L. The case $\dim L = 1$ is trivial: let $\dim L \geq 2$. If g is zero, there is nothing to prove. If g is non-zero, then in the symplectic case there exists a pair of vectors $l_1, l_2 \in L$ with $g(l_1, l_2) \neq 0$. According to §2.10, the subspace L_0 spanned by them is non-degenerate. According to Proposition 3.2, $L = L_0 \bigoplus L_0^{\perp}$, and we can further decompose $L_0^{\perp}$, as formulated in the theorem, by the induction hypothesis. This will give the required decomposition of L.

In the orthogonal and Hermitian case we shall show that the existence of a non-degenerate one-dimensional subspace L_0 follows from the non-triviality of g. After checking this, we shall be able to set $L = L_0 \bigoplus L_0^{\perp}$ and apply the previous arguments, that is, induction on the dimension of L.

Indeed, we assume that $g(l, l) = 0$ for all $l \in L$, and we shall show that then $g \equiv 0$. In fact, for all $l_1, l_2 \in L$ we have

$$0 = g(l_1 + l_2, l_1 + l_2) = g(l_1, l_1) + 2g(l_1, l_2) + g(l_2, l_2) = 2g(l_1, l_2)$$

or

$$0 = g(l_1 + l_2, l_1 + l_2) = g(l_1, l_1) + 2\,\mathrm{Re}\,g(l_1, l_2) + g(l_2, l_2) = 2\,\mathrm{Re}\,g(l_1, l_2).$$

In the orthogonal case it follows immediately from here that $g(l_1, l_2) = 0$. In the Hermitian case, we obtain only that $\mathrm{Re}\,g(l_1, l_2) = 0$, that is, $g(l_1, l_2) = ia$, $a \in \mathbf{R}$. But if $a \neq 0$, then also

$$0 = \mathrm{Re}\,g\left((ia)^{-1} l_1, l_2\right) = \mathrm{Re}(ia)^{-1} g(l_1, l_2) = 1$$

which is a contradiction.

This completes the proof.

We now proceed to the problem of uniqueness. In itself, the decomposition into an orthogonal direct sum, whose existence is asserted in Theorem 3.3, is by no means unique, except for the trivial cases of dimension 1 (or 2 in the symplectic case). Over general fields, in the case of an orthogonal geometry, the collection of invariants $a_i \in \mathcal{K}^*/\mathcal{K}^{*2}$, which characterizes the restriction of g to the one-dimensional subspaces L_i, is also not unique. The exact answer to the question of the classification of orthogonal spaces depends strongly on the properties of the main field, and for $\mathcal{K} = \mathbf{Q}$, for example, is related to subtle number-theoretic facts, such as the quadratic law of reciprocity. In the orthogonal case, therefore, we shall restrict our attention to the description of the result for $\mathcal{K} = \mathbf{R}$ and $\mathbf{C}$ (for further details see §14).

3.4. Invariants of metric spaces. Let (L, g) be a space with an inner product. We set $n = \dim L$, $r_0 = \dim L_0$, where L_0 is the kernel of the form g. In addition, we introduce two additional invariants, referring only to the orthogonal, for $\mathcal{K} = \mathbf{R}$, and Hermitian geometries: r_+ and r_-, the number of positive and negative one-dimensional subspaces L_i in an orthogonal decomposition of L into a direct sum, as in Theorem 3.3.

Obviously, $r_0 \leq n$ and $n = r_0 + r_+ + r_-$ for Hermitian and orthogonal geometries over $\mathbf{R}$. The set (r_0, r_+, r_-) is called the signature of the space. When $r_0 = 0$, (r_+, r_-) or $r_+ - r_-$ are also sometimes called the signature (provided that $n = r_+ + r_-$ is known).

We can now formulate the uniqueness theorem.

3.5. Theorem. a) *Symplectic spaces over an arbitrary field, as well as orthogonal spaces over* $\mathbf{C}$ *are determined up to isometry by two integers* n, r_0, *that is, the dimensions of the space and of the kernel of the inner product.*

b) *Orthogonal spaces over* $\mathbf{R}$ *and Hermitian spaces over* $\mathbf{C}$ *are determined, up to isometry, by the signature* (r_0, r_+, r_-), *which does not depend on the choice of orthogonal decomposition (this assertion is called the inertia theorem).*

Proof. a) Let (L, g) be a symplectic or orthogonal space over $\mathbf{C}$. We shall study its direct decomposition $L = \bigoplus_{i=1}^n L_i$, as in Theorem 3.3, and we shall show that r_0 equals the number of one-dimensional spaces in this decomposition that are degenerate for g. Indeed, the sum of these spaces L_0 coincides with the kernel of g. Indeed, it is obvious that it is contained in this kernel, because the elements of L_0 are orthogonal both to L_0 and to the remaining terms. On the other hand, if $L_0 = \bigoplus_{i=1}^{r_0} L_i$ and

$$l = \sum_{j=1}^{n} l_j, \; l_j \in L_j, \; \exists j > r_0, \; l_j \neq 0,$$

then

$$g(l, l_j) = g(l_j, l_j) \neq 0$$

in the orthogonal case and there exists a vector $l'_j \in L_j$ with

$$g(l, l'_j) = g(l_j, l'_j) \neq 0$$

in the symplectic case, because otherwise the kernel of the restriction of g to L_j would be non-trivial and the restriction of g to L_j would be null according to §2.10, contradicting the fact that $j > r_0$. Therefore $l \notin$ (kernel of g), and $L_0 =$ (kernel of g). If now (L, g) and (L', g') are two such spaces with identical n and r_0, then, after constructing their orthogonal direct decompositions $L = \bigoplus_{i=1}^n L_i$ and $L' = \bigoplus_{i=1}^n L'_i$, for which (kernel of g)$= \bigoplus_{i=1}^{r_0} L_i$ and (kernel of g')$= \bigoplus_{i=1}^{r_0} L'_i$, we can define the isometry of (L,g) to (L,g') as the direct sum of the isometries $\bigoplus f_i$, $f_i : L_i \to L_i'$ which exist by virtue of the results of §2.7 and §2.10.

b) Now let (L,g) and (L',g') be a pair of orthogonal spaces over $\mathbf{R}$ or Hermitian spaces over $\mathbf{C}$ with the signatures (r_0, r_+, r_-) and (r_0', r_+', r_-'), defined with the help of some orthogonal decompositions $L = \bigoplus L_i$, $L' = \bigoplus L_i'$ as in Theorem 3.3. Assume that an isometry exists between them. Then, first of all, $\dim L = \dim L'$ so that $r_0 + r_+ + r_- = r_0' + r_+' + r_-'$. Furthermore just as in the preceding section, we can verify that r_0 equals the dimension of the kernel of g, and r_0' equals the dimension of the kernel of g', and these kernels are the sums of the null spaces L_i and L_i' in the corresponding decompositions. Since the isometry determines a linear isomorphism between the kernels, we have $r_0 = r_0'$ and $r_+ + r_- = r_+' + r_-'$.

It remains to verify that $r_+ = r_+'$, $r_- = r_-'$. We set $L = L_0 \bigoplus L_+ \bigoplus L_-$, $L' = L_0' \bigoplus L_+' \bigoplus L_-'$, where L_0, L_+, L_- are the sums of the null, positive, and negative subspaces of the starting decomposition of L, and correspondingly for L'. We assume that $r_+ = \dim L_+ > r_+' = \dim L_+'$, and we arrive at a contradiction; the possibility $r_+ < r_+'$ is analysed analogously. We restrict the isometry $f : L \to L'$ to $L_+ \subset L$. Each vector $f(l)$ is uniquely represented as a sum

$$f(l) = f(l)_0 + f(l)_+ + f(l)_-,$$

where $f(l)_+ \in L_+'$ and so on. The mapping $L_+ \to L_+'$, $l \mapsto f(l)_+$ is linear. Since by assumption $\dim L_+ > \dim L_+'$, there exists a non-zero vector $l \in L_+$ for which $f(l)_+ = 0$, so that

$$f(l) = f(l)_0 + f(l)_-.$$

But $g(l,l) > 0$ so that $l \in L_+$ and L_+ is the orthogonal direct sum of positive one-dimensional spaces. Since f is an isometry, we must also have $g'(f(l), f(l)) > 0$. On the other hand,

$$g'(f(l), f(l)) = g'(f(l)_0 + f(l)_-,\ f(l)_0 + f(l)_-) = g'(f(l)_-, f(l)_-) \leq 0.$$

This contradiction completes the proof of the fact that the signatures of isometric spaces, calculated from any orthogonal decompositions, are identical.

Conversely, if (L,g) and (L',g') are two subspaces with identical signatures, then it is possible to establish between subspaces, from their orthogonal decompositions $L = \bigoplus L_i$ and $L' = \bigoplus L_i'$, a one-to-one correspondence $L_i \leftrightarrow L_i'$, preserving the sign of the restriction of g to L_i and of g' to L_i', respectively. According to the results of §2.7 and §2.8, the isometries $f_i : L_i \to L_i'$ exist, and their direct sum $\bigoplus f_i$ will be an isometry between L and L'.

We shall now derive several corollaries and reformulations of Theorems 3.3 and 3.5, which underscore the different aspects of the situation.

3.6. Bases. Let (L, g) be a space with a scalar product. The basis $\{e_1, \dots, e_n\}$ of L is called *orthogonal* if $g(e_i, e_j) = 0$ for all $i \neq j$. It follows from Theorem 3.5 that *any orthogonal or Hermitian space has an orthogonal basis.* Indeed, it is sufficient to construct the decomposition $L = \bigoplus L_i$ into orthogonal one-dimensional subspaces and then choose $e_i \in L_i, e_i \neq 0$.

An orthogonal basis $\{e_i\}$ is called *orthonormal* if $g(e_i, e_i) = 0$ or ± 1 for all i. The discussion at the end of §3.2 shows that *any orthogonal space over* **R** *or* **C** *and any Hermitian space has an orthonormal basis.* Theorem 3.5 shows that the number of elements e of the orthonormal basis with $g(e, e) = 0, 1$ or -1 does not depend on the basis for $\mathcal{K} = \mathbf{R}$ (orthogonal case) and $\mathcal{K} = \mathbf{C}$ (Hermitian case). In the orthogonal case over **C** it is always possible to make $g(e_i, e_i) = 0$ or 1, and the number of such vectors in the basis does not depend on the basis itself. The Gram matrix of an orthonormal basis has the form

$$\left(\begin{array}{c|c|c} E_{r_+} & 0 & 0 \\ \hline 0 & -E_{r_-} & 0 \\ \hline 0 & 0 & 0 \end{array}\right)$$

(with suitable ordering). The concept of an orthonormal basis is most often used in the non-degenerate case, when there are no vectors e_i with $g(e_i, e_i) = 0$. The next simple, but important, formula makes it possible to write explicitly the coefficients in the decomposition of any vector $e \in L$ with respect to an orthogonal basis (in the non-degenerate case):

$$e = \sum_{i=1}^{n} \frac{g(e, e_i)}{g(e_i, e_i)} e_i.$$

Indeed, the inner products on the left and right sides with all e_i are equal, and non-degeneracy implies that if $g(e, e_i) = g(e', e_i)$ for all i, then $e = e'$, because $e - e'$ lies in the kernel of the form g.

In a symplectic space an orthogonal basis can evidently exist only if $g = 0$. Theorem 3.3, on the other hand, guarantees the existence of a *symplectic basis* $\{e_1, e_2, \dots, e_r; e_{r+1}, \dots, e_{2r}; e_{2r+1}, \dots, e_n\}$, which is characterized by the fact that

$$g(e_i, e_{r+i}) = -g(e_{r+1}, e_i) = 1, \; i = 1, \dots, r,$$

while all remaining inner products of pairs of vectors equal zero. Indeed, we must decompose L into an orthogonal direct sum of two-dimensional non-singular subspaces L_i, $1 \leq i \leq r$, and one-dimensional singular subspaces L_j, $2r + 1 \leq j \leq n$, and choose for $\{e_i, e_{r+1}\}$ $(1 \leq i \leq r)$ the basis of L_i constructed in §2.10, and for e_j $(2r + 1 \leq j \leq n)$ any non-zero vector in L_j.

The Gram matrix of a symplectic basis has the form

$$\left(\begin{array}{c|c|c} 0 & E_r & 0 \\ \hline -E_r & 0 & 0 \\ \hline 0 & 0 & 0 \end{array}\right).$$

The rank of the symplectic form, according to Theorem 3.5, equals $2r$. In particular, *the dimension of a non-singular symplectic space is necessarily even.*

Let L be a non-singular symplectic space and $\{e_1, \dots, e_r; e_{r+1}, \dots, e_{2r}\}$ a symplectic basis of it. Set L_1 equal to the linear span of $\{e_1, \dots, e_r\}$ and L_2 equal the linear span of $\{e_{r+1}, \dots, e_{2r}\}$. Evidently the spaces L_1 and L_2 are isotropic, their dimensions equal one-half the dimension of L, and $L = L_1 \bigoplus L_2$. The canonical mapping $\tilde{g} : | \to L^*$ determines the mapping

$$\tilde{g}_1 : L_2 \to L_1^*; \quad \tilde{g}_1(l_2)(l_1) = g(l_2, l_1).$$

This mapping is an isomorphism, because $\dim L_2 = \dim L_1 = \dim L_1^*$ and $\ker \tilde{g}_1 = 0$: the vector from $\ker \tilde{g}_1$ is orthogonal to L_2, because L_2 is isotropic, and to L_1 by definition, while L is non-degenerate.

It follows that *any non-degenerate symplectic space is isometric to a space of the form $L = L_1^* \bigoplus L_1$ with symplectic form*

$$g((f,l),(f',l')) = f(l') - f'(l); \quad f, f' \in L_1^*, \quad l, l' \in L_1.$$

Further details are given in §12.

3.7. Matrices. Describing inner products by their Gram matrices and transforming from the accidental basis to the orthogonal or symplectic basis, we obtain, by virtue of the results of §§2.2 and 3.6, the following facts:

a) any quadratic symmetric matrix G over the field $\mathcal{K}$ can be reduced to diagonal form via the transformation $G \mapsto A^t G A$, where A is non-singular. If $\mathcal{K} = \mathbf{R}$, it is possible to achieve a situation in which only 0 and ± 1 appear on the diagonal, and if $\mathcal{K} = \mathbf{C}$ only 0 and 1 appear on the diagonal; the numbers of 0 and ± 1 (correpondingly 0 and 1) will depend only on G, and not on A.

b) Any quadratic antisymmetric matrix G over the field $\mathcal{K}$ with characteristic $\neq 2$ can be reduced by the transformation $G \mapsto A^t G A$, where A is non-singular, to the form

$$\left(\begin{array}{c|c|c} 0 & E_r & 0 \\ \hline -E_r & 0 & 0 \\ \hline 0 & 0 & 0 \end{array}\right).$$

The number $2r$ equals the rank of G.

c) Any Hermitian matrix G over $\mathbf{C}$ can be reduced to diagonal form with the numbers 0, ± 1 on the diagonal by the transformation $G \mapsto A^t G \bar{A}$, where A is non-singular. The numbers of 0 and ± 1 depend only on G.

3.8. Bilinear forms. If the vectors in the space (L, g) with a fixed basis are written in terms of the coordinates in this basis, then the expression of g in terms of the coordinates is a bilinear form of $2n$ variables, $n = \dim L$:

$$g = (x_1, \ldots, x_n; y_1, \ldots, y_n) - \sum_{i,j=1}^{n} g_{ij} x_i y_j = \vec{x}^t G \vec{y},$$

where G is the Gram matrix of the basis. A substitution of the basis reduces to a linear transformation of the variables $x_1, \ldots, x_n$ and $y_1, \ldots, y_n$ with the help of the same non-singular matrix A in the bilinear case (or the matrix A for $\vec{x}$, $\bar{A}$ for $\vec{y}$ in the sesquilinear case). The preceding results show that depending on the symmetry properties of the matrix G the form can be reduced by such a transformation to one of the following forms, called canonical forms.

Orthogonal case over any field:

$$g(\vec{x}, \vec{y}) = \sum_{i=1}^{n} a_i x_i y_i;$$

it is possible to achieve $a_i = 0, \pm 1$ over the field $\mathbf{R}$ and $a_i = 0$ or 1 over the field $\mathbf{C}$.

Hermitian case (sesquilinear form):

$$g(\vec{x}, \vec{y}) = \sum_{i=1}^{n} a_i x_i \bar{y}_i;$$

$a_i = 0$ or 1.

Symplectic case: $n = 2r + r_0$, in which case

$$g(\vec{x}, \vec{y}) = \sum_{i=1}^{r} (x_i y_{r+i} - y_i x_{r+i}).$$

3.9. Quadratic forms. A *quadratic form* q on the space L is a mapping $q : L \to \mathcal{K}$ for which there exists a bilinear form $h : L \times L \to \mathcal{K}$ with the property

$$q(l) = h(l, l) \quad \text{for all} \quad l \in L.$$

We shall show that if the characteristic of the field $\mathcal{K}$ is not equal to 2, then for any quadratic form q there exists a *unique symmetric* bilinear form g with the property $q(l) = g(l, l)$, called the *polarization of* q.

To prove existence, we set $q(l) = h(l, l)$, where h is the starting bilinear form, and

$$g(l, m) = \frac{1}{2}[h(l, m) + h(m, l)].$$

Obviously, g is symmetric, that is, $g(l, m) = g(m, l)$. In addition,

$$g(l, l) = \frac{1}{2}[h(l, l) + h(l, l)] = q(l).$$

The bilinearity of g follows immediately from the bilinearity of h.

To prove uniqueness, we note that if $q(l) = g_1(l,l) = g_2(l,l)$, where g_1, g_2 are symmetric and bilinear, then the form $g = g_1 - g_2$ is also symmetric and bilinear, and $g(l,l) = 0$ for all $l \in L$. But according to the arguments in the proof of Theorem 3.3, it follows that $g(l,m) = 0$ for all $l, m \in L$, which completes the proof. We note that if $q(l) = g(l,l)$, and g is symmetric, then

$$g(l,m) = \frac{1}{2}[q(l+m) - q(l) - q(m)].$$

We have thus established that orthogonal geometries (over fields with characteristic $\neq 2$) can be interpreted as geometries of pairs (L,q), where $q : L \to \mathcal{K}$ is a quadratic form. In terms of coordinates, the quadratic form is written in the form

$$q(\vec{x}) = \sum_{i,j=1}^{n} a_{ij} x_i x_j,$$

where the matrix (a_{ij}) is determined uniquely if it is symmetric: $a_{ij} = a_{ji}$. For example,

$$q(x_1, x_2) = a_{11}x_1^2 + 2a_{12}x_1x_2 + a_{22}x_2^2.$$

Classification theorems indicate that a quadratic form can be reduced by a non-degenerate linear substitution of variables to a sum of squares with coefficients:

$$q(\vec{x}) = \sum_{i=1}^{n} a_i x_i^2.$$

If $\mathcal{K} = \mathbf{R}$, it may be assumed that $a_i = 0, \pm 1$; the numbers r_0, r_+, r_- of zeros, pluses and minuses are determined uniquely and comprise the signature of the starting quadratic form; $r_+ + r_-$ is its rank. If $\mathcal{K} = \mathbf{C}$, it may be assumed that $a_i = 0, 1$; the number of 1's is the rank of the form; the rank is also determined uniquely.

§4. The Orthogonalization Algorithm and Orthogonal Polynomials

In this section we shall describe the classical algorithms for constructing orthogonal bases and we shall present important examples of such bases in function spaces.

1. Reduction of a quadratic form to a sum of squares. Let

$$q(x_1, \ldots, x_n) = \sum_{i,j=1}^{n} a_{ij} x_i x_j, \quad a_{ij} = a_{ji},$$

be a quadratic form over the field $\mathcal{K}$ with characteristic $\neq 2$. The following procedure is a convenient practical method for finding a linear substitution of variables x_i that reduces q to the sum of squares (with coefficients).

Case 1. *There exists a non-zero diagonal coefficient.* Renumbering the variables, we can assume that $a_{11} \neq 0$. Then

$$q(x_1, \dots, x_n) = a_{11}x_1^2 + x_1(2a_{12}x_2 + \dots + 2a_{1n}x_n) + q'(x_2, \dots, x_n),$$

where q' is a quadratic form of $\leq n-1$ variables. Separating off the complete square, we find

$$q(x_1, \dots, x_n) = a_{11}\left(x_1 + \frac{a_{12}}{a_{11}}x_2 + \dots + \frac{q_{1n}}{a_{11}}x_n\right)^2 + q''(x_2, \dots, x_n),$$

where q'' is a new quadratic form of $\leq n-1$ variables. Setting

$$y_1 = x_1 + a_{11}^{-1}(a_{12}x_2 + \dots + a_{1n}x_n), \quad y_2 = x_2, \dots, \quad y_n = x_n,$$

we obtain in the new variables the form

$$a_{11}y_1^2 + q''(y_2, \dots, y_n),$$

and the next step of the algorithm consists in applying it to q''.

Case 2. *All diagonal coefficients equal zero.* If, in general, $q = 0$, then there is nothing to do: $q = \sum_{i=1}^n 0 \cdot x_i^2$. Otherwise, renumbering the variables, we can assume that $a_{12} \neq 0$. Then

$$q(x_1, \dots x_n) =$$
$$= 2a_{12}x_1x_2 + x_1l_1(x_3, \dots, x_n) + x_2l_2(x_3, \dots, x_n) + q'(x_3, \dots, x_n),$$

where l_1, l_2 are linear forms and q' is a quadratic form. We set

$$x_1 = y_1 + y_2, \ x_2 = y_1 - y_2, \ x_i = y_i, \ i \geq 3.$$

In the new variables the form q becomes

$$2a_{12}(y_1^2 - y_2^2) + q''(y_1, y_2, \dots, y_n),$$

where q'' does not contain terms with y_1^2, y_2^2. We can therefore apply to it the method of separating the complete square and again reduce the problem to the term of lowest degree. Successive application of these steps yields a form of the type $\sum_{i=1}^n a_i z_i^2$. The final substitution of variables will be non-degenerate, because all the intermediate substitutions are non-degenerate.

The final substitution of variables $u_i = \sqrt{|a_i|}z_i$ with $a_i \neq 0$ in the case $\mathcal{K} = \mathbf{R}$ and $u_i = \sqrt{a_i}z_i$ with $a_i \neq 0$ in the case $\mathcal{K} = \mathbf{C}$, will reduce the form to a sum of squares with coefficients $0, \pm 1$, or $0, 1$.

4.2. The Gram-Schmidt orthogonalization algorithm. This algorithm is very similar to the one described in the preceding section, but it is formulated in more geometric terms. We shall examine simultaneously the orthogonal and Hermitian case.

The space (L, g) with an orthogonal or Hermitian matrix, defined in the basis $\{e'_1, \dots, e'_n\}$ is given. Let L_i be the subspace spanned by $e'_1, \dots, e'_i$, $i = 1, \dots, n$. The orthogonalization process, applied to the basis $\{e'_1, \dots, e'_n\}$, can be viewed as a constructive proof of the following result.

4.3. Proposition. *Suppose that in the above notation, all subspaces $L_1, \dots, L_n$ are non-degenerate. Then there exists an orthogonal basis $\{e_1, \dots, e_n\}$ of the space L such that the linear span of $\{e_1, \dots, e_i\}$ coincides with L_i for all $i = 1, \dots, n$. It is called the result of the orthogonalization of the starting basis $\{e'_1, \dots, e'_n\}$. Each vector e_i is determined uniquely up to a non-zero scalar factor.*

Proof. We construct e_i by induction over i. For e_1, we can take e'_1. If $e_1, \dots, e_{i-1}$ have already been constructed, then we seek e_i in the form

$$e_i = e'_i - \sum_{j=1}^{i-i} x_j e'_j, \ x_j \in \mathcal{K}.$$

Since $\{e'_1, \dots, e'_i\}$ generate L_i, and $\{e'_1, \dots, e'_{i-1}\}$ and $\{e_1, \dots, e_{i-1}\}$ generate L_{i-1}, any such vector e_i together with $e_1, \dots, e_{i-1}$ will generate L_i. It is therefore sufficient to achieve a situation such that e_i is orthogonal to $e_1, \dots, e_{i-1}$ or, which is the same thing, to $e'_1, \dots, e'_{i-1}$. These conditions mean that $g(e_i, e'_k) = 0$, $k = 1, \dots, i-1$, or

$$\sum_{j=1}^{i-1} x_j g(e'_j, e'_k) = g(e'_i, e'_k) \quad k = 1, \dots, i-1.$$

This is a system of $i-1$ linear equations for $i-1$ unknowns x_j. Its matrix of coefficients is the Gram matrix of the basis $\{e'_1, \dots, e'_{i-1}\}$ of the space L_{i-1}. It is non-singular by definition, so that the x_j exist and are determined uniquely. Any non-zero vector $\tilde{e}_i$ orthogonal to L_{i-1} must be proportional to e_i.

A simpler and immediately solvable system of equations is obtained if e_i is sought in the form

$$e_i = e'_i - \sum_{j=1}^{i-1} y_j e_j, \ y_j \in \mathcal{K},$$

assuming that $e_1, \dots, e_{i-1}$ have already been found. Since $e_1, \dots, e_{i-1}$ are pairwise orthogonal, from the condition $g(e_i, e_j) = 0$, $1 \le j \le i-1$, we find that

$$y_j = \frac{g(e'_i, e_j)}{g(e_j, e_j)}, \ j = 1, \dots, i-1.$$

The whole point of this proof is to write out explicitly the systems of linear equations whose successive solution determines e_i. We note that the matrices of the coefficients of the first system are the successive diagonal minors of the Gram matrix of the starting basis:

$$G_i = (g(e'_j, e'_k)),\ 1 \le j,\ k \le i.$$

If we were not striving to construct an algorithm it would have been simpler to argue as follows: Proposition 3.2 and the non-degeneracy of L_{i-1} imply that

$$L_i = L_{i-1} \oplus L_{i-1}^{\perp};\ \dim L_{i-1}^{\perp} = \dim L_i - \dim L_{i-1} = 1.$$

We now take for e_i any non-zero vector from $L_{i-1}^{\perp}$.

4.4. Remarks and corollaries. a) The Gram–Schmidt orthogonalization process is most often used in a situation when $g(l,l) > 0$ for all $l \in L$, $l \neq 0$, that is, it is applied to *Euclidean* and *unitary* spaces, which we shall study in detail later. In this case *all subspaces* of L are automatically *non-degenerate*, and any starting basis can be orthogonalized. The form g with this property and its Gram matrices are said to be *positive definite*.

b) If $\mathcal{K} = \mathbf{R}$ or $\mathbf{C}$, an orthogonalized basis can be constructed immediately. To do this, after the vector e_i is determined, as in the proof of the proposition, it must be replaced here by $|g(e_i, e_i)|^{-1/2} e_i$ or $g(e_i, e_i)^{-1/2} e_i$ (for orthogonal spaces over $\mathbf{C}$).

c) *Any orthogonal basis of a non-degenerate space $L_0 \subset L$ can be extended to an orthogonal basis of the entire space L.*

Indeed, $L = L_0 \bigoplus L_0^{\perp}$, and the orthogonal basis of $L_0^{\perp}$ can be taken as the extension. It can be found by the Gram–Schmidt method, if the basis of L_0 is somehow extended to a basis of L, taking care that the intermediate subspaces are non-degenerate.

d) Let $\{e'_1, \ldots, e'_n\}$ be a basis of (L, g) and let $\{e_1, \ldots, e_n\}$ be its orthogonalized form. We set $a_i = g(e_i, e_i)$; these are the only non-zero elements of the Gram matrix of the basis $\{e_i\}$. We shall assume that g is Hermitian or g is orthogonal over $\mathbf{R}$. Then all numbers a_i are real and the signature of g is determined by the number of positive and negative numbers a_i. We shall show how to reconstruct it from the minors of the starting Gram matrix $G = \{g(e'_i, e'_k)\}$. Let G_i be the ith diagonal minor, that is, the Gram matrix of $\{e'_1, \ldots, e'_i\}$. If A_i is the matrix of the transformation to the basis $\{e_1, \ldots, e_i\}$ then

$$\det(g(e_k, e_j))_{1 \le k, j \le i} = a_1 \ldots a_i = \det(A_i^t G_i A_i) = \det G_i (\det A_i)^2$$

in the orthogonal case or

$$a_1 \ldots a_i = \det(A_i^t G_i \bar{A}_i) = \det G_i |\det A_i|^2$$

in the Hermitian case. Therefore, it is always true that

$$\operatorname{sign} a_1 \dots a_i = \operatorname{sign} \det G_i.$$

Thus, *the signature of the form g is determined by the number of positive and negative elements of the sequence*

$$\det G_1, \ \frac{\det G_2}{\det G_1}, \dots, \ \frac{\det G_n}{\det G_{n-1}}.$$

In particular, *the form g (and its matrix G) is positive definite, if and only if all minors of* $\det G_i$ *are positive* (we recall that G is either real and symmetric or complex and Hermitian symmetric). This result is called *Sylvester's criterion.*

More generally, for a non-degenerate quadratic form over any field the identity

$$a_1 \dots a_i = \det G_i (\det A_i)^2$$

shows that the starting *form with a symmetric matrix G and non-singular diagonal minors G_i can be reduced by a linear transformation of variables to the form*

$$\sum_{i=1}^{n} \frac{\det G_i}{\det G_{i-1}} y_i^2, \ \det G_0 = 1,$$

because the squares $(\det A_i)^2$, which prevent the expression of a_i directly in terms of $\det G_j$, can be incorporated by cofactors into the variables. This result is called *Jacobi's theorem.*

4.5. Bilinear forms in function space. We shall study the functions f_1, f_2 defined on the interval (a, b) of the real axis (a can be $-\infty$ and b can be ∞) and assuming real or complex values. Let $G(x)$ be a fixed function of $x \in (a, b)$. Bilinear forms on function spaces in analysis are often defined by expressions of the type

$$g(f_1, f_2) = \int_a^b G(x) f_1(x) f_2(x) dx$$

or (the sesquilinear case)

$$g(f_1, f_2) = \int_a^b G(x) f_1(x) \overline{f_2(x)} dx.$$

Of course, G, f_1 and f_2 must satisfy some integrability conditions; in the following examples they will be automatically satisfied.

The function G is called the *weight* of the form g. The value

$$g(f, f) = \int_a^b G(x) f(x)^2 dx \quad \text{or} \quad \int_a^b G(x) |f(x)|^2 dx$$

is the *weighted mean-square* of the function f (with weight G); if $G \geq 0$, then it can be viewed as some integral measure of the deviation of f from zero. The typical problem of the approximation of the function f by linear combinations of some given collection of functions $f_1, \ldots, f_n, \ldots$ consists in searching for coefficients $a_1, \ldots, a_n, \ldots$, which for fixed n minimize the weighted mean-square value of the function

$$f - \sum a_i f_i.$$

It will later become evident that the coefficients a_i are especially simply found in the case when $\{f_i\}$ form an orthogonal or orthonormal system relative to the inner product g. In this section we shall restrict ourselves to an explicit description of several important orthogonal systems.

4.6. Trigonometric polynomials. Here $G = 1$, $(a, b) = (0, 2\pi)$. *Trigonometric polynomials* (or *Fourier polynomials*) are finite linear combinations of the functions $\cos nx$ and $\sin nx$ or finite linear combinations of the functions e^{inx}, $n \in \mathbf{Z}$. The former are usually used in the theory of real-valued functions and the latter are used in the theory of complex-valued functions. Since $e^{inx} = \cos nx + i \sin nx$, over $\mathbf{C}$ both spaces of Fourier polynomials coincide. A bilinear metric is used over $\mathbf{R}$ and a sesquilinear metric is used over $\mathbf{C}$. The functions $\{1, \cos nx, \sin nx | n \geq 1\}$ and $\{e^{inx} | n \in \mathbf{Z}\}$ are linearly independent (over both $\mathbf{R}$ and $\mathbf{C}$). In addition they form an orthogonal system, as follows from the easily verifiable formulas:

$$\int_0^{2\pi} \cos mx \cos nx \, dx = \int_0^{2\pi} \sin mx \sin nx \, dx = \begin{cases} \pi & \text{for } m = n > 0, \\ 0 & \text{for } m \neq n, \end{cases}$$

$$\int_0^{2\pi} \cos mx \sin nx \, dx = 0$$

$$\int_0^{2\pi} e^{imx} \overline{e^{inx}} \, dx = \int_0^{2\pi} e^{i(m-n)x} \, dx = \begin{cases} 2\pi & \text{for } m = n, \\ 0 & \text{for } m \neq n. \end{cases}$$

The systems

$$\left\{ \frac{1}{\sqrt{2\pi}}, \frac{1}{\sqrt{\pi}} \cos nx, \frac{1}{\sqrt{\pi}} \sin nx | n \geq 1 \right\} \quad \text{and} \quad \left\{ \frac{1}{\sqrt{2\pi}} e^{inx} | n \in \mathbf{Z} \right\}$$

are therefore orthonormalized. The inner products of any function f in $[0, 2\pi]$ with the elements of these orthonormal systems are called the *Fourier coefficients* of this function:

$$a_0 = \frac{1}{\sqrt{2\pi}} \int_0^{2\pi} f(x) \, dx,$$

$$a_n = \frac{1}{\sqrt{\pi}} \int_0^{2\pi} f(x) \cos nx \, dx, \; n \geq 1,$$

$$b_n = \frac{1}{\sqrt{\pi}} \int_0^{2\pi} f(x) \sin nx \, dx, \ n \geq 1,$$

for real functions f and

$$a_n = \frac{1}{\sqrt{2\pi}} \int_0^{2\pi} f(x) e^{-inx} dx, \ n \in \mathbf{Z},$$

for complex functions. If the function f is itself a Fourier polynomial, then according to the expansion theorem of §3.6, we have

$$f(x) = \frac{1}{\sqrt{2\pi}} a_0 + \frac{1}{\sqrt{\pi}} \sum_{n \geq 1} (a_n \cos nx + b_n \sin nx)$$

for real functions f, and

$$f(x) = \frac{1}{\sqrt{2\pi}} \sum_{n=-\infty}^{\infty} a_n e^{inx}$$

for complex functions f. The sums on the right, of course, are finite in the case under study. Infinite series with this structure are called *Fourier series*. The question of their convergence in general and the convergence to the function f whose Fourier coefficients are a_n and b_n, in particular, is studied in one of the most important sections of analysis.

4.7. Legendre polynomials. Here $G = 1$, $(a, b) = (-1, 1)$. The Legendre polynomials $P_0(x), P_1(x), P_2(x), \ldots$ are obtained as the result of the orthogonalization process applied to the basis $\{1, x, x^2, \ldots, \}$ of the space of real polynomials. They are usually normalized by the condition $P_n(1) = 1$. With this normalization their explicit form is given by the following result.

4.8. Proposition. $P_0(x) = 1$, $P_n(x) = \frac{1}{2^n n!} \frac{d^n}{dx^n} (x^2 - 1)^n$, $n \geq 1$.

Proof. Since the degree of the polynomial $(x^2 - 1)^n$ equals $2n$, the degree of the polynomial $\frac{d^n}{dx^n}(x^2 - 1)^n$ equals n, so that $P_1, \ldots, P_i$ generate the same space over $\mathbf{R}$ as do $1, x, \ldots, x^i$. Therefore, in order to check the orthogonality of P_i, P_j, $i \neq j$ it is sufficient to verify that

$$\int_{-1}^{1} x^k P_n(x) \, dx = 0 \quad \text{for } k < n.$$

Integrating by parts we obtain

$$\int_{-1}^{1} x^k \frac{d^n}{dx^n} (x^2 - 1)^n \, dx =$$

$$= x^k \frac{d^{n-1}}{dx^{n-1}} (x^2 - 1)^n \Big|_{-1}^{1} - k \int_{-1}^{1} x^{k-1} \frac{d^{n-1}}{dx^{n-1}} (x^2 - 1)^n \, dx.$$

The first term vanishes, because $(x^2-1)^n$ has a zero of order n at the points ± 1, and every differentiation lowers the order of the zero by one. An analogous procedure can be applied to the second term; after k steps, we obtain an integral proportional to

$$\int_{-1}^{1} \frac{d^{n-k}}{dx^{n-k}}(x^2-1)^n\,dx = \frac{d^{n-k-1}}{dx^{n-k-1}}(x^2-1)^n\Big|_{-1}^{1} = 0.$$

Then, according to Leibnitz's formula

$$\frac{d^n}{dx^n}[(x-1)^n(x+1)^n] = \sum_{k=0}^{n}\binom{n}{k}\frac{d^k}{dx^k}(x-1)^n\,\frac{d^{n-k}}{dx^{n-k}}(x+1)^n.$$

At the point $x=1$ the term corresponding to $k=n$ is the only term that does not vanish, so that

$$P_n(1) = \frac{1}{2^n n!}\binom{n}{n}\left[\frac{d^n}{dx^n}(x-1)^n\right](x+1)^n\Big|_{x=1} = \frac{1}{2^n n!}\cdot 1\cdot n!\cdot 2^n = 1,$$

which completes the proof.

4.9. Chebyshev polynomials. $G = \frac{1}{\sqrt{1-x^2}}$, $(a,b) = (-1,1)$. The polynomials $T_n(x)$, $n \geq 0$ are the result of orthogonalization of the basis $\{1, x, x^2, \ldots\}$. The explicit formulas are:

$$T_n(x) = \frac{(-2)^n\,n!}{(2n)!}\sqrt{1-x^2}\,\frac{d^n}{dx^n}(1-x^2)^{n-1/2} = \cos(n\cos^{-1}x).$$

They are normalized as follows:

$$\int_{-1}^{1}\frac{T_m(x)\,T_n(x)\,dx}{\sqrt{1-x^2}} = \begin{cases} 0 & \text{for } m \neq n, \\ \pi/2 & \text{for } m = n \neq 0, \\ \pi & \text{for } m = n = 0. \end{cases}$$

4.10. Hermite polynomials. $G = e^{-x^2}$, $(a,b) = (-\infty,\infty)$. The polynomials $H_n(x)$ are the result of the orthogonalization of the basis $\{1, x, x^2, \ldots\}$. The explicit formulas are

$$H_n(x) = (-1)^n e^{x^2}\frac{d^n}{dx^n}(e^{-x^2}).$$

They are normalized as follows:

$$\int_{-\infty}^{\infty} e^{-x^2}H_m(x)\,H_n(x)\,dx = \begin{cases} 0 & \text{for } m \neq n, \\ 2^n n!\sqrt{\pi} & \text{for } m = n. \end{cases}$$

The proof is left to the reader as an exercise.

EXERCISES

1. Prove that the Hermitian or orthogonal form g is positive definite, that is, $g(l,l) \geq 0$ for all $l \in L$, if and only if all diagonal minors of its Gram matrix are non-negative.

2. Prove the assertions of §§4.9 and 4.10.

§5. Euclidean Spaces

5.1. Definition. *A Euclidean space is a finite-dimensional, real, linear space L with a symmetric positive-definite inner product.*

We shall write (l, m) instead of $g(l, m)$ and $\|l\|$ instead of $(l, l)^{1/2}$; we shall call the number $\|l\|$ the *length* of the vector l.

It follows from the results proved in §§3–4 that:

a) Any Euclidean space has an orthonormal basis, all vectors of which have unit length.

b) Therefore, it is isometric to the coordinate Euclidean space $\mathbf{R}^n (n = \dim L)$, where

$$(\vec{x}, \vec{y}) = \sum_{i=1}^{n} x_i y_i, \quad |\vec{x}| = \left(\sum_{i=1}^{n} x_i^2\right)^{1/2}.$$

The key to many properties of Euclidean space is the repeatedly rediscovered Cauchy–Bunyakovskii–Schwarz inequality:

5.2. Proposition. *For any $l_1, l_2 \in L$ we have*

$$(l_1, l_2) \leq \|l_1\| \, \|l_2\|.$$

The equality holds if and only if the vectors l_1, l_2 are linearly independent.

Proof. If $l_1 = 0$, the equality holds and l_1, l_2 are linearly independent. We shall assume that $l_1 \neq 0$. For any real number t we have

$$\|tl_1 + l_2\|^2 = (tl_1 + l_2, tl_1 + l_2) = t^2\|l_1\|^2 + 2t(l_1, l_2) + \|l_2\|^2 \geq 0$$

by virtue of the positive-definiteness of the inner product. Therefore the discriminant of the quadratic trinomial on the right is non-positive, that is,

$$(l_1, l_2)^2 - \|l_1\|^2\|l_2\|^2 \leq 0.$$

It equals zero if and only if this trinomial has a real root t_0. Then

$$\|t_0 l_1 + l_2\|^2 = 0 \Leftrightarrow l_2 = -t_0 l_1,$$

which completes the proof.

5.3. Corollary (triangle inequality). *For any $l_1, l_2, l_3 \in L$*

$$\|l_1 + l_2\| \leq \|l_1\| + \|l_2\|, \ \|l_1 - l_3\| \leq \|l_1 - l_2\| + \|l_2 - l_3\|.$$

Proof. We have

$$\|l_1 + l_2\|^2 = \|l_1\|^2 + 2(l_1, l_2) + \|l_2\|^2 \leq \|l_1\|^2 + 2\|l_1\| \, \|l_2\| + \|l_2\|^2 = (\|l_1\| + \|l_2\|)^2.$$

Replacing here l_1 by $l_1 - l_2$ and l_2 by $l_2 - l_3$, we obtain the second inequality.

5.4. Corollary. *The Euclidean length of the vector $\|l\|$ is a norm on L in the sense of Definition 10.4 of Chapter 1, and the function $d(l, m) = \|l - m\|$ is the metric in the sense of Definition 10.1 of Chapter 1.*

Proof. It remains to verify only that $\|al\| = |a|\,\|l\|$ for all $a \in \mathbf{R}$; but

$$\|al\| = (al, al)^{1/2} = (a^2\|l\|^2)^{1/2} = |a|\,\|l\|.$$

5.5. Angles and distances. Let $l_1, l_2 \in L$ be non-zero vectors. Proposition 5.2 implies that

$$-1 \leq \frac{(l_1, l_2)}{\|l_1\|\,\|l_2\|} \leq 1.$$

Therefore there exists a unique angle ϕ, $0 \leq \phi \leq \pi$ for which

$$\cos\phi = \frac{(l_1, l_2)}{\|l_1\|\,\|l_2\|}.$$

This angle is called the angle between the vectors l_1, l_2. Since the inner product is symmetric, this is an "unoriented angle", which explains also the range of its values. In accordance with high-school geometry, the angle between orthogonal vectors equals $\pi/2$. Euclidean geometry can be developed systematically based on these definitions of length and angle, and it can be verified that in spaces of two and three dimensions it coincides with the classical geometry.

For example, the multidimensional Pythagoras theorem is a trivial consequence of the definitions. If the vectors $l_1, \ldots, l_n$ are pairwise orthogonal, then

$$\left|\sum_{i=1}^{n} l_i\right|^2 = \sum_{i=1}^{n} \|l_i\|^2.$$

The usual formula for cosines in plane geometry, applied to a triangle with sides l_1, l_2, l_3, asserts that

$$\|l_3\|^2 = \|l_1\|^2 + \|l_2\|^2 - 2\|l_1\|\,\|l_2\|\cos\phi,$$

where ϕ is the angle between l_1 and l_2. In the vector variant $l_3 = l_1 - l_2$, and this formula transforms into the identity

$$\|l_1 - l_2\|^2 = \|l_1\|^2 + \|l_2\|^2 - 2(l_1, l_2)$$

in accordance with our definition of angle.

Let $U, V \subset L$ be two sets in a Euclidean space. The non-negative number

$$d(U, V) = \inf\{\|l_1 - l_2\| \,|\, l_1 \in U,\ l_2 \in V\}$$

is called the *distance* between them. Consider the following particular case: $U = \{l\}$ (one vector), $V = L_0 \subset L$ is a linear space. Proposition 3.2 implies that $L = L_0 \oplus L_0^\perp$ and $l = l_0 + l_0'$ where $l_0 \in L_0$, $l_0' \in L_0^\perp$. The vectors l_0, l_0' are *orthogonal projections* of l onto $L_0, L_0^\perp$ respectively.

5.6. Proposition. *The distance from l to L_0 equals the length of the orthogonal projection of l onto $L_0^\perp$.*

Proof. Pythagoras's theorem implies that for any vector $m \in L_0$,

$$\|l - m\|^2 = \|l_0 + l_0' - m\|^2 = \|l_0 - m\|^2 + \|l_0'\|^2$$

because the vectors $l_0 - m \in L_0$ and $l_0' \in L_0^\perp$ are orthogonal. Therefore,

$$\|l - m\|^2 \geq \|l_0'\|^2,$$

and the equality holds only in the case $m = l_0$, which proves the assertion.

If an orthonormal basis $\{e_1, \ldots, e_m\}$ is selected in L_0, then the projection of l onto L_0 is determined by the formula

$$l_0 = \sum_{i=1}^{m} (l, e_i) e_i.$$

Indeed, the left and right sides have the same inner products with all e_i, and therefore their difference lies in $L_0^\perp$. Finally

$$d(l, L_0) = \left| l - \sum_{i=1}^{m} (l, e_i) e_i \right|$$

is the smallest value of $\|l - m\|$, when m runs through L_0. Since $\|l_0\|^2 \leq \|l\|^2$, according to the same Pythagoras's theorem we have

$$\sum_{i=1}^{m} (l, e_i)^2 \leq \|l\|^2.$$

5.7. Applications to function spaces. As an example, consider the space of continuous real functions on $[a, b] \subset \mathbf{R}$ with the inner product

$$(f, g) = \int_a^b fg \, dx.$$

It is finite-dimensional, but all our inequalities will refer to a finite number of such functions, so that in every case we shall be able to assume that we are working with a finite-dimensional Euclidean space: $\int_a^b f^2(x)\,dx \geq 0$ and if $\int_a^b f^2(x)\,dx = 0$, then $f(x) \equiv 0$.

The Cauchy-Bunyakovskii-Schwarz inequality assumes the form

$$\left(\int_a^b f(x)g(x)\,dx\right)^2 \leq \int_a^b f(x)^2\,dx \int_a^b g(x)^2\,dx.$$

The triangle inequality assumes the form

$$\left(\int_a^b (f(x)+g(x))^2\,dx\right)^{1/2} \leq \left(\int_a^b f(x)^2\,dx\right)^{1/2} + \left(\int_a^b g(x)^2\,dx\right)^{1/2}.$$

If $(a,b) = (o, 2\pi)$ and a_i, b_i are the Fourier coefficients of the function $f(x)$, as in §4.6, then the Fourier polynomial

$$f_N(x) = \frac{1}{\sqrt{2\pi}}a_0 + \frac{1}{\sqrt{\pi}}\sum_{n=1}^{N}(a_n \cos nx + b_n \sin nx)$$

is the orthogonal projection of $f(x)$ on the linear span of $\{1, \cos nx, \sin nx | 1 \leq \leq n \leq N\}$. Therefore, the Fourier coefficients of $f(x)$ for each N minimize the mean-square deviation of $f(x)$ from the Fourier polynomials of "degree" $\leq N$. The inequality $|f_N|^2 \leq |f|^2$ assumes the form

$$a_0^2 + \sum_{i=1}^{N}(a_i^2 + b_i^2) \leq \int_0^{2\pi} f(x)^2\,dx.$$

Since the right side does not depend on N and $a_i^2, b_i^2 \geq 0$, the series

$$a_0^2 + \sum_{i=1}^{\infty}(a_i^2 + b_i^2)$$

converges for any continuous function $f(x)$ in $[0, 2\pi]$. It can be shown that it converges exactly to $\int_0^{2\pi} f(x)^2\,dx$.

Entirely analogous arguments are applicable to the Legendre, Chebyshev, and Hermite polynomials. We leave them as an exercise to the reader.

5.8. Method of least squares. We shall study the system of m linear equations for n unknowns with real coefficients

$$\sum_{j=1}^{n} a_{ij}x_j = b_i,\ i = 1, \ldots, m.$$

Assume that this system is "overdetermined", that is, $m > n$ and the rank of the matrix of coefficients equals n. Then, in general, it does not have any solutions.

But we can try to find values of the unknowns $x_1^0, \dots x_n^0$, such that the total squared deviation of the left sides from the right sides

$$\sum_{i=1}^{m}\Big(\sum_{j=1}^{n} a_{ij}x_j^0 - b_i\Big)^2$$

assumes the smallest possible value. Then this problem has important practical applications. For example, in geodesic work the site is partitioned into a grid of triangles, some elements of which are measured while the others are calculated using trigonometric formulas. Since all measurements are approximate, it is recommended that more measurements than the number strictly necessary for calculating the remaining elements be made. Then, for the same reason, the equations for these elements will almost certainly be incompatible. The method of least squares permits obtaining an "approximate solution" that is more reliable due to the large quantity of information incorporated into the system.

We shall show that our problem can be solved using the results of §5.7. We interpret the columns of the matrix of coefficients $e_i = (a_{1i}, \dots, a_{mi})$ and columns of free terms $f = (b_1, \dots, b_m)$ as vectors of a coordinate Euclidean space $\mathbf{R}^m$ with the standard inner product. Setting

$$e = \sum_{i=1}^{n} x_i e_i,$$

we obtain

$$\sum_{i=1}^{m}\Big(\sum_{j=1}^{n} a_{ij}x_j - b_i\Big)^2 = \Big|\sum_{i=1}^{n} x_i e_i - f\Big|^2.$$

Therefore the minimum squared deviation is achieved when $\sum_{i=1}^{n} x_i^0 e_i$ is the orthogonal projection of f on the subspace spanned by the e_i. This means that the coefficients x_i^0 must be found from a system of n equations with n unknowns

$$\left(\sum_{i=1}^{n} x_i^0 e_i, e_j\right) = (f, e_j),\ j = 1, \dots, n,$$

the so-called "*normal system*". Its determinant is the determinant of the Gram matrix $((e_i, e_j))$, where

$$(e_i, e_j) = \sum_{k=1}^{m} a_{ki}a_{kj}.$$

It differs from zero, because it was assumed that the rank of the starting system, that is, the system of vectors (e_i), equals n (see the exercise for §2.5). Therefore the solution exists and is unique.

We now return to the subject of "measure in Euclidean space".

5.9. n-dimensional volume. In a one-dimensional Euclidean space one can associate with the simplest figures — segments and their finite unions — lengths and sums of lengths. In the Euclidean plane, high-school geometry teaches us how to measure the areas of figures such as rectangles, triangles, and, with some difficulty, circles. The generalization of these concepts is provided by the profound general *theory of measure*, whose natural place is not here. We shall restrict ourselves to a list of the basic properties and elementary calculations, associated with the special measure of figures in n-dimensional Euclidean space — their *n-dimensional volume*.

The n-dimensional volume is a function vol^n which is defined on *some* subsets of an n-dimensional Euclidean space L, called *measurable*, and which assumes non-negative real values or ∞ (on bounded measurable sets - only finite values). The collection of measurable sets is very rich. We shall simply postulate the following list of properties of vol^n and the measurability of sets entering into them, without proving the existence of the function with these properties and without indicating its natural domain of definition.

a) The function vol^n is countably additive, that is,

$$\mathrm{vol}^n\left(\bigcup_{i=1}^{\infty} U_i\right) = \sum_{i=1}^{\infty} \mathrm{vol}^n\, U_i, \text{ if } U_i \cap U_j = \emptyset \text{ for } i \neq j;$$

vol^1 (point)=0; vol^1 (segment)=length of segment. (A segment in a one-dimensional Euclidean space is a set of vectors of the type $tl_1 + (1-t)l_2$, $0 \leq t \leq 1$; its length equals $\|l_1 - l_2\|$.)

b) If $U \subseteq V$, then $\mathrm{vol}^n\, U \leq \mathrm{vol}^n\, V$.

c) If $L = L_1 \oplus L_2$ (orthogonal direct sum), $\dim L_1 = m$, $\dim L_2 = n$, $U \subset L_1$, $V \subset L_2$, then for $U \times V = \{(l_1, l_2) | l_1 \in U,\ l_2 \in V\} \in L_1 \oplus L_2$ we have

$$\mathrm{vol}^{m+n}(U \times V) = \mathrm{vol}^m\, U \cdot \mathrm{vol}^n\, V.$$

d) If $f : L \to L$ is an arbitrary linear operator, then

$$\mathrm{vol}^n(f(U)) = |\det f|\, \mathrm{vol}^n\, U,\ n = \dim L.$$

The properties a) and b) hardly need explanation. Property c) is a strong generalization of the formula for the area of a rectangle (product of the lengths of its sides) or the volume of a straight cylinder (product of the area of the base by the length of the generatrix). We note that from property c) it follows that the $(m+n)$-dimensional volume of a bounded set W in L, lying in the subspace L_1 of dimension $m < m+n$, equals zero. Indeed, then $L = L_1 \oplus L_1^{\perp}$ and $W = V \times \{0\}$, and finally $\mathrm{vol}^n(\{0\}) = 0$ for $n > 0$ by virtue of b) and c).

The meaning of property d) is less obvious. It is the main contribution of linear algebra to the theory of Euclidean volumes, and it is responsible for the appearance of Jacobians in the formalism of multidimensional integration. It could be explained more intuitively by remarking that the operator of stretching by a factor of $a \in \mathbf{R}$ along one of the vectors of the orthogonal basis must multiply the volume by $|a|$ by virtue of the properties b) and c). But any non-zero vector can be extended to an orthogonal basis, and therefore the diagonalizable operator f with the eigenvalues $a_1, \ldots, a_n \in \mathbf{R}$ must multiply the volumes by $|a_1| \ldots |a_n| = |\det f|$. Finally, isometries must preserve volume and, as we shall convince ourselves later, any operator is a composition of a diagonalizable operator and an isometry (see Exercise 11 of §8).

Using these axioms, we shall now present a list of volumes of the simplest and most important n-dimensional figures.

5.10. Unit cube. This is the set $\{t_1 e_1 + \ldots + t_n e_n | 0 \leq t_i \leq 1\}$ where $\{e_1, \ldots, e_n\}$ is some orthonormal basis of L. From the axioms of §5.9a) and b) it follows immediately that its volume equals unity.

A cube with side $a > 0$ is obtained if t_i is allowed to run through the values $0 \leq t_i \leq a$. Since it is the image of a unit cube relative to homothety — multiplication by a — its volume equals a^n.

5.11. Parallelepiped with sides $\{l_1, \ldots, l_n\}$. This is the set $\{t_1 l_1 + \ldots + t_n l_n | 0 \leq \leq t_i \leq 1\}$. We shall show that its volume equals $\sqrt{|\det G|}$ where $G = ((l_i, l_j))$ is the Gram matrix of the sides. Indeed, if $\{l_1, \ldots, l_n\}$ are linearly dependent, then the corresponding parallelepiped lies in a space of dimension $< \dim L$ and its n-dimensional volume equals zero according to the remark in §5.9. At the same time, G is singular.

Therefore it remains to analyse the case when $\{l_1, \ldots, l_n\}$ are linearly independent. Let $\{e_1, \ldots, e_n\}$ be an orthonormal basis in L, and f a linear mapping $L \to L$, transforming e_i into l_i, $i = 1, \ldots, n$. If A is the matrix of this mapping in the basis $\{e_i\}$:

$$(l_1, \ldots, l_n) = (e_1, \ldots, e_n)A,$$

then the Gram matrix of $\{l_i\}$ equals $A^t A$, because the Gram matrix of $\{e_i\}$ is the unit matrix. Therefore,

$$\sqrt{|\det G|} = \sqrt{|\det(A^t A)|} = |\det A|.$$

On the other hand, $|\det A| = |\det f|$ and f transforms the unit cube into our parallelepiped. According to the axiom d) of §5.9 the volume of the parallelepiped equals $|\det f|$, which completes the proof.

5.12. n-dimensional ball with radius r. This is the set of vectors

$$B^n(r) = \{l \mid \|l\| \leq r\},$$

or, in orthogonal coordinates,

$$B^n(r) = \left\{(x_1 \dots, x_n) \Big| \sum_{i=1}^{n} x_i^2 \leq r^2\right\}.$$

Since $B^n(r)$ is obtained from $B^n(1)$ by stretching by a factor of r, we have

$$\text{vol}^n\ B^n(r) = \text{vol}^n\ B^n(1) r^n.$$

The constant $\text{vol}^n\ B^n(1) = b_n$ can be calculated only by analytic means. Partitioning the $(n+1)$-dimensional ball into n-dimensional linear submanifolds, orthogonal to some direction, we obtain the induction formula

$$b_{n+1} = \left[2\int_0^1 \left(\sqrt{1 - x_{n+1}^2}\right)^n dx_{n+1}\right] b_n, \quad n \geq 1.$$

Of course, $b_1 = 2$, $b_2 = \pi$, $b_3 = \frac{4}{3}\pi$.

5.13. n-dimensional ellipsoid with semi-axes $r_1, \dots, r_n$. It is defined in orthogonal coordinates by the equations

$$\sum_{i=1}^{n} \left(\frac{x_i}{r_i}\right)^2 \leq 1.$$

Since it is obtained from $B^n(1)$ by stretching by a factor of r_i along the ith semi-axis, its volume equals $b_n r_1 \dots r_n$.

5.14. A property of the n-dimensional volume. It consists in the fact that for very large n the "*volume of an n-dimensional figure is concentrated near its surface.*" For example, the volume of the spherical ring between spheres of radius 1 and $1-\epsilon$ equals $b_n[1-(1-\epsilon)^n]$, which, for fixed arbitrarily small ϵ, but increasing n approaches b_n. A 20-dimensional watermelon with a radius of 20 cm. and a skin with a thickness of 1 cm. is nearly two-thirds skin:

$$1 - \left((1 - \frac{1}{20}\right)^{20} \approx 1 - e^{-1}.$$

This circumstance plays an important role in statistical mechanics. Consider, for example, the simplest model of a gas in a reservoir consisting of n atoms, which we shall assume are material points with mass 2 (in an appropriate system of units). We represent the instantaneous state of the gas by n three-dimensional vectors

$(\vec{v}_1, \ldots, \vec{v}_n)$ of the velocities of all molecules in the physical Euclidean space, that is, by a point in the three n-dimensional coordinate space $\mathbf{R}^{3n}$. The square of the lengths of the vectors in $\mathbf{R}^{3n}$ has a direct physical interpretation as the *energy* of the system (the sum of the kinetic energies of the atoms):

$$E = \sum_{i=1}^{n} |v_i|^2.$$

For a macroscopic volume of gas under normal conditions n is of the order of 10^{23} (Avogadro's number), so that the state of the gas is described only on a sphere of an enormous dimension, whose radius is the square root of its energy.

Let two such reservoirs be connected so that they can exchange energy, but not atoms, and let the sum of their energies $E_1 + E_2 = E$ remain constant. Then the energies E_1 and E_2 most of the time will be close to the values that maximize the "volume of the state space" accessible to the combined system, that is, the product

$$\mathrm{vol}^{n_1}\, B(E_1^{1/2})\, \mathrm{vol}^{n_2}\, B(E_2^{1/2})$$

(we replace the areas of the spheres by the volumes of the spheres, which is not literally correct, but has practically no effect on the result). Since as E_1 increases and E_2 decreases ($E_1 + E_2 = \text{const}$), the first volume increases extremely rapidly while the second volume decreases, there is a sharp peak in the volume of this product for some values of E_1, E_2 corresponding to the "most probable" state of the combined system. Evidently this occurs where

$$\frac{d}{dE_1} \log \mathrm{vol}^{n_1}\, B(E_1^{1/2}) = \frac{d}{dE_2} \log \mathrm{vol}^{n_2}\, B(E_2^{1/2}).$$

The inverses of these quantities are (to within a proportionality factor) the *temperatures* of the reservoirs and the most probable state corresponds to the situation when the temperatures are equal.

EXERCISES

1. Prove that the angle ϕ of inclination of the straight line in the plane $\mathbf{R}^2$, passing from the origin of coordinates as close as possible, in the mean-square, to m given points (a_i, b_i) $i = 1, \ldots, m$, is determined by the formula

$$\tan\phi = \Big(\sum_{i=1}^{m} a_i b_i\Big) \Big/ \Big(\sum_{i=1}^{m} a_i^2\Big).$$

(Hint: find the best "approximate solution" of the system of equations $a_i x = b_i$.)

2. Let $P_n(x)$ be the nth order Legendre polynomial. Prove that the leading coefficient of the polynomial $u_n(x) = \frac{2^n (n!)^2}{(2n)!} P_n(x)$ equals one and that the minimum of the integral $I(u) = \int_{-1}^{1} u(x)^2\,dx$ on the set of polynomials $u(x)$ of degree n with a leading coefficient equal to 1 is attained when $u = u_n$. (Hint: expand u with respect to the Legendre polynomials of degree $\leq n$.)

3. Let (S, μ) be a pair consisting of a finite set S and a real function $\mu : S \to \mathbf{R}$, satisfying two conditions: $\mu(s) \geq 0$ for all $s \in S$ and $\sum_{s \in S} \mu(s) = 1$. Consider on the space of real functions $F(S)$ on S (with real values in $\mathbf{R}$) the linear functional $E : F(S) \to \mathbf{R}$:

$$E(f) = \sum_{s \in S} \mu(s) f(s).$$

We denote the kernel of E by $F_0(S)$.

(S, μ) is called a finite probability space, the elements of $F(S)$ are random quantities in it, the elements of $F_0(S)$ are normalized random quantities, and the number $E(f)$ is the mathematical expectation of the quantity f. Random quantities form a ring under ordinary multiplication of functions.

Prove the following facts.

a) $F(S)$ and $F_0(S)$ have the structure of an orthogonal space in which the squared length of the vector f equals $E(f^2)$. The space $F(S)$ is Euclidean if and only if $\mu(s) > 0$ for all $s \in S$.

b) For any random quantities $f, g \in F(S)$ and $a, b \in \mathbf{R}$, we set

$$P(f = a) = \sum_{f(s)=a} \mu(s); \quad P(f = a;\ g = b) = \sum_{\substack{f(s)=a \\ g(s)=b}} \mu(s)$$

("the probabilities that f assumes the value a or $f = a$ and $g = b$ simultaneously"). Two random quantities are said to be independent if

$$P(f = a;\ g = b) = P(f = a) P(g = b)$$

for all $a, b \in \mathbf{R}$. Prove that if the normalized random quantities $f, g \in F_0(S)$ are independent, then they are orthogonal.

Construct an example showing that the converse is not true.

The inner product of the quantities $f, g \in F_0(S)$ is called their covariation, and the cosine of the angle between them is called the correlation coefficient.

§6. Unitary Spaces

6.1. Definition. *A unitary space is a complex linear space L with a Hermitian positive-definite inner product.*

As in §5, we shall write (l, m) instead of $g(l, m)$ and $\|l\|$ instead of $(l, l)^{1/2}$. We shall show below that $\|l\|$ is a norm in L in the sense of §10 of Chapter I. Unitary spaces, which are complete with respect to this norm, are also called *Hilbert spaces.* In particular, finite-dimensional unitary spaces are Hilbert spaces.

It follows from the results proved in §3 and §4 that

a) any finite-dimensional unitary space has an orthonormal basis, all vectors of which have unit length;

b) therefore, it is isomorphic to a coordinate unitary space $\mathbf{C}^n (n = \dim L)$ with the inner product

$$(\vec{x}, \vec{y}) = \sum_{i=1}^{n} x_i \bar{y}_i, \quad \|\vec{x}\| = \left(\sum_{i=1}^{n} |x_i|^2\right)^{1/2}.$$

A number of properties of unitary spaces are close to the properties of Euclidean spaces, primarily for the following reason: if L is a finite-dimensional unitary space, then its decomplexification $L_{\mathbf{R}}$ has the (unique) structure of a Euclidean space in which the norm $\|l\|$ of a vector is the same as in L. The existence is evident from the preceding item: if $\{e_1, \ldots, e_n\}$ is an orthonormal basis in L and $\{e_1, ie_1, e_2, ie_2, \ldots, e_n, ie_n\}$ is the corresponding basis of $L_{\mathbf{R}}$, then

$$\left|\sum_{j=1}^{n} x_j e_j\right|^2 = \sum_{j=1}^{n} |x_j|^2 = \sum_{j=1}^{n} ((\operatorname{Re} x_j)^2 + (\operatorname{Im} x_j)^2),$$

and the expression on the right is the Euclidean squared norm of the vector

$$\sum_{j=1}^{n} \operatorname{Re} x_j e_j + \sum_{j=1}^{n} \operatorname{Im} x_j (ie_j)$$

in the orthonormal basis $\{e_j, ie_j\}$. Uniqueness follows from §3.9.

Inner products in a unitary space L *do not*, however, *coincide* with inner products in the Euclidean space $L_{\mathbf{R}}$: the second assumes only real values, whereas the first assumes complex values. In reality, a Hermitian inner product in a complex space leads not only to an orthogonal, but also to a symplectic structure on $L_{\mathbf{R}}$ with the help of the following construction.

We temporarily return to the notation $g(l, m)$ for a Hermitian inner product in L and we set

$$a(l, m) = \operatorname{Re} g(l, m),$$

$$b(l, m) = \operatorname{Im} g(l, m)$$

Then the following facts hold:

6.2. Proposition. a) *$a(l,m)$ is a symmetric and $b(l,m)$ is an antisymmetric inner product in $L_{\mathbf{R}}$; both products are invariant under multiplication by i, that is, the canonical complex structure on $L_{\mathbf{R}}$:*

$$a(il,im) = a(l,m), \quad b(il,im) = b(l,m);$$

b) *a and b are related by the following relations:*

$$a(l,m) = b(il,m), \quad b(l,m) = -a(il,m);$$

c) *any pair of i-invariant forms a, b in $L_{\mathbf{R}}$, antisymmetric and symmetric, related by relations* b), *defines a Hermitian inner product in L according to the formula*

$$g(l,m) = a(l,m) + ib(l,m);$$

d) *the form g is positive-definite if and only if the form a is positive definite.*

Proof. The condition of Hermitian symmetry $g(l,m) = \overline{g(m,l)}$ is equivalent to the fact that

$$a(l,m) + ib(l,m) = a(m,l) - ib(m,l),$$

that is, symmetry of a and antisymmetry of b. The condition $g(il,im) = i\bar{i}g(l,m) = g(l,m)$ is equivalent to i-invariance of a and b. The condition of $\mathbf{C}$-linearity of g according to the first argument indicates $\mathbf{R}$-linearity and linearity relative to multiplication by i, that is,

$$a(il,m) + ib(il,m) = g(il,m) = ig(l,m) = -b(l,m) + ia(l,m),$$

whence follow relations b) and assertion c). Finally, $g(l,l) = a(l,l)$ by virtue of the antisymmetry of b, whence follows d).

6.3. Corollary. *In the previous notation, if g is positive-definite and $u\{e_1,\ldots,e_n\}$ is an orthonormal basis for g, then $\{e_1,\ldots,e_n, ie_1,\ldots,ie_n\}$ is an orthonormal basis for a and a symplectic basis for b.*

Conversely, if L is a $2n$-dimensional real space with a Euclidean form a and a symplectic form b as well as with a basis $\{e_1,\ldots,e_n,e_{n+1},\ldots,e_{2n}\}$ which is orthonormal for a and symplectic for b, then, introducing on L a complex structure with the help of the operator

$$J(e_j) = e_{n+j}, \ 1 \le j \le n; \quad J(e_j) = -e_{j-n}, \ n+1 \le j \le 2n,$$

and an inner product $g(l,m) = a(l,m) + ib(l,m)$, we obtain a complex space with a positive-definite Hermitian form, for which $\{e_1,\ldots,e_n\}$ is an orthonormal basis over **C**.

The proof is obtained by a simple verification with the help of Proposition 6.2, which we leave to the reader.

We now turn to unitary spaces L. The complex Cauchy–Bunyakovskii–Schwarz inequality has the following form:

6.4. Proposition. *For any $l_1, l_2 \in L$*

$$|(l_1,l_2)|^2 \le \|l_1\|^2\|l_2\|^2,$$

and, in addition, equality holds if and only if the vectors l_1, l_2 are proportional.

Proof. As in §5.2, for any real t we have

$$|tl_1 + l_2|^2 = t^2\|l_1\|^2 + 2t\ \mathrm{Re}(l_1,l_2) + \|l_2\|^2 \ge 0.$$

The case $l_1 = 0$ is trivial. Assuming that $l_1 \ne 0$, we deduce that

$$(\mathrm{Re}(l_1,l_2))^2 \le \|l_1\|^2\|l_2\|^2.$$

But, if $(l_1,l_2) = |(l_1,l_2)|e^{i\phi}$, $\phi \in \mathbf{R}$, then $\mathrm{Re}(e^{-i\phi}l_1,l_2) = |(l_1,l_2)|$. Therefore,

$$|(l_1,l_2)|^2 \le \|e^{-i\phi}l_1|^2\|l_2\|^2 = \|l_1\|^2\|l_2\|^2.$$

Strict equality holds here if and only if $\|t_0e^{-i\phi}l_1 + l_2\| = 0$ for an appropriate $t_0 \in \mathbf{R}$, which completes the proof.

In exactly the same manner as in the Euclidean case, the following corollaries are derived from here:

6.5. Corollary (triangle inequality). *For any $l_1, l_2, l_3 \in L$*

$$\|l_1 + l_2\| \le \|l_1\| + \|l_2\|,$$

$$\|l_1 - l_3\| \le \|l_1 - l_2\| + \|l_2 - l_3\|.$$

6.6. Corollary. *The unitary length of the vector $\|l\|$ is the norm in L in the sense of* Definition 10.4 *of* Chapter I.

(Here the property $\|al\| = |a|\,\|l\|$ is verified somewhat differently:

$$\|al\| = (al,al)^{1/2} = (a\bar{a}(l,l))^{1/2} = |a|\,\|l\|.)$$

6.7. Angles. Let $l_1, l_2 \in L$ be non-zero vectors. Proposition 6.4 implies that

$$0 \le \frac{|(l_1, l_2)|}{\|l_1\|\,\|l_2\|} \le 1.$$

Therefore there exists a unique angle ϕ, $0 \le \phi \le \pi/2$, for which

$$\cos\phi = \frac{|(l_1, l_2)|}{\|l_1\|\,\|l_2\|}.$$

However, in the very important models in the natural sciences, which make use of unitary spaces, the same quantity $\frac{|(l_1,l_2)|}{\|l_1\|\|l_2\|}$ (more precisely, its square) is interpreted not only as the cosine of an angle, but also as a *probability*. We shall briefly describe the postulates of quantum mechanics, which include this interpretation.

6.8. State space of a quantum system. In quantum mechanics it is postulated that physical systems such as an electron, a hydrogen atom, and so on, can be associated (not uniquely !) with a mathematical model consisting of the following data.

a) A unitary space $\mathcal{H}$, called the *state space* of the system. Such spaces, which are studied in standard textbooks, for the most part are infinite-dimensional Hilbert spaces, which are realized as a space of functions in models of "physical" space or space-time. Finite-dimensional spaces $\mathcal{H}$ arise, roughly speaking, as spaces of the *internal degrees of freedom* of a system, if the system is viewed as being localized or if its motion in physical space can in one way or another be neglected. The two-dimensional unitary space of the "spin states" of an electron, to which we shall return again, is such a space.

b) Rays, that is, one-dimensional complex subspaces in $\mathcal{H}$, are called (pure) *states* of the system.

All information on the state of a system at a fixed moment in time is determined by specifying the ray $L \subset \mathcal{H}$ or the non-zero vector $\psi \in L$, which is sometimes called the ψ *function*, corresponding to this state, or the *state vector*.

The fundamental postulate that the ψ functions form a complex linear space is called the *principle of superposition*, and the linear combination $\sum_{j=1}^{n} a_j\psi_j$, $a_j \in \mathbf{C}$ describes the superposition of the states $\psi_1, \ldots, \psi_n$. We note that since only the rays $\mathbf{C}\psi_j$ and not the vectors ψ_j themselves have physical meaning, the coefficients a_j cannot be assigned a uniquely defined meaning. If, however, the ψ_j are chosen to be normalized, $|\psi_j|^2 = 1$, and linearly independent and $\sum_{j=1}^{n} a_j\psi_j$ is also normalized, then the arbitrariness in the choice of the vectors ψ_j in its ray reduces to multiplications by the numbers $e^{i\phi_j}$, which are called *phase factors*; the same arbitrariness exists in the choice of the coefficients a_j, which we can then make real and non-negative, which together with the normalization condition $\left|\sum_{j=1}^{n} a_j\psi_j\right| = 1$ makes it possible to define them uniquely.

The strongly idealized assumptions about the connection between this scheme and reality consist of the fact that we have physical instruments A_ψ ("furnaces"), capable of preparing many samples of our system in instantaneous states ψ (more precisely, $\mathbf{C}\psi$) for different $\psi \in \mathcal{H}$. In addition, there exist physical instruments B_χ ("filters") into whose input systems in some state ψ are fed, and the same systems are observed at the output in some (possibly different) state χ, or nothing is observed at all (the system "does not pass" through the filter B_χ).

The second basic (after the principle of superposition) postulate of quantum mechanics is as follows:

A system prepared in the state $\psi \in \mathcal{H}$ can be observed immediately after preparation in the state $\chi \in \mathcal{H}$ with the probability

$$\frac{|(\psi,\chi)|^2}{|\psi|^2|\chi|^2} = \cos^2\theta, \text{ where } \theta \text{ is the angle between } \psi \text{ and } \chi.$$

In what follows, as additional geometric concepts are introduced, we shall refine the mathematical description of "furnaces" and "filters". Moreover, we shall explain what will happen if a system prepared in the state ψ is not immediately introduced into the filter, but rather after some time t: it turns out that during that interval the state ψ, also the scalar product (ψ,χ) , will change, and this change is also accurately described in terms of linear algebra.

If ψ and χ are normalized, then the probability indicated above equals $|(\psi,\chi)|^2$, and the inner product (ψ,χ) itself, which is a complex number, is called the *probability amplitude* (of a transition from ψ to χ). We note that physicists, following Dirac, usually study inner products which are antilinear with respect to the *first* argument, and write our (ψ,χ) in the form $\langle\chi|\psi\rangle$, so that the initial and final states of the system are arranged from *right to left*. The symbol $\langle\ \rangle$ is called "bracket". Correspondingly, Dirac calls the symbol $|\psi\rangle$ a "ket vector", and the symbol $\langle\chi|$ the corresponding "bra vector". From the mathematical viewpoint $|\psi\rangle$ is an element of $\mathcal{H}$, and $\langle\psi|$ is the corresponding element of the space of antilinear functionals $\overline{\mathcal{H}}^*$, and $\langle\chi|\psi\rangle$ is the value of χ on ψ.

If ψ, χ are orthogonal, that is, $(\psi,\chi) = 0$, then the system prepared in the state ψ cannot be observed (immediately after preparation) in the state χ, that is, it will not pass through the filter B_χ (conversely, it will pass through the filter B_ψ with certainty). In all other cases, there is a non-zero probability for a transition from ψ to χ.

The elements of any orthonormalized basis $\{\psi_1,\ldots,\psi_n\}$ form the set of *basis states* of the system. Assume that we have filters $B_{\psi_1},\ldots,B_{\psi_n}$. By repeatedly passing systems through them prepared in the state $\psi = \sum_{i=1}^n a_i\psi_i$, $0 \le a_i \le 1$ (the vector is assumed to be normalized), we observe ψ_i with probability a_i^2. Thus the coefficients in this linear combination can be measured experimentally, but in a fundamentally statistical test. This is one of the reasons that quantum mechanical measurements require the processing of a large statistical sample. Moreover, systems in the state ψ often enter a filter with a "flux" and at the output the probabilities a_i^2 are obtained in the form of intensities, something like "spectral lines"; these intensities in themselves are already the result of statistical averaging. In what follows, we shall make more precise the connection between this scheme and the theory of the spectra of linear operators.

6.9. Feynman rules. Let an orthonormal basis $\{\psi_1, \ldots, \psi_n\}$ of $\mathcal{H}$ be given. For any state vector $\psi \in \mathcal{H}$ we have

$$\psi = \sum_{i=1}^{n} (\psi, \psi_i)\psi_i,$$

whence

$$(\psi, \chi) = \sum_{i=1}^{n} (\psi, \psi_i)(\psi_i, \chi).$$

Analogously, $(\psi, \psi_i) = \sum_{j=1}(\psi, \psi_j)(\psi_j, \psi_i)$; substituting this equation into the preceding one, we obtain

$$(\psi, \chi) = \sum_{i_1, i_2=1}^{n} (\psi, \psi_{i_1})(\psi_{i_1}, \psi_{i_2})(\psi_{i_2}, \chi)$$

and, in general, for any $m \geq 1$

$$(\psi, \chi) = \sum_{i_1, \ldots, i_m=1}^{n} (\psi, \psi_{i_1})(\psi_{i_1}, \psi_{i_2}) \ldots (\psi_{i_m}, \chi).$$

These simple equations of linear algebra can be interpreted, according to Feynman, as laws of the "complex theory of probability", referring to amplitudes instead of probabilities. Namely, we shall study sequences of the type $(\psi, \psi_{i_1}, \psi_{i_2}, \ldots, \psi_{i_m}, \chi)$ as "classical trajectories" of the system, passing successively through the states in the parenthesis, and the number $(\psi, \psi_{i_1})(\psi_{i_1}, \psi_{i_2}) \ldots (\psi_{i_m}, \chi)$ as the probability amplitude of the transition from ψ to χ along the corresponding classical trajectory. This amplitude is the product of the transition amplitudes along successive segments of the trajectory.

Then the formula presented above for (ψ, χ) means that *this transition amplitude is the sum of the transition amplitudes from ψ to χ along all possible classical trajectories* ("of equal length").

R. Feynman placed the infinite-dimensional and more refined variant of this remark, in which the space-time (or energy-momentum) observables play the main role, at the foundation of his semi-heuristic technique for expressing amplitudes in terms of "continuum integrals over classical trajectories". The space of trajectories is an infinite-dimensional functional space, and mathematicians have still not been able to construct a general theory in which all of the remarkable calculations of physicists would be justified.

6.10. Distances. The distance between subsets in a unitary space L can be defined in the same way as in a Euclidean space:

$$d(U,V) = \inf\{\|l_1 - l_2\| \,|\, l_1 \in U,\ l_2 \in V\}.$$

The distance from the vector l to the subspace L_0 also equals the length of the orthogonal projection of l on $L_0^{\perp}$. The proof is identical in every respect to the Euclidean case. In particular, if $\{e_1, \ldots, e_m\}$ is an orthonormal basis of L_0, then

$$d(l, L_0) = \left| l - \sum_{i=1}^{m} (l, e_i) e_i \right|,$$

as in the Euclidean case, and

$$\left| \sum_{i=1}^{m} (l, e_i) e_i \right|^2 = \sum_{i=1}^{m} |(l, e_i)|^2 \leq \|l\|^2$$

according to Pythagoras's theorem.

6.11. Application to function spaces. As in §§4 and 5, we can introduce the inequalities for complex-valued functions:

$$\left| \int_a^b f(x)\bar{g}(x)\,dx \right|^2 \leq \int_a^b |f(x)|^2\,dx \int_a^b |g(x)|^2\,dx,$$

$$\left(\int_a^b |f(x) + g(x)|^2\,dx \right)^{1/2} \leq \left(\int_a^b |f(x)|^2\,dx \right)^{1/2} + \left(\int_a^b |g(x)|^2\,dx \right)^{1/2}$$

as well as for their Fourier coefficients. Studying functions in the interval $[0, 2\pi]$ and setting

$$a_n = \frac{1}{\sqrt{2\pi}} \int_0^{2\pi} f(x) e^{-inx}\,dx,$$

we find that in a space with the inner product $\int_0^{2\pi} f(x)\bar{g}(x)\,dx$, the sum

$$f_N(x) = \frac{1}{\sqrt{2\pi}} \sum_{n=-N}^{N} a_n e^{inx}$$

is the orthogonal projection of f on the space of Fourier polynomials "of degree $\leq N$" and minimizes the mean square deviation of f from this space. In particular,

$$\sum_{n=-N}^{N} |a_n|^2 \leq \int_0^{2\pi} |f(x)|^2\,dx,$$

so that the series $\sum_{n=-\infty}^{\infty} |a_n|^2$ converges.

§7. Orthogonal and Unitary Operators

7.1. Let L be a linear space with the scalar product g. The set of all isometries $f : L \to L$, that is invertible linear operators with the condition

$$g(f(l_1), f(l_2)) = g(l_1, l_2)$$

for all $l_1, l_2 \in L$, evidently, forms a group. If L is a Euclidean space, such operators are called *orthogonal*; if L is a Hermitian space, they are called *unitary*. Symplectic isometries will be examined later.

7.2. Proposition. *Let L be a finite-dimensional linear space with a symmetric or Hermitian non-degenerate inner product $(\, , \,)$. In order for the operator $f : L \to L$ to be an isometry it is necessary and sufficient that any of the following conditions hold:*

a) *$(f(l), f(l)) = (l, l)$ for all $l \in L$ (it is assumed here that the characteristic of the field of scalars does not equal 2).*

b) *Let $\{e_1, \ldots, e_n\}$ be a basis of L with the Gram matrix G and let A be the matrix of f in this basis. Then*

$$A^t G A = G, \quad \text{or} \quad A^t G \bar{A} = G;$$

c) *f transforms an orthonormal basis into an orthonormal basis.*

d) *If the signature of the inner product equals (p, q), then the matrix of f in any orthonormal basis $\{e_1, \ldots, e_p, e_{p+1}, \ldots, e_{p+q}\}$ with $(e_i, e_i) = +1$ for $i \le p$ and $(e_i, e_i) = -1$ for $p + 1 \le i \le p + q$ satisfies the condition*

$$A^t \begin{pmatrix} E_p & 0 \\ 0 & -E_q \end{pmatrix} A = \begin{pmatrix} E_p & 0 \\ 0 & -E_q \end{pmatrix}$$

or

$$A^t \begin{pmatrix} E_p & 0 \\ 0 & -E_q \end{pmatrix} \bar{A} = \begin{pmatrix} E_p & 0 \\ 0 & -E_q \end{pmatrix}$$

in the symmetric and Hermitian cases respectively.

Proof. a) In the symmetric case this assertion follows from §3.9: if f preserves the quadratic form $(l, l) = q(l)$, then f also preserves its polarization

$$(l, m) = \frac{1}{2}[q(l + m) - q(l) - q(m)].$$

In the Hermitian case we have, analogously,

$$\mathrm{Re}(l, m) = \frac{1}{2}[q(l + m) - q(l) - q(m)]$$

and Proposition 6.2 shows that (l,m) is uniquely reconstructed from $\mathrm{Re}(l,m)$ according to the formula

$$(l,m) = \mathrm{Re}(l,m) - i\,\mathrm{Re}(il,m)$$

and therefore f preserves (l,m).

b) If f is an isometry, then the Gram matrices of the bases $\{e_1,\dots,e_n\}$ and $\{f(e_1),\dots,f(e_n)\}$ are equal to one another. But the last Gram matrix equals A^tGA in the symmetric case and $A^tG\bar{A}$ in the Hermitian case. Conversely, if f transforms the basis $\{e_1,\dots,e_n\}$ into $\{e'_1,\dots,e'_n\}$ and the Gram matrices of the bases $\{e_i\}$ and $\{e'_i\}$ are equal to one another, then f is an isometry by virtue of the formulas from §2.2 for expressing the inner product in coordinate form.

c), d) These assertions are particular cases of the preceding assertions.

It follows from Proposition 7.2 that orthogonal (or unitary) operators are operators which in one (and therefore in any) *orthonormal* basis are specified by *orthogonal* (or *unitary*) matrices, that is, matrices U satisfying the relations

$$UU^t = E_n \quad \text{or} \quad U\bar{U}^t = E_n.$$

Collections of $n \times n$ matrices of this type were first introduced in §4 of Chapter 1. They were denoted by $O(n)$ and $U(n)$ respectively. Analogously, the matrices of the isometries in the orthonormal bases with signature (p,q), satisfying the conditons of Proposition 7.2d are denoted by $O(p,q)$ and $U(p,q)$; for $p,q \neq 0$ they are sometimes called *pseudo-orthogonal* and *pseudo-unitary*, respectively. In this section we shall be concerned only with the groups $O(n)$ and $U(n)$. We shall study the *Lorentz group* $O(1,3)$, which is fundamental for physics, in §10.

7.3. The groups $U(1)$, $O(1)$, and $O(2)$. It follows immediately from the definition that

$$U(1) = \{a \in \mathbf{C} \,|\, |a| = 1\} = \{e^{i\phi} \,|\, \phi \in \mathbf{R}\},$$

$$O(1) = \{\pm 1\} = U(1) \cap \mathbf{R}.$$

Further, if $U \in O(n)$, then $UU^t = E_n$, whence $(\det U)^2 = 1$ and $\det U = \pm 1$. If $U = \begin{pmatrix} a & b \\ c & d \end{pmatrix}$ is an orthogonal matrix whose determinant equals -1, then $\begin{pmatrix} a & b \\ -c & -d \end{pmatrix}$ is an orthogonal matrix (belonging to $SO(2)$) whose determinant equals $+1$. Matrices from $SO(2)$ have the form

$$\left\{\begin{pmatrix} a & b \\ c & d \end{pmatrix} \Big| ad - bc = a^2 + b^2 = c^2 + d^2 = 1,\ ac + bd = 0\right\}.$$

Any such matrix can obviously be represented in the form

$$\begin{pmatrix} \cos\phi & -\sin\phi \\ \sin\phi & \cos\phi \end{pmatrix},\ \phi \in [0, 2\pi),$$

that is, it represents a Euclidean rotation by an angle ϕ. The mapping

$$U(1) \to SO(2) : e^{i\phi} \mapsto \begin{pmatrix} \cos\phi & -\sin\phi \\ \sin\phi & \cos\phi \end{pmatrix}$$

is an isomorphism. Its geometric meaning is explained by the following remark: the decomplexification of the one-dimensional unitary space $(\mathbf{C}^1, |z|^2)$ is a two-dimensional Euclidean space $(\mathbf{R}^2, x_1^2+x_2^2)$, and the decomplexification of the unitary transformation $z \mapsto e^{i\phi}z$ is given by the matrix representing a rotation by an angle ϕ.

In §9 we shall construct the much less trivial epimorphism $SU(2) \to SO(3)$ with the kernel $\{\pm 1\}$.

The rotations $\begin{pmatrix} \cos\phi & -\sin\phi \\ \sin\phi & \cos\phi \end{pmatrix}$ by an angle $\phi \neq 0, \pi$ do not have eigenvectors in $\mathbf{R}^2$ and are therefore not diagonalizable. By contrast, all matrices $U \in O(2)$ with $\det U = -1$ are diagonalizable. More precisely, the characteristic polynomial of the matrix

$$\begin{pmatrix} \cos\phi & -\sin\phi \\ -\sin\phi & -\cos\phi \end{pmatrix}$$

equals t^2-1 and its roots equal ± 1. It is easy to check directly that the characteristic subspaces corresponding to these roots are orthogonal; this will be proved below with much greater generality. Therefore, any operator from $O(2)$ with $\det U = -1$ is a *reflection* relative to a straight line: it acts as an identity on this line and changes the sign of vectors orthogonal to it.

With this information we can now establish the structure of general orthogonal and unitary operators.

7.4. Theorem. a) *In order for an operator f in a unitary space to be unitary it is necessary and sufficient that f be diagonalizable in an orthonormal basis and have its spectrum situated on the unit circle in* $\mathbf{C}$.

b) *In order for an operator f in a Euclidean space to be orthogonal it is necessary and sufficient that its matrix in an appropriate orthonormal basis have the form*

$$\begin{pmatrix} A(\phi_1) & & & & & & & & \\ & \ddots & & & & & & & \\ & & A(\phi_m) & & & & 0 & & \\ & & & 1 & & & & & \\ & & & & \ddots & & & & \\ & & & & & 1 & & & \\ & & 0 & & & & -1 & & \\ & & & & & & & \ddots & \\ & & & & & & & & -1 \end{pmatrix}, \quad A(\phi) = \begin{pmatrix} \cos\phi & -\sin\phi \\ \sin\phi & \cos\phi \end{pmatrix}$$

where the empty spaces contain zeros.

c) *The eigenvectors of an orthogonal or unitary operator, corresponding to different eigenvalues, are orthogonal.*

Proof. a) The sufficiency of the assertion is obvious: if $U = \mathrm{diag}(\lambda_1, \dots \lambda_n)$, $|\lambda_i|^2 = 1$, then $U\overline{U}^t = E_n$, so that U is the matrix of a unitary operator. Conversely, let f be a unitary operator, λ an eigenvalue of it, and L_λ the corresponding characteristic subspace. According to Proposition 3.2, we have $L = L_\lambda \oplus L_\lambda^\perp$. The subspace L_λ is one-dimensional and f-invariant, and the restriction of f to L_λ is a one-dimensional unitary operator. Therefore $\lambda \in U(1)$, that is, $|\lambda|^2 = 1$. If we show that the subspace $L_\lambda^\perp$ is also f-invariant, then by induction on $\dim L$ we can deduce that L can be decomposed into a direct sum of f-invariant, pairwise orthogonal, one-dimensional subspaces, which will prove the required result.

Indeed, if $l_0 \in L_\lambda, l_0 \neq 0$ and $(l_0, l) = 0$, then

$$(l_0, f(l)) = (f(\lambda^{-1} l_0), f(l)) = (\lambda^{-1} l_0, l) = \lambda^{-1}(l_0, l) = 0,$$

so that $f(l) \in L_\lambda^\perp$.

b) In the orthogonal case the arguments are analogous: the sufficiency of the conditions is checked directly and induction on $\dim L$ is then performed. The cases $\dim L = 1, 2$ have been analysed in the preceding section. If $\dim L \geq 3$ and f has a real eigenvalue λ, then we must again set $L = L_\lambda \oplus L_\lambda^\perp$ and argue as above (we note that here necessarily $\lambda = \pm 1$). Finally, if f does not have real eigenvalues, then we must select a two-dimensional f-invariant subspace $L_0 \subset L$, which exists according to Proposition 12.16 of Chapter 1. The discussion in the preceding section implies that in this subspace the matrix of the restriction of f in any orthonormal basis will have the form $A(\phi)$. Therefore, it remains to verify that the subspace $L_0^\perp$ is likewise f-invariant. Indeed, if $(l_0, l) = 0$ for all $l_0 \in L_0$, then

$$(l_0, f(l)) = (f(f^{-1}(l_0)), f(l)) = (f^{-1}(l_0), l) = 0,$$

because $f^{-1}(l_0) \in L_0$ for all $l_0 \in L_0$. This completes the proof.

c) Let $f(l_i) = \lambda_i l_i$, $i = 1, 2$, Then

$$(l_1, l_2) = (f(l_1), f(l_2)) = \lambda_1 \bar{\lambda}_2 (l_1, l_2).$$

Since $|\lambda_i|^2 = 1$, for $\lambda_1 \neq \lambda_2$ we have $\lambda_1 \bar{\lambda}_2 \neq 1$. Therefore, $(l_1, l_2) = 0$. This argument is applicable simultaneously to the unitary and orthogonal cases. This completes the proof.

7.5. Corollary ("Euler's theorem"). *In a three-dimensional Euclidean space, any orthogonal mapping f which does not change the orientation (that is, an element of the group $SO(3)$) is a rotation with respect to some axis.*

Proof. Since the characteristic polynomial of f is of degree 3, it necessarily has a real root. If this is the only root, then it must equal one, because $\det f = 1$. If there is more than one real root, then all the roots are real, and the combinations $(1,1,1)$ or $(1,-1,-1)$ are possible. In either case we have the eigenvalue 1. The corresponding characteristic subspace is an axis of rotation, and induces an element of $SO(2)$, that is, a rotation by some angle, in a plane perpendicular to it.

§8. Self-Adjoint Operators

8.1. In Chapter I we saw that diagonalizable operators form the simplest and most important class of linear operators. It turns out that in Euclidean and unitary spaces operators with a real spectrum, that are diagonalizable in some orthonormal basis, play a very special role. In other words, these operators realize real stretching of the space along a system of pairwise orthogonal directions.

Let $\{e_1, \dots e_n\}$ be an orthonormal basis in L and let $f : L \to L$ be an operator, for which $f(e_i) = \lambda_i e_i$, $\lambda_i \in \mathbf{R}$, $i = 1, \dots, n$. It is not difficult to verify that it has the following simple property:

$$(f(l_1), l_2) = (l_1, f(l_2)) \quad \text{for all} \quad l_1, l_2 \in L. \tag{1}$$

Indeed,

$$\left(f\left(\sum x_i e_i\right), \sum y_j e_j\right) = \sum \lambda_i x_i y_i \quad \text{for} \quad \sum \lambda_i x_i \bar{y}_i,$$

$$\left(\sum x_i e_i, f\left(\sum y_j e_j\right)\right) = \sum \lambda_i x_i y_i \quad \text{for} \quad \sum \lambda_i x_i \bar{y}_i$$

(in the unitary case the fact that λ_i are real was used in the second formula). Operators with the property (1) are called *self-adjoint operators*, and we have established the fact that *operators with a real spectrum, that are diagonalizable in an orthonormal basis, are self-adjoint.* We shall soon prove the converse assertion as well, but we shall first study the property of self-adjointness more systematically.

8.2. Adjoint operators in spaces with a bilinear form. In the first part of this course we showed that for any linear mapping $f : L \to M$ there exists a unique linear mapping $f^* : M^* \to L^*$ for which

$$(f^*(m^*), l) = (m^*, f(l)),$$

where $m^* \in M^*$, $l \in L$ and the parentheses indicate the canonical bilinear mappings $L^* \times L \to \mathcal{K}$, $M^* \times M \to \mathcal{K}$.

In particular, if $M = L$, then the operator $f : L \to L$ corresponds to an operator $f^* : L^* \to L^*$. We shall now assume that a non-degenerate bilinear form

$g : L \times L \to \mathcal{K}$, determining the isomorphism $\tilde{g} : L \to L^*$, exists on L. Then, identifying L^* with L by means of $\tilde{g}^{-1}$, we can study f^*, more precisely $\tilde{g}^{-1} \circ f^* \circ \tilde{g}$, as an operator on L. We shall denote it as before by f^* (it would be more accurate to write, for example, f_q^*, but f^* in the old sense will no longer be used in this section). Evidently, the new operator f^* is uniquely defined by the formula

$$g(f^*(l), m) = g(l, f(m)).$$

It is called, as before, the adjoint of f (with respect to the inner product g).

In the sesquilinear case $\tilde{g}$ defines an isomorphism of L to $\bar{L}^*$, and not to L^*. Therefore, the operator $\bar{f}^* : \bar{L}^* \to \bar{L}^*$, which is defined as $\bar{f}^*(m) = \overline{f^*(m)}$ should transfer to L with the help of this isomorphism. The transferred operator $\tilde{g}^{-1} \circ \bar{f}^* \circ \tilde{g} : L \to L$ is linear. It should be denoted by f^+, but we shall retain the more traditional notation f^*. Then, the following formula will also hold in the sesquilinear case:

$$g(f^*(l), m) = g(l, f(m)).$$

The operation $f \mapsto f^*$ is linear if g is bilinear, and antilinear if g is sesquilinear.

The operators $f : L \to L$ with the property $f^* = f$ in Euclidean and finite-dimensional unitary spaces are called *self-adjoint*; also *symmetric* in the Euclidean case and *Hermitian* in the unitary case. This terminilogy is explained by the following simple remark.

8.3. Proposition. *If the operator $f : L \to L$ in an orthonormal basis is defined by the matrix A, then the operator f^* is defined in the same basis by the matrix A^t (Euclidean case) or $\bar{A}^t$ (unitary case).*

In particular, an operator is self-adjoint if and only if its matrix in an orthonormal basis is symmetric or Hermitian.

Proof. Denoting the inner product in L by parentheses and vectors by columns of their coordinates in an orthonormal basis, we have

$$(f(\vec{x}), y) = (A\vec{x})^t y = (\vec{x}^t A^t)\vec{y} = \vec{x}^t (A^t \vec{y}) = (\vec{x}, f^*(\vec{y}))$$

(Euclidean case). It follows that the matrix f^* equals A^t. The unitary case is analysed analogously.

8.4. Self-adjoint operators and scalar products. Let L be a space with a symmetric or Hermitian inner product (,). For any linear operator $f : L \to L$ we can determine a new inner product $(\, , \,)_f$ on L by setting

$$(l_1, l_2)_f = (f(l_1), l_2).$$

Suppose that L is non-degenerate, so that we can use the concept of an adjoint operator. Then

$$(l_2, l_1)_f = (f(l_2), l_1) = (l_2, f^*(l_1)) = (f^*(l_1), l_2) = (l_1, l_2)_{f^*},$$

in the Euclidean case, and analogously

$$\overline{(l_2, l_1)_f} = \overline{(f(l_2), l_1)} = \overline{(l_2, f^*(l_1))} = (f^*(l_1), l_2) = (l_1, l_2)_{f^*}$$

in the unitary case. Therefore, *if the operator f is self-adjoint, then the new metric $(l_1, l_2)_f$ constructed according to it will be, as before, symmetric or Hermitian.* The converse is also true, as is easily verified directly or with the help of Proposition 8.3.

We have thus established a bijection between sets of self-adjoint operators on the one hand and between symmetric inner products in a space in which one non-degenerate inner product is given on the other. In the Euclidean and unitary cases, after selecting an orthonormal basis, the correspondence is easily described in matrix language: the Gram matrix $(\ ,\ )_f$ is the transpose of the matrix of the mapping f.

We shall now prove the main theorem about self-adjoint operators, which is parallel to Theorem 7.4 on orthogonal and unitary operators and is closely related to it.

8.5. Theorem. a) *In order that the operator f in a finite-dimensional Euclidean or unitary space be self-adjoint it is necessary and sufficient that it be diagonalizable in an orthonormal basis and have a real spectrum.*

b) *The eigenvectors of a self-adjoint operator corresponding to different eigenvalues, are orthogonal.*

Proof. a) We verified sufficiency at the beginning of this section. The fact that the spectrum is real in the unitary case is easily established. Let λ be an eigenvalue of f, and let $l \in L$ be the corresponding eigenvector. Then

$$\lambda(l, l) = (f(l), l) = (l, f(l)) = \bar{\lambda}(l, l),$$

whence $\lambda = \bar{\lambda}$, because $(l, l) \neq 0$. The orthogonal case is reduced to the unitary case by the following device: we examine the complexified space $L^{\mathbf{C}}$ and introduce on it a sesquilinear inner product according to the formula

$$(l_1 + il_2, l_3 + il_4) = (l_1, l_3) + (l_2, l_4) + i(l_2, l_3) - i(l_1, l_4).$$

A simple direct verification shows that $L^{\mathbf{C}}$ transforms into a unitary space and $f^{\mathbf{C}}$ transforms into an Hermitian operator on it. The spectrum of $f^{\mathbf{C}}$ coincides with the spectrum of f, because in any $\mathbf{R}$-basis of L, which is also a $\mathbf{C}$-basis of $L^{\mathbf{C}}$, f and $f^{\mathbf{C}}$ are specified by identical matrices. Therefore the spectrum of f is real.

Further, both cases can be studied in parallel and induction on $\dim L$ can be performed. The case $\dim L = 1$ is trivial. For $\dim L > 1$ we select an eigenvalue λ and the corresponding characteristic subspace L_0, and then set $L_1 = L_0^\perp$. Proposition 3.2 implies that $L = L_0 \bigoplus L_1$. The subspace L_1 is invariant with respect to f, so that if $l_0 \in L_0$, $l_0 \neq 0$ and $l \in L_1$, that is, $(l_0, l) = 0$, then

$$(l_0, f(l)) = (f(l_0), l) = \lambda(l_0, l) = 0,$$

so that $f(l) \in L$. By the induction hypothesis the restriction of f to L_1 is diagonalized in the orthonormal basis of L_1. Adding to it the vector $l_0 \in L_0$, $|l_0| = 1$, we obtain the required basis of L.

b) Let $f(l_1) = \lambda_1 l_1, f(l_2) = \lambda_2 l_2$. Then

$$\lambda_1(l_1, l_2) = (f(l_1), l_2) = (l_1, f(l_2)) = \lambda_2(l_1, l_2),$$

whence it follows that if $\lambda_1 \neq \lambda_2$, then $(l_1, l_2) = 0$.

8.6. Corollary. *Any real symmetric or complex Hermitian matrix has a real spectrum and is diagonalizable.*

Proof. We construct, in terms of the matrix A, a self-adjoint operator in the coordinate space $\mathbf{R}^n$ or $\mathbf{C}^n$ with a canonical Euclidean or unitary metric and apply Theorem 8.5. Even more can be gleaned from it: a matrix X, such that $X^{-1}AX$ is diagonal, exists in $O(n)$ or $U(n)$, respectively.

8.7. Corollary. *The mapping* $\exp : u(n) \to U(n)$ *is surjective.*

Proof. The Lie algebra $u(n)$ consists of anti-Hermitian matrices (see §4 of Chapter 1), and any anti-Hermitian matrix has the form iA, where A is a Hermitian matrix. In order to solve the equation $\exp(iA) = U$ for A, where $U \in U(n)$, we realize U as a unitary operator f in a Hermitian coordinate space $\mathbf{C}^n$. Then, according to Theorem 7.4, we find in $\mathbf{C}^n$ a new orthonormal basis $\{e_1, \ldots, e_n\}$ in which the matrix f acquires the form $\operatorname{diag}(e^{i\phi_1}, \ldots, e^{i\phi_n})$, define in this basis the operator g by the matrix $\operatorname{diag}(\phi_1, \ldots, \phi_n)$, and denote by A the matrix of g in the starting basis. Evidently, $\exp(ig) = f$ and $\exp(iA) = U$.

8.8. Corollary. a) *Let g_1, g_2 be two orthogonal or Hermitian forms in a finite-dimensional space L, and let one of them, say g_1, be positive-definite. Then in the space L there exists a basis whose Gram matrix with respect to g_1 is a unit matrix, and is diagonal and real with respect to g_2.*

b) *Let g_1, g_2 be two real symmetric or complex Hermitian symmetric forms with respect to the variables $x_1, \ldots, x_n$; $y_1, \ldots, y_n$, and let g_1 be positive-definite.*

Then, with the help of a non-singular linear substitution of variables (common to both $\vec{x}$ and $\vec{y}$) these two forms can be written as

$$g_1(\vec{x},\vec{y}) = \sum_{i=1}^{n} x_i y_i; \quad g_2(\vec{x},\vec{y}) = \sum_{i=1}^{n} \lambda_i x_i y_i, \quad \lambda_i \in \mathbf{R},$$

or

$$g_1(\vec{x},\vec{y}) = \sum_{i=1}^{n} x_i \bar{y}_i; \quad g_2(\vec{x},\vec{y}) = \sum_{i=1}^{n} \lambda_i x_i \bar{y}_i, \quad \lambda_i \in \mathbf{R}.$$

Proof. Both formulations are obviously equivalent. To prove them we examine (L, g_1) as an orthogonal or unitary space, rewrite $g_1(l_1, l_2)$ as (l_1, l_2) and represent $g_2(l_1, l_2)$ in the form $(l_1, l_2)_f$, where $f : L \to L$ is some self-adjoint operator, as was done in §8.4. Next we find the orthonormal basis of L in which f is diagonalized. The remark at the end of §8.4 implies that this basis will satisfy the requirements of the corollary (more precisely, the assertion a)).

8.9. Orthogonal projection operators. Let L be a linear space over $\mathcal{K}$, and let its decomposition into the direct sum $L = L_1 \oplus L_2$ be given. As was demonstrated in Chapter 1, it determines two projection operators $p_i : L \to L$ such that $\operatorname{im} p_i = L_i$, $\operatorname{id}_L = p_1 + p_2$, $p_1 p_2 = p_2 p_1 = 0$, $p_i^2 = p_i$. The eigenvalues of the projection operators equal 0 or 1.

If L is a Euclidean or unitary space and $L_2 = L_1^{\perp}$, then the corresponding orthogonal projection operators are diagonalized in an orthonormal basis of L which is the union of orthonormal bases of L_1 and L_2, and are therefore self-adjoint. Conversely, *any self-adjoint projection operator p is the operator of an orthogonal projection onto a subspace.* Indeed $\ker p$ and $\operatorname{im} p$ are spanned by the eigenvectors of p corresponding to eigenvalues 0 and 1 respectively, so that $\ker p$ and $\operatorname{im} p$ are orthogonal according to Theorem 8.5 and $L = \ker p \oplus \operatorname{im} p$.

Further, if the self-adjoint operator f is diagonalized in the orthonormal basis $\{e_i\}$, $f(e_i) = \lambda_i e_i$ and p_i is the orthogonal projection operator of L on the subspace spanned by e_i, then

$$f = \sum_{i=1}^{n} \lambda_i p_i. \tag{2}$$

This formula is called the *spectral decomposition* of the operator f.

It may be assumed that λ_i runs through only pairwise different eigenvalues, and p_i is the operator of an orthogonal projection on the complete root subspace $L(\lambda_i)$; equation (2) remains correct.

Theorem 8.5 can also be extended to (norm) bounded (and with certain complications to unbounded) self-adjoint operators in infinite-dimensional Hilbert spaces. This extension, however, requires a highly non-trivial change in some basic concepts. The main problems are associated with the structure of the spectrum: in the finite-dimensional case λ is the eigenvalue of f, if and only the operator $\lambda\,\mathrm{id} - f$ is invertible, while in the infinite case the set of points of non-invertibility of the operator $\lambda\,\mathrm{id} - f$ can be larger than the set of eigenvalues of f: for points λ_0 not isolated in the spectrum, there are, generally speaking, no eigenvectors. On the other hand, it is precisely the set of points of non-invertibility of the operator $\lambda\,\mathrm{id} - f$ that serves as the correct extension of the spectrum in the infinite-dimensional case. This shortage of eigenvectors requires that many formulations be modified. The main result is an extension of equation (2) where, however, the summation is replaced by integration.

We shall confine our attention to a description of several important principles for examples where these difficulties do not arise.

8.10. Formally adjoint differential operators. Consider the space of real functions on the interval $[a, b]$ with the inner product

$$(f, g) = \int_a^b f(x)g(x)\,dx.$$

Suppose that the operator $\frac{d}{dx}$ transforms it into itself. According to the formula for integration by parts,

$$\left(\frac{df}{dx}, g\right) + \left(f, \frac{dg}{dx}\right) = fg\Big|_a^b = f(b)g(b) - f(a)g(a).$$

Therefore, if the space consists only of functions that assume identical values at the ends of the interval, then

$$\left(\frac{df}{dx}, g\right) = \left(f, -\frac{dg}{dx}\right),$$

that is, in such a space the operator $-\frac{d}{dx}$ is the adjoint of the operator $\frac{d}{dx}$.

Applying the formula for integration by parts several times or making use of the formal operator relation $(f\circ\ldots\circ f_n)^* = f_n^*\circ\ldots\circ f_1^*$, we find that on such spaces

$$\left[\sum_{i=0}^{n} a_i(x)\frac{d^i}{dx^i}\right] = \sum_{i=0}^{n}(-1)^i\frac{d^i}{dx^i}\circ a_i(x), \tag{3}$$

where the notation $\frac{d^i}{dx^i}\circ a_i$ for the operator means that on applying it to the function $f(x)$, we first multiply it by $a_i(x)$ and then differentiate it i times with respect to x. Equation (3) defines the operation of taking the *(formal) adjoint* of differential operators: $D \mapsto D^*$. The operator D is called formally self-adjoint, if $D = D^*$. The word "formally" here reminds us of the fact that the definition does not indicate explicitly the space on which D is realized as a linear operator.

If the inner product is determined with the help of the weight $G(x)$

$$(f, g)_G = \int_a^b G(x)f(x)g(x)\,dx,$$

then obvious calculations show that instead of D^* we must study the operator $G^{-1} \circ D^* \circ G$ (assuming that G does not vanish); it is precisely this operator that is the candidate for the role of an operator adjoint to D with respect to $(f,g)_G$.

We shall show that the orthogonal systems of functions studied in §4 consist of the eigenfunctions of self-adjoint differential operators.

a) *Real Fourier polynomials of degree* $\leq N$. The operator $\frac{d^2}{dx^2}$, which is formally self-adjoint, transforms this space into itself and is self-adjoint in it. In addition, its eigenvalues equal zero (multiplicity 1) and $-1^2, -2^2, \dots, -N^2$ (multiplicity 2). The corresponding eigenvectors are 1 and $\{\cos nx, \sin nx\}$, $1 \leq n \leq N$.

b) *Legendre polynomials.* The operator

$$(x^2-1)\frac{d^2}{dx^2} + 2x\frac{d}{dx} = \frac{d}{dx} \circ \left[(x^2-1)\frac{d}{dx}\right]$$

is formally self-adjoint and transforms the space of polynomials of degree $\leq N$ into itself. The obvious identity

$$(x^2-1)\frac{d}{dx}(x^2-1)^n = 2nx(x^2-1)^n,$$

holds whence, according to Leibnitz's formula applied to both sides,

$$\frac{d^{n+1}}{dx^{n+1}}\left[(x^2-1)\frac{d}{dx}(x^2-1)^n\right] =$$

$$= (x^2-1)\frac{d^{n+2}}{dx^{n+2}}(x^2-1)^n + 2(n+1)x\frac{d^{n+1}}{dx^{n+1}}(x^2-1)^n +$$

$$+n(n+1)\frac{d^n}{dx^n}(x^2-1)^n = 2nx\frac{d^{n+1}}{dx^{n+1}}(x^2-1)^n + 2n(n+1)\frac{d^n}{dx^n}(x^2-1)^n.$$

Dividing the last equality by $2^n n!$ and recalling the definiton of Legendre polynomials we obtain:

$$\left[(x^2-1)\frac{d^2}{dx^2} + 2x\frac{d}{dx}\right] P_n(x) = n(n+1)P_n(x).$$

Thus *the operator* $(x^2-1)\frac{d^2}{dx^2} + 2x\frac{d}{dx}$ *in the space of polynomials of degree* $\leq N$ *is diagonalized in an orthogonal basis consisting of Legendre polynomials and has a simple real spectrum. Therefore it is self-adjoint.*

Of course, self-adjointness in this space could have been verified also by direct integration by parts: a term of the type $fg\big|_{-1}^{1}$ vanishes here due to the factor x^2-1 in the coefficients of the operator. Then from the results of this section and Theorem 8.4, a different proof is obtained for the pairwise orthogonality of the Legendre polynomials.

We leave to the reader the verification and interpretation in terms of linear algebra of the corresponding factors for Hermite and Chebyshev polynomials (don't forget the weight factors $G(x)$!).

c) *The Hermite polynomial* $H_n(x) = (-1)^n e^{x^2} \frac{d^n}{dx^n}(e^{-x^2})$ *is the eigenvector with the eigenvalue* $-2n$ *of the operator*

$$K = \frac{d^2}{dx^2} - 2x\frac{d}{dx}.$$

The function $e^{-x^2/2}H_n(x) = (-1)^n e^{x^2/2}\frac{d^n}{dx^n}(e^{-x^2})$ *is the eigenvector of the operator*

$$H = \frac{d^2}{dx^2} - x^2$$

with the eigenvalue $-(2n+1)$.

The first assertion is verified by direct induction on n, which we omit. To prove the second assertion we examine the auxiliary operator

$$M = \frac{d}{dx} - x.$$

It is easy to verify that

$$[H, M] = HM - MH = -2\left(\frac{d}{dx} - x\right) = -2M.$$

From here it follows that if f is an eigenfunction of the operator H with eigenvalue λ, then Mf is an eigenfunction of the operator H with eigenvalue $\lambda - 2$:

$$HMf = [H, M]f + MHf = -2Mf + \lambda Mf = (\lambda - 2)Mf.$$

Since $H(e^{-x^2/2}) = -e^{-x^2/2}$, we obtain that $M^n(e^{-x^2/2})$ is an eigenfunction for H with eigenvalue $-(2n+1)$ for all $n \geq 0$. On the other hand a direct check shows that

$$e^{x^2/2}M(e^{-x^2/2}f(x)) = e^{x^2}\frac{d}{dx}(e^{-x^2}f(x)),$$

whence it follows that $e^{-x^2/2}H_n(x) = (-1)^n M^n(e^{-x^2/2})$, as required.

d) *Chebyshev polynomial*

$$T_n(x) = \frac{(-2)^n n!}{(2n)!}\sqrt{1-x^2}\,\frac{d^n}{dx^n}(1-x^2)^{n-\frac{1}{2}}$$

is the eigenvector with eigenvalue $-n^2$ *of the operator*

$$(1-x^2)\frac{d^2}{dx^2} - x\frac{d}{dx}.$$

11. Normal operators. Both unitary and self-adjoint operators in a unitary space are a particular case of *normal operators*, which can be described by two equivalent properties:

a) *These are operators that are diagonalizable in an orthonormal basis.*

b) *These are operators that commute with their own conjugate operator.*

Let us verify equivalence.

If $\{e_i\}$ is an orthonormal basis with $f(e_i) = \lambda e_i$, then $f^*(e_i) = \bar{\lambda}_i e_i$, so that $[f, f^*] = 0$ and b) follows from a).

To prove the converse implication we choose the eigenvalue λ of the operator f and set

$$L_\lambda = \{l \in L | f(l) = \lambda l\}.$$

We verify that $f^*(L_\lambda) \subset L_\lambda$. Indeed, if $l \in L$, then

$$f(f^*(l)) = f^*(f(l)) = f^*(\lambda l) = \lambda f^*(l),$$

since $ff^* = f^*f$. From here it follows that the space $L_\lambda^\perp$ is f-invariant: if $(l, l_0) = 0$ for all $l_0 \in L$, then

$$(f(l), l_0) = (l, f^*(l_0)) = 0.$$

The same argument shows that $L_\lambda^\perp$ is f^*-invariant. The restrictions of f and f^* to $L_\lambda^\perp$ evidently commute. Applying induction on the dimension of L, we can assume that on $L_\lambda^\perp$ f is diagonalized in an orthonormal basis. Since the same is true for L_λ, this completes the proof.

EXERCISES

1. Let $f : L \to L$ be an operator in a unitary space. Prove that if $|(f(l), l)| \le c|l|^2$ for all $l \in L$ and some $c > 0$, then

$$|(f(l), m)| + |(l, f(m))| \le 2c|l|\,|m|$$

for all $l, m \in L$.

2. Let $f : L \to L$ be a self-adjoint operator. Prove that

$$|(f(l), l)| \le |f|\,|l|^2$$

for all $l \in L$, where $|f|$ is the induced norm of f, and if $c < |f|$, then there exists a vector $l \in L$ with

$$|(f(l), l)| > c|l|^2.$$

3. The self-adjoint operator f is said to be non-negative, $f \geq 0$ if $(f(l)l) \geq 0$ for all l. Prove that this condition is equivalent to non-negativity for all points of the spectrum f.

4. Prove that the relation $f \geq g : f - g \geq 0$ is an order relation on the set of self-adjoint operators.

5. Prove that the product of two commuting non-negative self-adjoint operators is non-negative.

6. Prove that from each non-negative self-adjoint operator it is possible to extract a unique non-negative square root.

7. Calculate explicitly the second-order correction to the eigenvector and eigenvalue of the operator $H_0 + \epsilon H_1$.

8. Let f be a self-adjoint operator, $\omega \in \mathbf{C}$, $\mathrm{im}\,\omega \neq 0$. Prove that the operator

$$g = (f - \bar{\omega}\,\mathrm{id})\,(f - \omega\,\mathrm{id})^{-1}$$

is unitary, its spectrum does not contain unity and

$$f = (\omega g - \bar{\omega}\,\mathrm{id})\,(g - \mathrm{id})^{-1}.$$

9. Conversely, let g be a unitary operator, whose spectrum does not contain unity. Prove that the operator

$$f = (\omega g - \bar{\omega}\,\mathrm{id})\,(g - \mathrm{id})^{-1}$$

is self-adjoint and

$$g = (f - \bar{\omega}\,\mathrm{id})\,(f - \omega\,\mathrm{id})^{-1}.$$

(The mappings described here, which relate the self-adjoint and unitary operators, are called the Cayley transformations. In the one-dimensional case they correspond to the mapping $a \mapsto \frac{a-\bar{\omega}}{a-\omega}$, which transforms the real axis into the unit circle.)

10. Let $f : L \to L$ be any linear operator in a unitary space. Prove that $f^* f$ is a non-negative self-adjoint operator and that it is positive if and only if f is invertible.

11. Let f be invertible and $r_1^2 = ff^*$, $r_2^2 = f^* f$, where r_1, r_2 are positive self-adjoint operators. Prove that

$$f = r_1 u_1 = u_2 r_2,$$

where u_1, u_2 are unitary. (These representations are called polar decompositions of the linear operator f, where u_1, u_2 are respectively the right and left phase factors of f. In the one-dimensional case, a representation of non-zero complex numbers in the form $re^{i\phi}$ is obtained.)

12. Prove that the polar decompositions $f = r_1u_1 = u_2r_2$ are unique.

13. Prove that the polar decompositions also exist for non-invertible operators f, but only r_1 and r_2 are determined uniquely, and not the unitary cofactors.

§9. Self-Adjoint Operators in Quantum Mechanics

9.1. In this section we shall continue the discussion of the basic postulates of quantum mechanics which we started in §6.8.

Let $\mathcal{H}$ be the unitary state space of a quantum system. In physics, specific states are characterized by determining ("measuring") the values of some *physical quantities*, such as the energy, spin, coordinate, momentum, etc. If the unit of measurement of each such quantity as well as the reference point ("zero") are chosen, then the possible values are *real numbers* (this is essentially the measurement of scalar quantities), and we shall always assume that this condition is satisfied.

The third (after the principle of superposition and interpretation of inner products as probability amplitudes) postulate of quantum mechanics consists of the following.

With every scalar physical quantity, whose value can be measured on the states of the system with the state space $\mathcal{H}$, there can be associated a self-adjoint operator $f : \mathcal{H} \to \mathcal{H}$ with the following properties:

a) *The spectrum of f is a complete set of values of the quantity which can be determined by measuring this quantity in different states of the system.*

b) *If $\psi \in \mathcal{H}$ is the eigenvector of the operator f with eigenvalue λ, then in measuring this quantity in a state ψ the value λ will be obtained with certainty.*

c) *More generally, by measuring the quantity f in a state ψ, $|\psi| = 1$, we can determine the value λ from the spectrum of f with a probability equal to the square of the norm of the orthogonal projection of ψ on the complete characteristic subspace $\mathcal{H}(\lambda)$ corresponding to λ.*

Since, according to Theorem 8.4, $\mathcal{H}$ can be decomposed into an orthogonal direct sum $\bigoplus_{i=1}^{m} \mathcal{H}(\lambda_i)$, $\lambda_i \neq \lambda_j$ for $i \neq j$, we can expand ψ in a corresponding sum of projections $\psi_i \in \mathcal{H}(\lambda_i)$, $i = 1, \dots, m$. Pythagoras's theorem

$$1 = |\psi|^2 = \sum_{i=1}^{m} |\psi_i|^2$$

is then interpreted as an assertion about the fact that by performing a measurement of f on any state ψ, we shall obtain with probability 1 one of the possible values of f.

Physical quantities, which we mentioned above, and the corresponding self-adjoint operators are also called *observables*. The postulate about observables is sometimes interpreted more broadly and it is assumed that any self-adjoint operator corresponds to some physical observable.

In infinite-dimensional spaces $\mathcal{H}$ these postulates are somewhat altered. In particular, instead of b) and c) one studies the probability that in measuring f in the state ψ the values will fall into some interval $(a,b) \subset \mathbf{R}$. With this interval one can also associate a subspace $\mathcal{H}_{(a,b)} \subset \mathcal{H}$ — the image of the orthogonal projector $p_{(a,b)}$ on $\bigoplus_{\lambda_i \in (a,b)} \mathcal{H}(\lambda_i)$ in the finite-dimensional case — and the probability sought equals

$$|p_{(a,b)}\psi|^2 = (\psi, p_{(a,b)}\psi).$$

In addition, in the infinite-dimensional case it can happen that the operators of observables are defined only on some subspace $\mathcal{H}_0 \subset \mathcal{H}$.

This terminology is related to the concepts introduced in §6.8 as follows. A *filter* B_χ is a device that measures the observable corresponding to the orthogonal projection operator on the subspace generated by χ. It is assigned the value 1, if the system passes through the filter and 0 otherwise. A *furnace* A_ψ is a combination of a device that transforms the system, generally speaking, into different states and a filter B_ψ which passes only systems which are in the state ψ. The recipe given in §6.8 for calculating probabilities, evidently, agrees with the recipe given above in §9.1b, c.

In this example it is evident that a device which measures an observable, say B_χ, in the state ψ, generally speaking *changes* this state: it transforms the system into the state χ with probability $|(\psi,\chi)|^2$ and "annihilates" the system with probability $1 - |(\psi,\chi)|^2$. Hence the term "measurement" as applied to this interaction of the system with the measurement device can lead to completely inadequate intuitive notions. Classical physics is based on the assumption that measurements can in principle be performed with as small a perturbation as desired of the system subjected to the measurement action. The term "measurement" is nevertheless generally used in physics texts, and we have found it necessary to introduce it here, beginning first with the less familiar, but intuitively more convenient, "furnaces" and "filters".

9.2. Average values and the uncertainty principle. Let f be an observable, $\{\lambda_i\}$ its spectrum, and $\mathcal{H} = \bigoplus_i \mathcal{H}(\lambda_i)$ the corresponding orthogonal decomposition. As already stated, in the state ψ, $|\psi| = 1$, f assumes the value λ_i with probability $(\psi, p_i\psi)$, where p_i is the orthogonal projection operator onto $\mathcal{H}(\lambda_i)$. Therefore, the average value $\hat{f}_\psi$ of the quantity f in the state ψ, obtained from many measurements, can be calculated as

$$\hat{f}_\psi = \sum_i \lambda_i(\psi, p_i\psi) = \sum_i(\psi, \lambda_i p_i\psi) = (\psi, f(\psi))$$

(we repeat that $|\psi| = 1$).

Our quantity $(\psi, f(\chi))$ is written in Dirac's notation as $\langle\chi|f|\psi\rangle$. The part $f|\psi\rangle$ of this symbol is the result of the action of the operator f on the ket vector $|\psi\rangle$, while $\langle\chi|f$ is the result of the action of the adjoint operator on the bra vector $\langle\chi|$.

We return to the average values. If the operators f and g are self-adjoint, then the operator fg, generally speaking, is not self-adjoint:

$$(fg)^* = g^* f^* = gf \neq fg,$$

if f and g do not commute. However, $f^2, f - \lambda(\lambda \in \mathbf{R})$ and the commutator $\frac{1}{i}[f,g] = \frac{1}{i}(fg - gf)$ are, as before, self-adjoint.

The average value $[(f - \hat{f}_\psi)^2]_{\hat{\psi}}$ of the observable $(f - \hat{f}_\psi)^2$ in the state ψ is the *mean-square deviation of the values of f from their average value*, or the variance (spread) of the values of f. We set

$$\widehat{\Delta f}_\psi = \sqrt{[(f - \hat{f}_\psi)^2]_{\hat{\psi}}}.$$

9.3. Proposition (Heisenberg's uncertainty principle). *For any self-adjoint operators f, g in a unitary space*

$$\widehat{\Delta f}_\psi \cdot \widehat{\Delta g}_\psi \geq \frac{1}{2}|([f,g]\psi, \psi)|.$$

Proof. Using the obvious formula

$$[f - \hat{f}_\psi, g - \hat{g}_\psi] = [f, g],$$

the self-adjointness of the operators f and g, and the Cauchy-Bunyakovskii-Schwarz inequality, we find ($f_1 = f - \hat{f}_\psi$, $g_1 = g - \hat{g}_\psi$)

$$|([f,g]\psi, \psi)| = |((f_1g_1 - g_1f_1)\psi, \psi)| = |(g_1\psi, f_1\psi) - (f_1\psi, g_1\psi)| =$$

$$= |2\mathrm{Im}(g_1\psi, f_1\psi)| \leq 2|(g_1\psi, f_1\psi)| \leq$$

$$\leq 2\sqrt{(f_1\psi, f_1\psi)}\sqrt{(g_1\psi, g_1\psi)} = 2\widehat{\Delta f_\psi}\widehat{\Delta g_\psi}.$$

This shows that the average spread in the values of the non-commuting observables f and g, generally speaking, cannot be made arbitrarily small at the same time. It is also said that the non-commuting observables are ***not simultaneously*** *measurable*; this formulation should be treated with the same precautions as the term "measurement".

The application of Heisenberg's inequality to the case of *canonically conjugate* pairs of observables, which by definition satisfy the relation $\frac{1}{i}[f,g] = \mathrm{id}$, plays a special role. For them

$$\widehat{\Delta f_\psi} \cdot \widehat{\Delta g_\psi} \geq \frac{1}{2},$$

irrespective of the state ψ. We note that in finite-dimensional spaces there are no such pairs, because $\mathrm{Tr}[f,g] = 0$, $\mathrm{Tr}\ \mathrm{id} = \dim \mathcal{H}$. They do exist, however, in infinite-dimensional spaces. The classical example is

$$\frac{1}{i}\left[x, \frac{1}{i}\frac{d}{dx}\right] = \mathrm{id}.$$

These operators appear in quantum models of physical systems, which in the classical language are called "particles moving in a one-dimensional potential".

We shall describe these and some other observables in greater detail.

9.4. a) *Coordinate observable.* This is the operator of multiplication by x in the space of complex functions on $\mathbf{R}$ (or some subsets of $\mathbf{R}$) with the inner product $\int f(x)\bar{g}(x)\,dx$. It is presumed that the quantum system is a "particle moving in a straight line in an external field".

b) *Momentum observable.* This is the operator $\frac{1}{i}\frac{d}{dx}$ in analogous function spaces. (It is usually multiplied by Planck's constant $\hbar$; this refers to the choice of the system of units, which we do not consider.)

c) *The energy observable of a quantum oscillator.* This is the operator $\frac{1}{2}\left[-\frac{d^2}{dx^2} + x^2\right]$, once again in appropriate units.

d) *Observable of the projection of the spin* for the system "particle with spin 1/2". This is any self-adjoint operator with the eigenvalues $\pm 1/2$ in a 2-dimensional unitary space. Further details concerning it will be given later.

In examples a)–c) we intentionally did not specify the unitary spaces in which our operators act. They are substantially infinite-dimensional and are constructed and studied by means of functional analysis. We shall have more to say about example d) below.

9.5. Energy observable and the temporal evolution of the system. The description of any quantum system, together with its spatial states $\mathcal{H}$, includes the specification of the fundamental observable $H : \mathcal{H} \to \mathcal{H}$, which is called the *energy observable* or the *Hamiltonian operator*, or the *Hamiltonian*.

The last of the basic postulates of quantum mechanics is formulated in terms of it.

If at time zero the system is in a state ψ and the system evolved over a time interval t as an isolated system, in particular, measurements were not performed on it, then at the time t it will be in a state $\exp(-iHt)(\psi)$, *where*

$$\exp(-iHt) = \sum_{n=0}^{\infty} \frac{(-iH)^n t^n}{n!} : \mathcal{H} \to \mathcal{H}$$

(see §11 of Chapter I).

The operator $\exp(-iHt) = U(t)$ is unitary. The evolution of an isolated system is completely determined by the one-parameter group of unitary operators $\{U(t)|t \in \mathbf{R}\}$.

The physical unit (energy) × (time) is called the "action". Many experiments permit determining the universal unit of action — the famous Planck's constant $\hbar = 1.055 \times 10^{-34}$ erg· sec. In our formula, it is presumed that Ht is measured in units of $\hbar$, and it is most often written in the form $\exp\left(\frac{Ht}{i\hbar}\right)\psi$. We shall drop $\hbar$ in order to simplify the notation.

We also note that because the operator e^{-iHt} is linear, it transforms rays in $\mathcal{H}$ into rays and indeed acts on the state of the system, and not simply on the vector ψ.

The law of evolution can be written in the differential form

$$\frac{d}{dt}(e^{-iHt}\psi) = -iH(e^{-iHt}\psi),$$

or, setting $\psi(t) = e^{-iHt}\psi$,

$$\frac{d\psi}{dt} = -iH\psi \quad \left(\frac{1}{i\hbar} H\psi, \text{ if we refer to the units}\right)$$

The last equation is called *Schrödinger's equation*. It was first written for the case when the states ψ are realized as functions in physical space and H is represented by a differential operator with respect to the coordinates.

In the following discussions, as usual, we shall confine our attention primarily to finite-dimensional state spaces $\mathcal{H}$.

9.6. Energy spectrum and stationary states of a system. The ***energy spectrum*** of a system is the spectrum of its Hamiltonian H. The ***stationary states*** are the states that do not change with time. The rays corresponding to them must be invariant with respect to the operator e^{itH}, that is, they must be one-dimensional characteristic subspaces of this operator. But these are the same subspaces as those of the operator H. The eigenvalue E_j of the Hamiltonian, or the ***energy level*** of the system, corresponds to the eigenvalue $e^{itE_j} = \cos tE_j + i \sin tE_j$ of the evolution operator, which varies with time.

If H has a simple spectrum, then the space $\mathcal{H}$ is equipped with a canonical orthonormal basis, consisting of the vectors of the stationary states (they are determined to within a phase factor $e^{i\phi}$). If the multiplicity of the energy level E is greater than one, then this level and the corresponding states are said to be *degenerate*, and the multiplicity of E is called the degree of degeneracy.

All states corresponding to the lowest level, that is, the lowest eigenvalue of H, are called ***ground states*** of the system; the ground state is unique, if the lowest level is not degenerate. This term is related to the idea that a quantum system can never be regarded as completely isolated from the external world: with some probability, it can emit or absorb some energy. Under some conditions it is much more probable that the energy will be lost rather than absorbed, and the system will have a tendency to "fall down" into its lowest state and will subsequently remain there. Therefore, states other than the ground state are sometimes called *excited* states.

In §9.4d we wrote down the Hamiltonian of the quantum oscillator:

$$\frac{1}{2}\left[-\frac{d^2}{dx^2}+x^2\right].$$

In §8.10c it was shown that the functions $e^{-x^2/2}H_n(x)$ form a system of stationary states of the harmonic oscillator with energy levels $E_n = n + \frac{1}{2}$, $n = 1,2,3,\ldots$. (A more detailed analysis shows that the energy is measured here in units of $\hbar\omega$, where the constant ω corresponds to the oscillation frequency of the corresponding classical oscillator.) Defining in a reasonable manner the unitary space in which one must work, it can be shown that this is a complete system of stationary states. For $n > 0$ the oscillator can emit energy $E_n - E_m = (n-m)\hbar\omega$ and make a transition out of the state ψ_n into the state ψ_m. In application to the quantum theory of the electromagnetic field this is referred to as "*emission of $n-m$ photons with frequency ω*". The inverse process will be the absorption of $n-m$ photons; in this case the oscillator will make a transition into a higher (excited) state. It is important that energy can be absorbed or transferred only in integer multiples of $\hbar\omega$.

In the ground state the oscillator has a non-zero energy $\frac{1}{2}\hbar\omega$ which, however, cannot be transferred in any way – the oscillator does not have lower energy levels. The electromagnetic field in quantum models is viewed as a superposition of infinitely many oscillators (corresponding, in particular, to different frequencies ω). In the ground state – the vacuum – the field therefore has an infinite energy, though from the classical viewpoint it has a zero energy – since energy cannot be taken away from it, it cannot act on anything ! This is a very simple model of the profound difficulties in the modern quantum theory of fields. Neither the mathematical apparatus nor the physical interpretation of the quantum theory of fields has achieved any degree of finality. This is an open and interesting science.

9.7. The formulas of perturbation theory. Situations in which the Hamiltonian H of the system can be viewed as the sum $H_0 + \epsilon H_1$, where H_0 is the "unperturbed" Hamiltonian while ϵH_1 is a small correction, a "perturbation", play an important role in the apparatus of quantum mechanics. From the physical viewpoint the perturbation often corresponds to the interaction of a system with the "external world" (for example, an external magnetic field) or the components of the system with one another (then H_0 corresponds to the idealized case of a system consisting of free, non-interacting components). From the mathematical viewpoint such a representation is justified when the spectral analysis of the unperturbed Hamiltonian H_0 is simpler than that of H, and it is convenient to represent the spectral characteristics of H by power series in powers of ϵ, whose first terms are determined in terms of H_0. We shall confine our attention to the following most widely used formulas and qualitative remarks about them.

a) *First order corrections.* Let $H_0 e_0 = \lambda_0 e_0$, $|e_0| = 1$. We shall find an eigenvector and eigenvalues of $H_0 + \epsilon H_1$ close to e_0 and λ_0 respectively, to within second-order infinitesimals with respect to ϵ, that is, we shall solve the equation

$$(H_0 + \epsilon H_1)(e_0 + \epsilon e_1) = (\lambda_0 + \epsilon\lambda_1)(e_0 + \epsilon e_1) + o(\epsilon^2).$$

Equating the coefficients in front of ϵ, we obtain

$$(H_0 - \lambda_0)e_1 = (\lambda_1 - H_1)e_0.$$

The unknowns here are the number λ_1 and the vector e_1. They can be found successively with the help of the following device. We examine the inner product of both parts of the last equality by e_0. On the left we obtain zero, by virtue of the self-adjointness of $H - \lambda_0$:

$$((H_0 - \lambda_0)e_1, e_0) = (e_1, (H_0 - \lambda_0)e_0) = 0.$$

Therefore $((\lambda_1 - H_1)e_0, e_0) = 0$ and since e_0 is normalized

$$\lambda_1 = (H_1 e_0, e_0).$$

This is the first-order correction to the eigenvalue λ_0: the "shift of the energy level" $\epsilon\lambda_1$ equals $(\epsilon H_1 e_0, e_0)$, that is, according to the results of §9.12 *it equals the average value of the "perturbation energy" ϵH_1 in the state e_0.*

To determine e_1 we must now invert the operator $H_0 - \lambda_0$. Of course, it is not invertible, because λ_0 is an eigenvalue of H_0 but, *the right side of the equation, $(\lambda_1 - H_1)e_0$, is orthogonal to e_0.* Therefore it is sufficient for $H_0 - \lambda_0$ to be invertible in the orthogonal complement to e_0, which we denote as $e_0^{\perp}$. This condition (in the finite-dimensional case) is, evidently, equivalent to the condition that the *multiplicity of the eigenvalue λ_0 of H_0 equals 1, that is, the energy level λ_0 must be non-degenerate.*

If this is so, then

$$e_1 = ((H_0 - \lambda_0)|_{e_0^\perp})^{-1}(\lambda_1 - H_1)e_0,$$

which gives the first-order correction to the eigenvector.

We select an orthonormal basis $\{e_0 = e^{(0)}, e^{(1)}, \ldots, e^{(n)}\}$, in which H_0 is diagonal with eigenvalues $\lambda_0 = \lambda^{(0)}, \lambda^{(1)}, \ldots, \lambda^{(n)}$. In the basis $\{e^{(1)}, \ldots, e^{(n)}\}$ of the space $e_0^\perp$, we have

$$(\lambda_1 - H_1)e_0 = \sum_{i=1}^{n}((\lambda_1 - H_1)e_0, e^{(i)})e^{(i)} = -\sum_{i=1}^{n}(H_1 e_0, e^{(i)})e^{(i)},$$

whence

$$e_1 = \sum_{i=1}^{n} \frac{(H_1 e_0, e^{(i)})}{\lambda_0 - \lambda^{(i)}} e^{(i)}.$$

It is intuitively clear that this first-order correction could be a good approximation if the perturbation energy is small compared to the distance between the level λ_0 and a neighbouring level: ϵ must compensate the denominator $\lambda_0 - \lambda^{(i)}$. Physicists usually assume that this is the case.

b) *Higher order corrections.* By analogy with the case analysed above we shall show that when the eigenvalue λ_0 is non-degenerate, the $(i+1)$st order correction to (λ_0, e_0), can be found inductively under the assumption that the corrections of orders $\leq i$ have already been found. Let $i \geq 1$. We solve the equation

$$(H_0 + \epsilon H_1)\left(\sum_{k=0}^{i+1} \epsilon^k e_k\right) = \left(\sum_{l=0}^{i+1} \epsilon^l \lambda_l\right)\left(\sum_{j=0}^{i+1} \epsilon^j e_j\right) + o(\epsilon^{i+2})$$

for e_{i+1}, λ_{i+1}. Equating the coefficients in front of ϵ^{i+1}, we obtain

$$(H_0 - \lambda_0)e_{i+1} = (\lambda_1 - H_1)e_i + \sum_{l=2}^{i} \lambda_l e_{i+1-l} + \lambda_{i+1} e_0.$$

As above, the left side is orthogonal to e_0, whence

$$\lambda_{i+1} = ((H_1 - \lambda_1)e_i, e_0) - \sum_{l=2}^{i} \lambda_l (e_{i+1-l}, e_0),$$

$$e_{i+1} = \left((H_0 - \lambda_0)|_{e_0^\perp}\right)^{-1}\left[(\lambda_1 - H_1)e_i + \sum_{l=2}^{i+1} \lambda_l e_{i+1-l}\right].$$

Thus all corrections exist and are unique.

c) *Series in perturbation theory.* The formal power series in powers of ϵ

$$\sum_{i=0}^{\infty} \lambda_i \epsilon^i, \sum_{i=0}^{\infty} e_i \epsilon^i,$$

where λ_i and e_i are found from recurrence relations, are called perturbation series. It can be shown that in the finite-dimensional case they converge for sufficiently small ϵ. In the infinite-dimensional case they can diverge; nevertheless, the first few terms often yield predictions which are in good agreement with experiment. Perturbation series play a very large physical role in the quantum theory of fields. Their mathematical study leads to many interesting and important problems.

d) *Multiple eigenvalues and geometry.* In our previous calculations, the requirement that the eigenvalue λ_0 be non-degenerate stemmed from our desire to invert $H_0 - \lambda_0$ in $e_0^{\perp}$ and was formally expressed as the appearance of the differences $\lambda_0 - \lambda^{(i)}$ in the denominators. It is also possible to obtain formulas in the general case by appropriately altering the arguments, but we shall confine our attention to the analysis of the geometric effects of multiplicity. They are evident from the typical case $H_0 = \mathrm{id}$: all eigenvalues equal one. A small change in H_0 leads to the following effects.

The eigenvalues become *different*, if this change is sufficiently general: this effect in physics is called "splitting of levels", or "removal of degeneracy". For example, one spectral line can be split into two or more lines either by increasing the resolution of the instrument or by placing the system into an external field. The mathematical model in both cases will consist of taking into account a small, previously ignored correction to H_0 (though sometimes the starting state space will also change).

Consider now what can happen with eigenvectors. In the completely degenerate case H_0 is diagonalized in *any* orthonormal basis. A small change in H_0, removing the degeneracy, corresponds to the selection of an orthonormal basis along whose axes stretching occurs, and the coefficients of these stretchings. The coefficients must not differ much from the starting value of λ_0, but the axes themselves can be oriented in any direction. Thus the characteristic directions near a degenerate eigenvalue now depend very strongly on the perturbation. Two arbitrarily small perturbations of the unit operator with a simple spectrum can be diagonalized in two fixed orthonormal bases rigidly rotated away from one another. This explains why the differences $\lambda_0 - \lambda^{(i)}$ appear in the denominators.

§10. The Geometry of Quadratic Forms and the Eigenvalues of Self-Adjoint Operators

10.1. This section is devoted to the study of the geometry of graphs of quadratic forms q in a real linear space, that is, sets of the form $x_{n+1} = q(x_1, \ldots, x_n)$ in $\mathbf{R}^{n+1}$, and the description of some applications. One of the most important reasons that these classical results are of great interest, even now, is that quadratic forms give the next (after the linear) approximation to twice-differentiable functions and are therefore the key to understanding the geometry of the "curvature" of any smooth multidimensional surface.

In §§3 and 8 we proved general theorems on the classification of quadratic forms with the help of any linear or only orthogonal transformations, so that here we shall start with a clarification of the geometric consequences. We shall assume that we are working in Euclidean space with the standard metric $\sum_{i=1}^{n} x_i^2$. The disregard of the Euclidean structure merely means the introduction of a rougher equivalence relation between graphs. The plan of the geometric analysis consists in analysing low-dimensional structures, where the form of the graph can be represented more conveniently, and then studying multidimensional graphs in a low-dimensional section in different directions. It is recommended that the reader draw pictures illustrating our text. We assume that the axis x_{n+1} is oriented upwards, and the space $\mathbf{R}^n$ is arranged horizontally.

10.2. One-dimensional case. The graph of the curve $x_2 = \lambda x_1^2$ in $\mathbf{R}^2$ has three basic forms: "a cup" (convex downwards) for $\lambda > 0$, "a dome" (convex upwards) for $\lambda < 0$, and a horizontal line for $\lambda = 0$. With regard to the linear classification, admitting an arbitrary change in scale along the x_1 axis, these three cases exhaust all possibilities: it may be assumed that $\lambda = \pm 1$ or 0. In an orthogonal classification λ is an invariant: $|\lambda|$ determines the slope of the walls of the cup or of the dome, which increases with $|\lambda|$. The other characterization of $|\lambda|$ consists of the fact that $\frac{1}{2|\lambda|}$ is the radius of curvature of the graph at the bottom or at the top $(0,0)$. Indeed, the equation of a circle with radius R, tangent to the x_1 axis at the origin, has the form $x_1^2 + (x_2 - R)^2 = R^2$, and near zero we have $x_2 \approx \frac{x_1^2}{2R}$.

10.3. Two-dimensional case. To analyse this case we use an orthonormal basis in $\mathbf{R}^2$, in which q is reduced to a sum of squares with coefficients $q(y_1, y_2) = \lambda_1 y_1^2 + \lambda_2 y_2^2$. The straight lines spanned by the elements of this basis are called the *principal axes* of the form q; generally speaking, they are rotated relative to the starting axes. The numbers λ_1 and λ_2 are determined uniquely, and are the eigenvalues of the self-adjoint operator A, for which $q(\vec{x}) = \vec{x}^t A \vec{x}$ (in the old coordinates). For $\lambda_1 \neq \lambda_2$ the axes themselves are also determined uniquely, but with $\lambda_1 = \lambda_2$ they can be chosen arbitrarily (provided they are orthogonal). For each coefficient λ_1 or λ_2, there are three basic possibilities ($\lambda_i > 0$, $\lambda_i < 0$, $\lambda_i = 0$), but symmetry considerations lead to four basic cases (of which only the first two are non-degenerate).

a) $x_3 = \lambda_1 y_1^2 + \lambda_2 y_2^2$, $\lambda_1, \lambda_2 > 0$. The graph is an *elliptic paraboloid* shaped like a cup. The adjective "elliptic" refers to the fact that the projection of the horizontal sections $\lambda_1 y_1^2 + \lambda_2 y_2^2 = c$ with $c > 0$ are ellipses with semi-axes $\sqrt{c\lambda_i^{-1}}$ oriented along the principal axes of the form q (circles if $\lambda_1 = \lambda_2$). (These projections are contour lines of the function q.) The noun "paraboloid" refers to the fact that the sections of the graph by the vertical planes $ay_1 + by_2 = 0$ are parabolas (for $\lambda_1 = \lambda_2$ the graph is a paraboloid of revolution).

The case $\lambda_1, \lambda_2 < 0$ is the same cup, but turned upside down.

b) $x_3 = \lambda_1 y_1^2 - \lambda_2 y_2^2$, $\lambda_1, \lambda_2 > 0$. The graph is a *hyperbolic paraboloid*. The contour lines of q are hyperbolas, non-empty for all values of x_3, so that the graph passes above and below the plane $x_3 = 0$. Sections by vertical planes are, as before, parabolas. The contour line $x_3 = 0$ is a "degenerate hyperbola", converging to its asymptotes — the two straight lines $\sqrt{\lambda_1} y_1 \pm \sqrt{\lambda_2} y_2 = 0$. These straight lines in $\mathbf{R}^2$ are called "*asymptotic directions*" of the form q. If q is viewed as an (indefinite) metric in $\mathbf{R}^2$, then the asymptotic straight lines consist of all vectors of zero length. The asymptotic straight lines divide $\mathbf{R}^2$ into four sectors. For $x_3 > 0$ the contour lines $q = x_3$ lie in a pair of opposite sectors; when $x_3 \to +0$ from above, they "adhere to" the asymptotes and transform into them when $x_3 = 0$; for $x_3 < 0$, "passing straight through", they end up in the other pair of opposite sectors. The vertical sections of the graph by planes passing through the asymptotic straight lines are themselves these straight lines — "straightened out parabolas".

The case $\lambda_1 y_1^2 + \lambda_2 y_2^2$, $\lambda_1, \lambda_2 > 0$, is obtained from the one analysed above by changing the sign of x_3.

c) $x_3 = \lambda y_1^2$, $\lambda > 0$. Since the function does not depend on y_2, the sections of the graph by the vertical planes $y_2 =$ const have the same form: the entire graph is swept by the parabola $x_3 = \lambda y_1^2$ in the (y_1, x_3) plane as it moves along the y_2 axis and is called a *parabolic cylinder*. The contour lines are pairs of straight lines $y_1 = \pm\sqrt{x_3 \lambda^{-1}}$; when $x_3 = 0$ they coalesce into a single straight line; the entire graph lies above the plane $x_3 = 0$.

The case $x_3 = \lambda y_1^2, \lambda < 0$ is obtained by "flipping".

d) $x_3 = 0$. This is a plane.

10.4. General case. We are now in a position to understand the geometry of the graph $x_{n+1} = q(x_1, \ldots, x_n)$ for arbitrary values of n.

We transfer to the principal axes in $\mathbf{R}^n$, that is, to the orthonormal basis in which $q(y_1, \ldots, y_n) = \sum_{i=1}^m \lambda_i y_i^2$, $\lambda_1 \ldots \lambda_m \neq 0$. As above, they are determined uniquely if $m = n$ and $\lambda_i \neq \lambda_j$ for $i \neq j$ or if $m = n - 1$ and $\lambda_i \neq \lambda_j$ for $i \neq j$. The form q does not depend on the coordinates $y_{m+1}, \ldots, y_n$, so that the entire graph is obtained from the graph $\sum_{i=1}^m \lambda_i y_i^2$ in $\mathbf{R}^{m+1}$ by translation along the subspace spanned by $\{e_{m+1}, \ldots, e_n\}$. In other words, along this subspace the graph is "cylindrical". It is easy to verify that it is precisely the kernel of the bilinear form that is polar with respect to q, and is trivial if and only if q is non-degenerate.

Let q be non-degenerate, that is, $m = n$. It may be assumed that $\lambda_1, \ldots, \lambda_r > 0$, $\lambda_{r+1}, \ldots, \lambda_{r+s} < 0$, that is (r, s) is the signature of the form q. If the form q is positive-definite, that is, $r = n$, $s = 0$, then the graph has the form of an n-dimensional cup: all of its sections by vertical planes are parabolas, and all contour lines $q = c > 0$ are ellipsoids with semi-axes $\sqrt{c\lambda_i^{-1}}$ oriented along the principal axes. The equation of this ellipsoid has the form

$$\sum_{i=1}^{n} \left(\frac{x_i}{\sqrt{c\lambda_i^{-1}}} \right)^2 = 1,$$

that is, it is obtained from the unit sphere by stretching along the orthogonal directions. In particular, it is bounded: it lies entirely within the rectangular parallelepiped $|x_i| \leq \sqrt{c\lambda_i^{-1}}$, $i = 1, \ldots, n$. We shall verify below that the study of the variations of the lengths of the semi-axes in different cross sections of the ellipsoid (also ellipsoids) gives useful information on the eigenvalues of self-adjoint operators.

When $r = 0$ and $s = n$, a dome is obtained. In both cases the graph is called an (n-dimensional) *elliptic paraboloid*.

The intermediate cases $rs \neq 0$ lead to multidimensional hyperbolic paraboloids with different signatures. The key to their geometry is once again the structure of the *cone of asymptotic directions* C in $\mathbf{R}^n$, that is, the zero contour of the form $q(y_1, \ldots, y_n) = 0$.

It is called a cone because it is swept out by its *generatrices*: the straight line containing a single vector in C lies wholly inside it. In order to form a picture of the base of this cone we shall examine its intersection, for example, with the linear manifold $y_n = 1$:

$$-\lambda_n^{-1} \sum_{i=1}^{n-1} \lambda_i y_i^2 = 1.$$

It is evident that the base is a level set of the quadratic form of the $(n-1)$st variable. The simplest case is obtained when it is positive-definite. Then this level set is an ellipsoid, in particular, it is bounded, and our cone looks like the three-dimensional cones studied in high school. This case corresponds to the signature $(n-1, 1)$ or $(1, n-1)$; for $n = 4$ the space $(\mathbf{R}^4, q)$ is the famous *Minkowski space*, which will be studied in detail below. For other signatures C is much more complicated, because its space"goes off to infinity". Sections of the graph of q by vertical planes passing through the generatrices C coincide with these generatrices. For any other planes, either "cups" or "domes" are obtained – the asymptotic directions separate these two cases. The cone C therefore divides the space $\mathbf{R}^n \backslash C$ into two parts, which are swept out by the straight lines along which q is positive or negative, respectively.

One of these regions is called the collection of internal folds of the cone C, the other is the exterior region. *The geometric meaning of the signature (r, s) can be roughly, but conveniently, described by the following phrase: the graph of the form q passes upwards along r directions and downwards along s directions.*

Although we have been working with real quadratic forms, the same results are applicable to complex Hermitian forms. Indeed, the decomplexification of $\mathbf{C}^n$ is $\mathbf{R}^{2n}$, and the decomplexification of the Hermitian form $\sum_{i=1}^{n} a_{ij}x_i\bar{x}_j$, $a_{ij} = \bar{a}_{ji}$, is the real quadratic form. Decomplexification doubles all dimensions: in particular, the complex signature (r, s) transforms into the real signature $(2r, 2s)$.

We shall now briefly describe, without proof, two applications of this theory to mechanics and topology.

10.5. Oscillations. Imagine first a ball, which rolls in the plane $\mathbf{R}^2$ under the action of gravity along a channel of the form $x_2 = \lambda x_1^2$. The point $(0,0)$ is in all cases one of the possible motions of the ball: the position of equilibrium. For $\lambda > 0$ this position is stable: a small initial displacement of the ball with respect to position or velocity will cause it to oscillate around the bottom of the cup. For $\lambda < 0$ it is unstable: the ball will fall off along one of the two branches of the parabola. For $\lambda = 0$ displacements with respect to position are of no significance, but not so with the velocity: the ball can remain at any point of the straight line $x_2 = 0$ or move uniformly in any direction with the initial momentum.

It turns out that the mathematical description of a large class of mechanical systems near their equilibrium positions is modelled well qualitatively by the multidimensional generalization of this picture: the motion of the ball near the origin of coordinates along a multidimensional surface $x_{n+1} = q(x_1, \ldots, x_n)$ under the action of gravity. If q is positive-definite, any "small" motion will be close to a superposition of small oscillations along the principal axes of the form q. Along the null space of the form the ball can escape to infinity with constant velocity. Along directions where q is negative the ball can fall off downwards. The presence of both the null space and a negative component of the signature indicates that the equilibrium position is unstable and casts doubt on the approximation of "small oscillations". It is important, however, that when this equilibrium is stable, small changes in the form of the cup along which the ball rolls (or, more technically, the potential of our system) do not destroy this stability.

To understand this we return to the remark made at the beginning of the section on the approximate distribution of any (say, thrice-differentiable) real function $f(x_1, \ldots, x_n)$. Near zero it has the form

$$f(x_1, \ldots, x_n) = f(0, \ldots, 0) + \sum_{i=1}^{n} a_i x_i + \sum_{i,j=1}^{n} b_{ij} x_i x_j + o\Big(\sum_{i=1}^{n} |x_i|^2\Big)$$

where

$$a_i = \frac{\partial f}{\partial x_i}(0,\ldots,0),\ b_{ij} = \frac{\partial^2 f}{\partial x_i \partial x_j}(0,\ldots,0).$$

Subtracting from f its value at zero and the linear part we find that the remainder is quadratic, apart from higher order terms. This subtraction means that we are studying the deviation of the graph f from the tangential hypersurface to this graph at the origin. Denoting this tangent plane by $\mathbf{R}^n$, we find that the *behaviour of f near zero is determined by a quadratic form with the matrix* $\left(\frac{\partial^2 f}{\partial x_i \partial x_j}(0,\ldots,0)\right)$, at least *when this form is non-degenerate; otherwise higher order terms must be taken into account.* (For example, the graph $x_2 = x_1^3$ diverges to $-\infty$ on the left and $+\infty$ on the right; graphs of quadratic functions do not behave in this manner. The two-dimensional graph $x_3 = x_1^2 + x_2^3$ is a "monkey's saddle", in one curvilinear sector $x_1^2 + x_2^3 < 0$, diverging downwards – "for the tail".)

A point at which the differential $df = \sum_{i=1}^n \frac{\partial f}{\partial x_i} dx_i$ vanishes (that is, $\frac{\partial f}{\partial x_i} = 0$ for all $i = 1,\ldots,n$), is called a *critical point* of the function f (in our examples this was the origin of coordinates). It is said to be *non-degenerate*, if at this point the quadratic form $\sum_{i,j=1}^n \frac{\partial^2 f}{\partial x_i \partial x_j} \Delta x_i \Delta x_j$ is non-degenerate. The preceding discussion can be summarized in one phrase: *near a non-degenerate critical point the graph of the function is arranged relative to the tangential hyperplane like the graph of its quadratic part.* It can now be shown that a small change in the function (together with its first and second derivatives) can only slightly displace the position of a non-degenerate critical point, but does not change the signature of the corresponding quadratic form and therefore of the general behaviour of the graph (in the small).

It can also be shown that near a non-degenerate critical point it is possible to make a smooth and smoothly invertible (although, generally speaking, non-linear) substitution of coordinates $y_i = y_i(x_1,\ldots x_n)$, $i = 1,\ldots,n$, such that in the new coordinates f will be precisely a quadratic function:

$$f(y_1,\ldots,y_n) = f(0,\ldots,0) + \sum_{i,j=1}^n b_{ij} y_i y_j.$$

A rigorous exposition of the theory of small oscillations is given in the book by V.I. Arnol'd entitled "Mathematical Methods of Classical Mechanics", Nauka, Moscow (1974),Chapter 5.

10.6. Morse theory. Imagine in an $(n+1)$-dimensional Euclidean space $\mathbf{R}^{n+1}$ an n-dimensional, smooth, bounded hyperplane V, like an egg or a doughnut (torus) in $\mathbf{R}^3$. We shall study the section of V by the hyperplane $x_{n+1} = \text{const}$. Suppose that there exists only a finite number of values of $c_1,\ldots,c_m$, such that the hyperplanes $x_{n+1} = c_i$ are tangent to V and moreover, at a single point $v_i \in V$. Near these tangent points V can be approximated by a graph of a quadratic form $x_{n+1} = c_i + q_i(x_1 - x_1(v_i),\ldots,x_n - x_n(v_i))$, provided that V is in sufficiently general position (for example, the doughnut should not be horizontal). It turns out that *the most important topological properties of* V, in particular, the so-called homotopy type of V, *are completely determined by the collection of signatures of the forms* q_i, that is, an indication of how many directions along which V diverges downwards and upwards near v_i. The most remarkable thing is that although the information about signatures of q_i is purely local, the homotopy type of V reconstructed in terms of it is a global characteristic of the form V. For example, if there are only two critical points c_1 and c_2 with signatures $(n, 0)$ and $(0, n)$, then V is topologically structured like an n-dimensional sphere.

Details can be found in "Morse Theory" by John Milnor, Princeton University Press,Princeton,N.J.(1963).

10.7. Self-adjoint operators and multidimensional quadrics. Now let L be a finite-dimensional Euclidean or unitary space, and $f : L \to L$ a self-adjoint operator. We are interested in the properties of its spectrum. We arrange the eigenvalues of f in decreasing order taking into account their multiplicities $\lambda_1 \geq \lambda_2 \geq \ldots \geq \lambda_n$ and we select a corresponding orthonormal basis $\{e_1, e_2, \ldots, e_n\}$. We return to the viewpoint of §8.4 according to which, a representation of f is equivalent to a representation of a new symmetric or Hermitian form $(f(l_1), l_2)$ or a quadratic form $q_f(l) = (f(l), l)$ (in the unitary case it is quadratic on the decomplexified space). In the basis $\{e_1, \ldots, e_n\}$ it acquires the form

$$q_f(x_1, \ldots, x_n) = \sum_{i=1}^{n} \lambda_i x_i^2 \quad \text{or} \quad \sum_{i=1}^{n} \lambda_i |x_i|^2$$

and thus the directions $\mathbf{R}e_i$ (or $\mathbf{C}e_i$) are the *principal axes of* q_f.

The simplest extremal property of the eigenvalues λ_i is expressed by the following fact.

10.8. Proposition. *Let* $S = \{l \in L \mid |l| = 1\}$ *be the unit sphere in the space* L. *Then*

$$\lambda_1 = \max_{l \in S} q_f(l), \quad \lambda_n = \min_{l \in S} q_f(l).$$

Proof. Since $|x_i|^2 \geq 0$ and $\lambda_1 \geq \ldots \geq \lambda_n$, obviously,

$$\lambda_n \left(\sum_{i=1}^{n} |x_i|^2 \right) \leq \sum_{i=1}^{n} \lambda_i |x_i|^2 \leq \lambda_1 \left(\sum_{i=1}^{n} |x_i|^2 \right).$$

On the unit sphere the left side is λ_n and the right side is λ_1. These values are achieved on the vectors $(0, \ldots, 0, 1)$ and $(1, 0, \ldots, 0)$ respectively (the coordinates are chosen in a basis $\{e_1, \ldots, e_n\}$ diagonalizing f).

10.9. Corollary. *Let L_k^- be the linear span of $\{e_1, \dots, e_k\}$ and let L_k^+ be the linear span of $\{e_k, \dots, e_n\}$. Then*

$$\lambda_k = \max\{q_f(l) | l \in S \cap L_k^+\} = \min\{q_f(l) | l \in S \cap L_k^-\}.$$

Proof. Indeed, in obvious coordinates, the restriction of q_f to L_k^+ has the form $\sum_{i=k}^n \lambda_i |x_i|^2$ and to L_k^- the form $\sum_{i=1}^k \lambda_i |x_i|^2$.

The following important extension of this result, in which *any* linear subspace of L with codimension $k-1$ is studied instead of L_k^+, is called the *Fisher-Courant theorem.* It gives the "minimax" characterization of the eigenvalues of differential operators.

10.10. Theorem. *For any subspace $L' \subset L$ with codimension $k-1$, the following inequalities hold:*

$$\lambda_k \le \max\{q_f(l) | l \in S \cap L'\} \; \lambda_{n-k+1} \ge \min\{q_f(l) | l \in S \cap L'\}.$$

These estimates are exact for some L' (for example, L_k^+ and L_{n-k+1}^- respectively) so that

$$\lambda_k = \min_{L'} \max\{q_f(l) | l \in S \cap L'\}, \; \lambda_{n-k+1} = \max_{L'} \min\{q_f(l) | l \in S \cap L'\}.$$

Proof. Since

$$\dim L' + \dim L_k^- = (n-k+1) + k = n+1,$$

and $\dim(L' + L_k^-) \le \dim L = n$, Theorem 5.3 of Chapter I implies that $\dim(L' \cap L_k^-) \ge 1$. Choose a vector $l_0 \in L' \cap L_k^- \cap S$. According to Corollary 10.9, $\lambda_k = \min\{q_f(l) | l \in S \cap L_k^-\}$, so that $\lambda_k \le q_f(l_0)$ and, moreover, $\lambda_k \le \max\{q_f(l) | l \in S \cap L'\}$. The second inequality of the theorem is most easily obtained by applying the first inequality to the operator $-f$ and noting that the signs and order of the eigenvalues, in this case, are reversed.

10.11. Corollary. *Let $\dim L/L_0 = 1$ and let p be the operator of the orthogonal projection $L \to L_0$. We denote by $\lambda'_1 \ge \lambda'_2 \ge \dots \ge \lambda'_{n-1}$ the eigenvalues of the self-adjoint operator $pf : L_0 \to L_0$. Then*

$$\lambda_1 \ge \lambda'_1 \ge \lambda_2 \ge \lambda'_2 \ge \dots \ge \lambda'_{n-1} \ge \lambda_n,$$

that is, the eigenvlaues of f and pf alternate.

Proof. The restriction of the form q_f to L_0 coincides with $q_{pf} : (f(l), l) = (pf(l), l)$ if $l \in L_0$. Therefore

$$\lambda'_k = \max\{q_{pf}(l) | l \in S \cap L'\} = \max\{q_f(l) | l \in S \cap L'\}$$

for an appropriate subspace $L' \subset L_0$, with codimension $k-1$ in L_0. This means that in L it has the codimension k, whence $\lambda_{k+1} \leq \lambda'_k$. Writing this inequality for $-f$ instead of f, we obtain $-\lambda_k \leq -\lambda'_k$, that is, $\lambda'_k \leq \lambda_k$. This completes the proof.

We leave to the reader to verify the following simple geometric interpretation of Corollary 10.11. Assume that $\lambda_1 \geq \lambda_2 \geq \ldots \geq \lambda_n > 0$ and instead of the functions $q_f(l)$ on S study the ellipsoid $\epsilon : q_f(l) = 1$. Then its section ϵ_0 by the subspace L_0 is also an ellipsoid, the lengths of whose semi-axes alternate with the lengths of the semi-axes of the ellipsoid ϵ. Imagine, for example, an ellipsoid ϵ in $\mathbf{R}^3$ and its section by the plane ϵ_0. The long semi-axis of ϵ_0 is not longer than the semi-axis of ϵ ("obviously"), but it is not shorter than the middle semi-axis of ϵ. The short semi-axis of ϵ_0 is not shorter than the short semi-axis of ϵ ("obviously"), but it is not longer than the middle semi-axis of ϵ. The key question is: how does one obtain a circle in a section ?

§11. Three-Dimensional Euclidean Space

11.1. The three-dimensional Euclidean space $\mathcal{E}$ is the basic model of the Newtonian and Galilean physical space. The four-dimensional Minkowski space $\mathcal{M}$, equipped with a symmetric metric with the signature $(r_+, r_-) = (1,3)$ is a model of the space-time of relativistic physics. For this reason at least, they deserve a more careful study. They also have special properties from the mathematical viewpoint, which are important for understanding the structure of the world in which we live: the relationship between rotations in $\mathcal{E}$ and quaternions and the existence of a vector product; the geometry of vectors of zero length in $\mathcal{M}$.

These special properties are conveniently expressed in terms of the relationship of the geometries of $\mathcal{E}$ and $\mathcal{M}$ to the geometry of the *auxiliary two-dimensional unitary space* $\mathcal{H}$ called the spinor space. This relationship also has a profound physical meaning, which became clear only after the appearance of quantum mechanics. We have chosen precisely this manner of exposition.

11.2. Thus we fix a two-dimensional unitary space $\mathcal{H}$. We denote by $\mathcal{E}$ the *real linear space of self-adjoint operators in* $\mathcal{H}$ *with zero trace*. Each operator $f \in \mathcal{E}$ has two real eigenvalues; they differ only in sign, because the trace, equal to their sum, vanishes. We set

$$|f| = \sqrt{|\det f|} = \text{ positive eigenvalue of } f.$$

11.3. Proposition. *$\mathcal{E}$ with the norm $|\ |$ is a three-dimensional Euclidean space.*

Proof. In an orthonormal basis of $\mathcal{H}$ the operators f are represented by Hermitian matrices of the form

$$\begin{pmatrix} a & \bar{b} \\ b & -a \end{pmatrix}, \ a \in \mathbf{R}, \ b \in \mathbf{C},$$

that is, by the linear combinations

$$\operatorname{Re} b \cdot \sigma_1 + \operatorname{Im} b \cdot \sigma_2 + a\sigma_3,$$

where $\sigma_1, \sigma_2, \sigma_3$ are the Pauli matrices (see Exercise 5 in §4 of Chapter III):

$$\sigma_1 = \begin{pmatrix} 0 & 1 \\ 1 & 0 \end{pmatrix}, \ \sigma_2 = \begin{pmatrix} 0 & -i \\ i & 0 \end{pmatrix}, \ \sigma_3 = \begin{pmatrix} 1 & 0 \\ 0 & -1 \end{pmatrix}.$$

Since $\sigma_1, \sigma_2, \sigma_3$ are linearly independent over $\mathbf{R}$, $\dim_R \mathcal{E} = 3$.

We now set

$$(f, g) = \frac{1}{2} \operatorname{Tr}(fg).$$

This is a bilinear symmetric inner product, and if the eigenvalues of f equal $\pm\lambda$,

$$|f|^2 = \frac{1}{2} \operatorname{Tr}(f^2) = \frac{1}{2}(\lambda^2 + \lambda^2) = |\det f|.$$

Obviously, $\lambda^2 = 0$ if and only if $f = 0$. This completes the proof.

We call a *direction in* $\mathcal{E}$ a set of vectors of the form

$$\mathbf{R}_+ f = \{af | a > 0\},$$

where f is a non-zero vector from $\mathcal{E}$. In other words, a direction is a half-line in $\mathcal{E}$. The direction opposite to $\mathbf{R}_+ f$ is $\mathbf{R}_+(-f)$.

11.4. Proposition. *There is a one-to-one correspondence between the directions in $\mathcal{E}$ and decompositions of $\mathcal{H}$ into a direct sum of two orthogonal one-dimensional subspaces $\mathcal{H}_+ \oplus \mathcal{H}_-$. Namely, $\mathcal{H}_+$ is the characteristic subspace of $\mathcal{H}$ for the positive eigenvalue of f corresponding to the direction $\mathbf{R}_+ f$ and $\mathcal{H}_-$ is the same for the negative eigenvalue.*

Proof. $\mathcal{H}_+$ and $\mathcal{H}_-$ are orthogonal according to Theorem 7.4. Substituting af for f, $a > 0$, does not change $\mathcal{H}_+$ and $\mathcal{H}_-$. Conversely, if the orthogonal decomposition $\mathcal{H} = \mathcal{H}_+ \oplus \mathcal{H}_-$ is given, then the set of operators $f \in \mathcal{E}$, stretching $\mathcal{H}$ by a factor of $\lambda > 0$ along $\mathcal{H}_+$ and by a factor $-\lambda < 0$ along $\mathcal{H}_-$, forms a direction in $\mathcal{E}$.

Physical interpretation. We identify $\mathcal{E}$ with physical space, for example, by selecting orthogonal coordinates in $\mathcal{E}$ and in space. We identify $\mathcal{H}$ with the space of the internal states of a quantum system "particle with spin 1/2, localized near the origin" (for example, an electron). Choosing the direction $\mathbf{R}_+ f \subset \mathcal{E}$, we turn on a magnetic field in physical space along this direction. In this field the system will have two stationary states, which are precisely $\mathcal{H}_+$ and $\mathcal{H}_-$.

If the direction $\mathbf{R}_+ f$ corresponds, for example, to the upper vertical semi-axis of the chosen coordinate system in physical space (the "z axis"), then the state $\mathcal{H}_+$ is called the state "with spin projection $+1/2$ along the z axis" (or "up spin"), while $\mathcal{H}_-$ is correspondingly called the state "with spin projection $-1/2$" (or "down spin"). This traditional terminology is a relic of prequantum ideas about the fact that the observed spin corresponds to the classical observable "angular momentum" — a characteristic of the internal rotation of the system and can therefore be itself represented by a vector in $\mathcal{E}$, which for this reason has a projection on the coordinate axes in $\mathcal{E}$. This is completely incorrect: the states of the system are rays in $\mathcal{H}$, not vectors in $\mathcal{E}$. The disagreement with the classical interpretation becomes even more evident for systems with spin $s/2$, $s > 1$, for which $\dim \mathcal{H} = s + 1$. The precise assertion is stated in Proposition 11.4.

We have given an idealized description of the classical Stern-Gerlach (1922) experiment. In this experiment silver ions which passed between the poles of an electromagnet were used instead of electrons. Due to the non-homogeneity of the magnetic field, the ions, which exist in states close to $\mathcal{H}_+$ and $\mathcal{H}_-$ respectively, were spatially separated into two beams, which made it possible to identify these states macroscopically. The silver was evaporated in an electric furnace, and the magnetic field between the poles played the role of a combination of two filters, separately passing the states $\mathcal{H}_+$ and $\mathcal{H}_-$.

We now continue the study of Euclidean space $\mathcal{E}$.

11.5. Proposition. $(f, g) = 0$, *if and only if* $fg + gf = 0$.

Proof. We have

$$(f, g) = \frac{1}{2}\operatorname{Tr}(fg) = \frac{1}{4}\operatorname{Tr}(fg + gf) = \frac{1}{4}\operatorname{Tr}[(f+g)^2 - f^2 - g^2].$$

But f^2 has one eigenvalue $|f|^2$, so that all squares of operators from $\mathcal{E}$ are scalars, and therefore $fg+gf$ is also a scalar operator, which vanishes if and only if its trace vanishes.

11.6. Orthonormal bases in $\mathcal{E}$. It is clear from the proof of Proposition 11.5 that the operators $\{e_1, e_2, e_3\}$ form an orthonormal basis if and only if

$$e_1^2 = e_2^2 = e_3^2 = \mathrm{id};\ e_i e_j + e_j e_i = 0,\ i \neq j.$$

In particular, if an orthonormal basis is chosen in $\mathcal{H}$, then the operators, represented in it by the Pauli matrices $\sigma_1, \sigma_2, \sigma_3$, form an orthonormal basis in $\mathcal{E}$:

$$\sigma_1^2 = \sigma_2^2 = \sigma_3^2 = \sigma_0 = \begin{pmatrix} 1 & 0 \\ 0 & 1 \end{pmatrix};\ \sigma_i\sigma_j + \sigma_j\sigma_i = 0,\ i \neq j.$$

We can now explain the mathematical meaning of the Pauli matrices, proving the converse assertion.

11.7. Proposition. *For every orthonormal basis $\{e_1, e_2, e_3\}$ of the space $\mathcal{E}$ there exists an orthonormal basis $\{h_1, h_2\}$ of the space $\mathcal{H}$ with the property*

$$A_{e_1} = \sigma_1,\ A_{e_2} = \sigma_2 \text{ or } -\sigma_2,\ A_{e_3} = \sigma_3,$$

where A_e is the matrix of the operator e in the basis $\{h_1, h_2\}$. It is determined to within a complex factor of unit modulus.

Proof. The eigenvalues of e_i are ± 1. Let $\mathcal{H} = \mathcal{H}_+ \oplus \mathcal{H}_-$, where e_3 acts on $\mathcal{H}_+$ identically and on $\mathcal{H}_-$ by changing the sign. We first choose the vectors $h_1' \in \mathcal{H}_+$, $h_2' \in \mathcal{H}_-$, $|h_1'| = |h_2'| = 1$. They are determined to within factors $e^{i\phi_1}, e^{i\phi_2}$; the matrix of e_3 in the basis $\{h_1', h_2'\}$ is σ_3.

Next we have

$$e_1(h_1') = e_1 e_3(h_1') = -e_3 e_1(h_1'),$$

so that $e_1(h_1')$ is a non-zero eigenvector of e_3 with eigenvalue -1. Therefore $e_1(h_1') = \alpha h_2'$. Analogously $e_1(h_2') = \beta h_1'$. The matrix of e_1 in the basis $\{h_1', h_2'\}$ is Hermitian, and therefore $\alpha = \bar{\beta}$. Finally, $e_1^2 = \mathrm{id}$, and therefore $\alpha\beta = 1 = |\alpha|^2 = |\beta|^2$. Replacing $\{h_1', h_2'\}$ by $\{h_1, h_2\} = \{xh_1', yh_2'\}$, where $|x| = |y| = 1$, in order to transform the matrix of e_1 in the new basis into σ_1, we obtain

$$e_1(h_1) = xe_1(h_1') = x\,\alpha h_2' = \alpha\, xy^{-1} h_2,$$

$$e_1(h_2) = ye_1(h_2') = y\,\beta h_1' = \beta\, yx^{-1} h_1.$$

For this reason x and y must satisfy the additional condition $xy^{-1} = \alpha^{-1}$; then $\alpha xy^{-1} = \beta yx^{-1} = 1$ automatically. We can set, for example $x = 1$ and $y = \alpha$.

So in the basis $\{h_1, h_2\}$ we have $A_{e_3} = \sigma_3, A_{e_1} = \sigma_1$, and this basis is determined to within a scalar factor of unit modulus. The same arguments as for e_1 show that in such a basis A_{e_2} has the form $\begin{pmatrix} 0 & \gamma \\ \bar{\gamma} & 0 \end{pmatrix}$, where $|\gamma|^2 = 1$. In addition, the condition of orthogonality $e_1 e_2 + e_2 e_1 = 0$ gives

$$\begin{pmatrix} 0 & 1 \\ 1 & 0 \end{pmatrix} \begin{pmatrix} 0 & \gamma \\ \bar{\gamma} & 0 \end{pmatrix} + \begin{pmatrix} 0 & \gamma \\ \bar{\gamma} & 0 \end{pmatrix} \begin{pmatrix} 0 & 1 \\ 1 & 0 \end{pmatrix} = 0,$$

that is, $\gamma + \bar{\gamma} = 0$, whence $\gamma = i$ or $\gamma = -i$. Therefore, $A_{e_2} = \sigma_2$ or $A_{e_2} = -\sigma_2$.

11.8. Corollary. *The space $\mathcal{E}$ is equipped with a distinguished orientation: the orthonormal basis $\{e_1, e_2, e_3\}$ belongs to the class corresponding to this orientation if and only if there exists an orthonormal basis $\{h_1, h_2\}$ in $\mathcal{H}$ in which $A_{e_a} = \sigma_a$, $a = 1, 2, 3$.*

Proof. We must verify that if $\{e_a\}$ in the basis $\{h_b\}$ and $\{e'_a\}$ in the basis $\{h'_b\}$ are represented exactly by the Pauli matrices, then the determinant of the matrix of the transformation from $\{e_a\}$ to $\{e'_a\}$ is positive, or that $\{e_a\}$ can be transformed into $\{e'_a\}$ by a continuous motion. We shall construct this motion, showing that $\{h_b\}$ is transformed into $\{h'_b\}$ by a unitary continuous motion: there exists a system of unitary operators $f_t : \mathcal{H} \to \mathcal{H}$, depending on the parameter $t \in [0,1]$, such that $f_0 = \mathrm{id}$, $f_1(h_b) = h'_b$ and $\{f_t(h_1), f_t(h_2)\}$ form an orthonormal basis of $\mathcal{H}$ for all t. Then denoting by $\{g_t(e_1), g_t(e_2), g_t(e_3)\}$ the orthonormal basis of $\mathcal{E}$ represented by the Pauli matrices in the basis $\{f_t(h_1), f_t(h_2)\}$, we construct the required motion in $\mathcal{E}$.

Let $\{h'_1, h'_2\} = \{h_1, h_2\}U$. Since both bases are orthonormal, the transition matrix U must be unitary. According to Corollary 8.7, it can be represented in the form $\exp(iA)$, where A is a Hermitian matrix. Then for all $t \in A$ the matrix tA is Hermitian, the operator $\exp(itA)$ is unitary, and we can set

$$f_t\{h_1, h_2\} = \{h_1, h_2\}\exp(itA), \quad 0 \le t \le 1.$$

This completes the proof.

The operators $\sigma_1/2, \sigma_2/2, \sigma_3/2$ in $\mathcal{H}$ are called *observables of the projections of the spin* on the corresponding axes in $\mathcal{E}$: this terminology is explained by the quantum mechanical interpretation in §11.5. The factor 1/2 is introduced so that their eigenvalues would be equal to $\pm 1/2$.

11.9. Vector products. Let $\{e_1, e_2, e_3\}$ be an orthonormal basis in $\mathcal{E}$, belonging to the noted orientation. The vector product in $\mathcal{E}$ is determined by the classical formula

$$(x_1e_1 + x_2e_2 + x_3e_3) \times (y_1e_1 + y_2e_2 + y_3e_3) =$$
$$= (x_2y_3 - x_3y_2)e_1 + (x_3y_1 - x_1y_3)e_2 + (x_1y_2 - x_2y_1)e_3.$$

A change in basis to another basis with the same orientation does not change the vector product; if, on the other hand, the new basis is oppositely oriented, then the sign of the product changes.

It is not difficult to give an invariant construction of the vector product in our terms. We recall that *anti-Hermitian* operators in $\mathcal{H}$ with zero trace form a Lie algebra $su(2)$ (see §4 of Part I). The space $\mathcal{E}$ can be identified with this Lie algebra, dividing each operator from $\mathcal{E}$ by i. Hence $\frac{1}{i}\mathcal{E}$ has the structure of a Lie algebra. We have

$$\left[\frac{1}{i}\sigma_a, \frac{1}{i}\sigma_b\right] = 2\epsilon_{abc}\frac{1}{i}\sigma_c,$$

where $\epsilon_{123} = 1$ and ϵ_{abc} are skew-symmetric with respect to all indices, or

$$[\sigma_a, \sigma_b] = 2i\epsilon_{abc}\sigma_c.$$

Therefore

$$\left[\sum_{a=1}^{3} x_a\sigma_a, \sum_{b=1}^{3} y_b\sigma_b\right] = 2i\left(\sum_{a=1}^{3} x_a\sigma_a\right) \times \left(\sum_{b=1}^{3} x_b\sigma_b\right),$$

so that *the vector product, to within a trivial factor, simply equals the commutator of operators.* This makes it possible to establish without any calculations the following classical identities:

$$\vec{x} \times \vec{y} = -\vec{y} \times \vec{x};$$

$$\vec{x} \times (\vec{y} \times \vec{z}) + \vec{z} \times (\vec{x} \times \vec{y}) + \vec{y} \times (\vec{z} \times \vec{x}) = 0.$$

There is one more method for introducing the vector product, simultaneously relating it to the inner product and quaternions.

11.10. Quaternions. Like commutation, multiplication of operators from $\mathcal{E}$, generally speaking, takes us out of $\mathcal{E}$: the Hermitian property and the condition that the trace vanish break down at the same time. Indeed, the product of operators from $\mathcal{E}$ lies in $\mathbf{R}\,\mathrm{id} + i\mathcal{E}$; in addition, the "real part" is precisely the inner product, while the "imaginary part" is the vector product. Indeed,

$$\sigma_a\sigma_b = i\epsilon_{abc}\sigma_c \quad \text{for} \quad a \neq b,\ \{a,b,c\} = \{1,2,3\},$$

$$\sigma_a^2\sigma_0 = \begin{pmatrix} 1 & 0 \\ 0 & 1 \end{pmatrix},\ a = 1,2,3,$$

so that

$$\left(\sum_{a=1}^{3} x_a\sigma_a\right)\left(\sum_{b=1}^{3} y_b\sigma_b\right) = \left(\sum_{a=1}^{3} x_a y_a\right)\sigma_0 + i\left(\sum_{a=1}^{3} x_a\sigma_a\right) \times \left(\sum_{b=1}^{3} y_b\sigma_b\right),$$

or, as physicists write,

$$(\vec{x}\cdot\vec{\sigma})(\vec{y}\cdot\vec{\sigma}) = (\vec{x}\cdot\vec{y})\sigma_0 + i(\vec{x}\times\vec{y})\cdot\vec{\sigma}, \quad \vec{\sigma} = (\sigma_1,\sigma_2,\sigma_3).$$

It is evident from here that the real space of operators $\mathbf{R}\,\mathrm{id} + i\mathcal{E}$ is closed under multiplication. Its basis consists of the following elements (in classical notation):

$$\mathbf{1} = \sigma_0,\ \mathbf{i} = -i\sigma_1,\ \mathbf{j} = -i\sigma_2,\ \mathbf{k} = -i\sigma_3$$

with the multiplication table

$$\mathbf{i}^2 = \mathbf{j}^2 = \mathbf{k}^2 = -1,\ \mathbf{ij} = -\mathbf{ji} = \mathbf{k},$$

$$\mathbf{ki} = -\mathbf{ik} = \mathbf{j};\ \mathbf{jk} = -\mathbf{kj} = \mathbf{i}.$$

In other words we obtain the *division ring of quaternions* in one of the traditional matrix representations (cf. "Introduction to Algebra", Ch.9,§4).

11.11. The homomorphism $SU(2) \to SO(3)$. We fix the orthonormal basis $\{h_1, h_2\}$ in $\mathcal{H}$ and the corresponding orthonormal basis $\{e_1, e_2, e_3\}$ in $\mathcal{E}$, for which $A_{e_i} = \sigma_i$. Any unitary operator $U : \mathcal{H} \to \mathcal{H}$ transforms $\{h_1, h_2\}$ into $\{h'_1, h'_2\}$. This last basis corresponds to the basis $\{e'_1, e'_2, e'_3\}$ and there exists an orthogonal operator $s(U) : \mathcal{E} \to \mathcal{E}$, which transforms $\{e_i\}$ into $\{e'_i\}$. By Corollary 11.8, $s(U) \in SO(3)$, because the determinant of $s(U)$ is positive.

Realizing $\mathcal{E}$ by means of matrices in the basis $\{h_1, h_2\}$ we can represent the action of $s(U)$ on $\mathcal{E}$ by the simple formula

$$s(U)(A) = UAU^{-1}$$

for any $A \in \mathcal{E}$. Indeed, this is a particular case of the general formula for changing the matrix of operators when the basis is changed. We can now prove the following important result.

11.12. Theorem. *The mapping s, restricted to $SU(2)$, defines a surjective homomorphism of groups $SU(2) \to SO(3)$ with the kernel $\{\pm E_2\}$.*

Proof. It is evident immediately from the formula $s(U)(A) = UAU^{-1}$ that $s(E) = \mathrm{id}$ and $s(UV) = s(U)s(V)$, so that s is a homomorphism of groups. Its surjectivity is verified thus.

We choose an element $g \in SO(3)$ and let g transform the basis $(\sigma_1, \sigma_2, \sigma_3)$ in $\mathcal{E}$ into a new basis $(\sigma'_1, \sigma'_2, \sigma'_3)$. We construct according to it the basis $\{h'_1, h'_2\}$ in $\mathcal{H}$, in which the operators σ'_i are represented by the matrices σ_i. By Proposition 11.7 $\{h'_1, h'_2\}$ exists, apart from the fact that the matrix σ'_2, possibly, equals $-\sigma_2$ and not σ_2. Actually, this possibility is excluded by Corollary 11.8; since g belongs to $SO(3)$ it preserves the orientation of $\mathcal{E}$. The operator U, transforming $\{h_1, h_2\}$ into $\{h'_1, h'_2\}$, satisfies the condition $s(U) = g$. It is true that it can belong only to $U(2)$ and not to $SU(2)$. If $\det U = e^{i\phi}$, then $e^{-i\phi/2}U \in SU(2)$. The matrix $e^{-i\phi/2}U$ transforms $\{h_1, h_2\}$ into $\{e^{-i\phi/2}h'_1, e^{-i\phi/2}h'_2\}$, and this basis in $\mathcal{H}$ corresponds, as before, to the basis $\{\sigma'_1, \sigma'_2, \sigma'_3\}$ in $\mathcal{E}$. Therefore $s(e^{-i\phi/2}U) = g$ also and we obtain that $s : SU(2) \to SO(3)$ is surjective.

The kernel of the homomorphism $s : U(2) \to SO(3)$ consists only of the scalar operators $\{e^{i\phi}\mathrm{id}\}$: this follows from Proposition 11.7, according to which the basis $\{h'_1, h'_2\}$ is reconstructed from $\{e'_1, e'_2, e'_3\}$ precisely, apart from a factor of $e^{i\phi}$. The intersection of the group $\{e^{i\phi}\ \mathrm{id}\}$ with $SU(2)$ equals precisely $\{\pm\ \mathrm{id}\}$, which completes the proof.

The meaning of the homomorphism constructed is clarified in topology: the group $SU(2)$ is *simply connected*, that is, any closed curve in it can be contracted by a continuous motion into a point, while for $SO(3)$ this is not true. Thus $SU(2)$ is the universal covering of the group $SO(3)$.

We shall use the theorem proved above to clarify the structure of the group $SO(3)$, exploiting the fact that $SU(2)$ has a simpler structure. Here it is appropriate to quote R. Feynman:

"It is rather strange, because we live in three dimensions, but it is hard for us to appreciate what happens if we turn this way and then that way. Perhaps if we were fish or birds and had a real appreciation of what happens when we turn somersaults in space, we could more easily appreciate such things." R.P. Feynman, R.B. Leighton and M.Sands, The Feynman Lectures on Physics, Addison-Wesley,New York,1965, Vol.3,p.6–11.

11.13. Structure of $SU(2)$. First of all, the elements of $SU(2)$ are 2×2 matrices with complex elements, for which $U^t = \bar{U}$ and $\det U = 1$. From this it follows immediately that

$$SU(2) = \left\{ \begin{pmatrix} a & b \\ -\bar{b} & \bar{a} \end{pmatrix} \middle| \, |a|^2 + |b|^2 = 1 \right\}.$$

The set of pairs $\{(a,b) \,|\, |a|^2 + |b|^2 = 1\}$ in $\mathbf{C}^2$ transforms into a sphere with unit radius in the realization of $\mathbf{C}^2$, that is $\mathbf{R}^4$:

$$(\operatorname{Re} a)^2 + (\operatorname{Im} a)^2 + (\operatorname{Re} b)^2 + (\operatorname{Im} b)^2 = 1.$$

Thus *the group $SU(2)$ is topologically arranged like the three-dimensional sphere in four-dimensional Euclidean space.*

We now write down a system of generators of the group $SU(2)$, taking inspiration from Corollary 8.7, according to which the mapping $\exp : u(2) \to U(2)$ is surjective. Direct calculation of the exponentials from the three generators of the space $su(2)$ gives:

$$\exp\left(\frac{1}{2} it\sigma_1\right) = \begin{pmatrix} \cos\frac{t}{2} & i\sin\frac{t}{2} \\ i\sin\frac{t}{2} & \cos\frac{t}{2} \end{pmatrix},$$

$$\exp\left(\frac{1}{2} it\sigma_2\right) = \begin{pmatrix} \cos\frac{t}{2} & \sin\frac{t}{2} \\ -\sin\frac{t}{2} & \cos\frac{t}{2} \end{pmatrix},$$

$$\exp\left(\frac{1}{2} it\sigma_3\right) = \begin{pmatrix} e^{it/2} & 0 \\ 0 & e^{-it/2} \end{pmatrix}.$$

Any element $\begin{pmatrix} a & b \\ -\bar{b} & \bar{a} \end{pmatrix} \in SU(2)$, for which $ab \neq 0$, can be represented in the form

$$\exp\left(\frac{1}{2} i\phi\sigma_3\right)\exp\left(\frac{1}{2} i\theta\sigma_1\right)\exp\left(\frac{1}{2} i\psi\sigma_3\right) =$$

$$= \begin{pmatrix} \cos\frac{\theta}{2} e^{i\frac{\phi+\psi}{2}} & i\sin\frac{\theta}{2} e^{i\frac{\phi-\psi}{2}} \\ i\sin\frac{\theta}{2} e^{i\frac{\psi-\phi}{2}} & \cos\frac{\theta}{2} e^{-i\frac{\phi+\psi}{2}} \end{pmatrix},$$

where $0 \le \phi < 2\pi$, $0 < \theta < \pi$, $-2\pi \le \psi < 2\pi$. For this it is sufficient to set $|a| = \cos\frac{\theta}{2}$, $\arg a = \frac{\phi+\psi}{2}$, $\arg b = \frac{\phi-\psi+\pi}{2}$. (The elements of $SU(2)$ with $b = 0$, evidently, have the form $(\frac{1}{2}\, i\phi\sigma_3)$; we leave it to the reader to determine the elements for which $a = 0$).

The angles ϕ, θ, ψ are called *Euler angles* in the group $SU(2)$.

11.14. Structure of $SO(3)$. We identified $SU(2)$ topologically with the three-dimensional sphere. The homomorphism $s : SU(2) \to SO(3)$ transforms pairs of elements $\pm U \in SU(2)$ into a single point of $SO(3)$. On the sphere they form the ends of one of the diameters. Therefore $SO(3)$ is topologically the result of *sewing the three-dimensional sphere at pairs of antipodal points.* On the other hand, pairs of antipodes of a sphere are in a one-to-one correspondence with straight lines, connecting the points of the pair, in four-dimensional real space. The set of such straight lines is called *three-dimensional real projective space* and is sometimes denoted by RP^3; later we shall study projective spaces in greater detail. Thus $SO(3)$ is topologically equivalent to RP^3.

We now consider what the generators of $SU(2)$, described in the preceding section, are transformed into by the homomorphism s. In the standard basis $\{\sigma_1, \sigma_2, \sigma_3\}$ of the space $\mathcal{E}$ we have

$$\exp\left(\frac{1}{2}\, it\sigma_1\right)\sigma_1 \exp\left(-\frac{1}{2}\, it\sigma_1\right) = \sigma_1,$$

$$\exp\left(\frac{1}{2}\, it\sigma_1\right)\sigma_2 \exp\left(-\frac{1}{2}\, it\sigma_1\right) = (\cos t)\sigma_2 - (\sin t)\sigma_3,$$

$$\exp\left(\frac{1}{2}\, it\sigma_1\right)\sigma_3 \exp\left(-\frac{1}{2}\, it\sigma_1\right) = (\sin t)\sigma_2 + (\cos t)\sigma_3.$$

Therefore

$$s\left(\exp\left(\frac{1}{2}\, it\sigma_1\right)\right) = \begin{pmatrix} 1 & 0 & 0 \\ 0 & \cos t & -\sin t \\ 0 & \sin t & \cos t \end{pmatrix}$$

is the rotation of $\mathcal{E}$ by an angle t around the axis $\mathbf{R}\sigma_1$. It can be verified in an analogous manner that $s\,(\exp\,(\frac{1}{2}\, it\sigma_k))$ is the rotation of $\mathcal{E}$ by an angle t around an axis σ_k for $k = 2$ and 3 also. In particular, any rotation from $SO(3)$ can be decomposed into a product of three rotations with respect to $\sigma_3, \sigma_1, \sigma_3$ by the Euler angles ψ, θ, ϕ, and in addition ψ may be assumed to vary from 0 to 2π.

EXERCISES

1. Prove the following identities:

$$(\vec{x} \times \vec{y}, \vec{z}) = (\vec{x}, \vec{y} \times \vec{z}),$$

$$(\vec{x} \times \vec{y}) \times \vec{z} - \vec{x} \times (\vec{y} \times \vec{z}) = (\vec{x}, \vec{y})\vec{z} - \vec{x}(\vec{y}, \vec{z}).$$

(Hint: use the associative property of multiplication in the algebra of quaternions.)

2. In a three-dimensional Euclidean space, two axes z and z', making an angle ϕ, are distinguished. A beam of electrons with spin projection $+1/2$ along the z axis is introduced into a filter, which transmits only electrons with spin projection $+1/2$ along the z' axis. Show that the relative fraction of electrons passing through the filter equals $\cos^2 \frac{\phi}{2}$.

§12. Minkowski Space

12.1. A *Minkowski space* $\mathcal{M}$ is a four-dimensional, real, linear space with a non-degenerate symmetric metric with the signature (1,3) (sometimes the signature (3,1) is used). Before undertaking the mathematical study of this space, we shall indicate the basic principles of its physical interpretation, which form the basis for Einstein's *special theory of relativity*.

a) *Points.* A point (or vector) in the space $\mathcal{M}$ is an idealization of a physical event, localized in space and time, such as "flashes", "emission of a photon by an atom", "collision of two elementary particles", etc. The origin of coordinates of $\mathcal{M}$ must be regarded as an event occurring "here and now" for some observer; it fixes simultaneously the origin of time and the origin of the spatial coordinates.

b) *Units of measurement.* In classical physics, length and time are measured in different units. Since $\mathcal{M}$ is a model of space-time, the special theory of relativity must have a method for transforming spatial units into time units and vice versa. The method used is equivalent to the principle that "the velocity of light c is constant": it consists of the fact that a selected unit of time t_0 is associated with a unit of length $l_0 = ct_0$, which is the distance traversed by light over the period of time t_0 (for example, a "light second"). One of the units l_0 or t_0 is then assumed to be chosen once and for all; after the second is fixed by the condition $l_0 = ct_0$, the velocity of light in these units equals 1.

c) *Space-time interval.* If $l_1, l_2 \in \mathcal{M}$ are two points in Minkowski space, the inner product $(l_1 - l_2, l_1 - l_2)$ is called the square of the space-time interval between them. It can be positive, zero, or negative; in physical terms, *time-*, *light-*, or *space-like*, respectively. (An explanation of these terms will be given below.) If $l_2 = 0$, the same terms are applied to the vector l_1, depending on the sign of (l_1, l_1).

d) *World lines of inertial observers.* If on the straight line $L \subset \mathcal{M}$ at least one vector is timelike, then all vectors are timelike. Such straight lines are called *the world lines of inertial observers.* A good approximation to a segment of such a line is the set of events occurring in a space ship, moving freely (with the engines off) far from celestial bodies (taking into account their gravitational force requires changing the mathematical scheme for describing space-time and the use of "curved" models of the general theory of relativity). We note that we have introduced into the analysis thus far only world lines emanating from the origin of coordinates. An inertial observer, who is not "here and now", moves along some translation $l+L$ of the timelike straight line L. Let l_1, l_2 be two points on the world line of an inertial observer. Then $(l_1-l_2, l_1-l_2)>0$, and the interval $|l_1-l_2|=(l_1-l_2, l_1-l_2)^{1/2}$ is the *proper time of this observer, passing between events l_1, l_2 and measured by clocks moving together with him.* The world line of the inertial observer is his proper "river of time".

The physical fact that *time has a direction* (from past to future) is expressed mathematically by fixing the *orientation* of each timelike straight line, so that the length $\|l\|$ of a timelike vector can be equipped with a sign distinguishing vectors oriented into the future and into the past. We shall see below that the notion of matching of two orientations, that is, the existence of a general direction of time ! — but not of the times themselves — for two inertial observers, makes sense.

e) *The physical space of an inertial observer.* The linear subvariety

$$\mathcal{E}_l = l + L^\perp \subset \mathcal{M}$$

is interpreted as a set of points of "instantaneous physical space" for an inertial observer located at the point l of his world line L. The orthogonal complement is taken, of course, relative to the Minkowski metric in $\mathcal{M}$. It is easy to verify that $\mathcal{M} = L \oplus L^\perp$ and that the structure of a three-dimensional Euclidean space is induced in $L^\perp$ (only with a negative-definite metric instead of the standard positive-definite metric). All events corresponding to the points in $L^\perp$ are interpreted by an observer as occurring "now"; for a different observer they will not be simultaneous, because $L_1^\perp \neq L_2^\perp$ for $L_1 \neq L_2$.

f) *Inertial coordinate systems.* Let L be an oriented timelike straight line, e_0 a positively oriented vector of unit length in L, and $\{e_1, e_2, e_3\}$ an orthonormal basis of $L^\perp$: $(e_i, e_i) = -1$ for $i = 1, 2, 3$. A system of coordinates in $\mathcal{M}$ corresponding to the basis $\{e_0, \dots, e_3\}$, is called an *inertial system*. In it

$$\left(\sum_{i=0}^{3} x_i e_i, \sum_{i=0}^{3} y_i e_i\right) = x_0 y_0 - \sum_{i=1}^{3} x_i y_i.$$

Since $x_0 = ct_0$, where t_0 is the proper time, the space-time interval from the origin to the point $\sum_{i=0}^{3} x_i e_i$ equals $\left(c^2 t_0^2 - \sum_{i=1}^{3} x_i^2\right)^{1/2}$. Every inertial coordinate system in $\mathcal{M}$ determines an identity between $\mathcal{M}$ and the coordinate Minkowski space $\left(\mathbf{R}^4, x_0^2 - \sum_{i=1}^{3} x_i^2\right)$. The *isometries* of $\mathcal{M}$ (or of the coordinate space) form the *Lorentz group*; the isometries which preserve the time orientation form its *orthochronous subgroup*.

g) *Light cone.* The set of points $l \in \mathcal{M}$ with $(l,l) = 0$ is called the *light cone* C (origin of coordinates). In any inertial coordinate system, C is given by the equation

$$x_0^2 = \sum_{i=1}^{3} x_i^2.$$

For $x_0 > 0$ the point on the light cone (x_0, x_1, x_2, x_3) is separated from the position of an observer $(x_0, 0, 0, 0)$ by a spacelike interval with the square $-\sum_{i=1}^{3} x_i^2 = -x_0^2$, that is, it is located at the distance that a quantum of light, emitted from the origin of coordinates at the initial moment in time, traverses within a time x_0. (For $x_0 < 0$ the set of such points corresponds to flashes which occurred at the proper time x_0 and could be observed at the origin of the reference system: "arriving radiation"). Correspondingly, the "zero straight lines", lying entirely in C, are world lines of particles emitted from the origin of coordinates and moving with the velocity of light, for example, photons or neutrinos. The reader can see the base of the "arriving fold" of the light cone, looking out the window — it is the celestial sphere.

Straight lines in $\mathcal{M}$, consisting of vectors with a negative squared length, do not have a physical interpretation. They should correspond to the world lines of particles moving faster than light, the hypothetical "tachyons", which have not been observed experimentally.

We now proceed to the mathematical study of $\mathcal{M}$.

12.2. *Realization of $\mathcal{M}$ as a metric space.* As in §9, we fix the two-dimensional complex space $\mathcal{H}$ and examine in it the set $\mathcal{M}$ of Hermitian symmetric inner products. It is a real linear space. If the basis $\{h_1, h_2\}$ in $\mathcal{H}$ is chosen, then the Gram matrix of these metrics will consist of all possible 2×2 Hermitian matrices. We associate with the metric $l \in \mathcal{M}$ the determinant of its Gram matrix G, which we shall denote by $\det L$. A transformation to the basis $\{h_1', h_2'\} = \{h_2, h_2\}V$ will replace G by $G' = V^t G \overline{V}$ and $\det G' = |\det V|^2 \det G$. In particular, if $V \in SL(2, \mathbf{C})$, then $\det G = \det G'$. Therefore, a calculation of $\det l$ in any of the bases of $\mathcal{H}$ belonging to the same class relative to the action of $SL(2, \mathbf{C})$, will lead to the same result. Henceforth, we shall *fix such a class of bases of* $\mathcal{H}$, and we shall calculate all determinants with respect to it. Changing the class merely multiplies the determinant by a positive scalar.

12.3. Proposition. a) *$\mathcal{M}$ is a four-dimensional real space.*

b) *$\mathcal{M}$ has a unique symmetric metric (l, m) for which $(l,l) = \det l$. Its signature equals* (1,3), *so that $\mathcal{M}$ is a Minkowski space.*

Proof. a) The space of 2×2 Hermitian matrices has the basis $\sigma_0 = \begin{pmatrix} 1 & 0 \\ 0 & 1 \end{pmatrix} = E_2$; $\sigma_1, \sigma_2, \sigma_3$ where $\sigma_i, i \geq 1$ are the Pauli matrices. Therefore $\dim \mathcal{M} = 4$.

b) We shall show that in the matrix realization of $\mathcal{M}$ the function $\det l$ is a quadratic form, whose polarization has the form

$$(l, m) = \frac{1}{2}(\operatorname{Tr} l \operatorname{Tr} m - \operatorname{Tr} lm),$$

which is clearly symmetric and bilinear. Indeed, if λ and μ are eigenvalues of l, then $\det l = \lambda\mu$, $\operatorname{Tr} l = \lambda + \mu$, $\operatorname{Tr} l^2 = \lambda^2 + \mu^2$, so that

$$\lambda\mu = \det l = \frac{1}{2}((\lambda+\mu)^2 - \lambda^2 - \mu^2) = \frac{1}{2}((\operatorname{Tr} l)^2 - \operatorname{Tr} l^2) = (l, l).$$

It is now evident that $\{\sigma_0, \sigma_1, \sigma_2, \sigma_3\}$ is an orthonormal basis of $\mathcal{M}$ with the Gram matrix $\operatorname{diag}(1, -1, -1, -1)$ so that the signature of our metric equals (1,3). This completes the proof.

12.4. Corollary. *Let $L \subset \mathcal{M}$ be a timelike straight line. Then $L^\perp$ with the metric $-(l, m)$ is a three-dimensional Euclidean space, and $\mathcal{M} = L \oplus L^\perp$.*

Proof. The assertion $\mathcal{M} = L \oplus L^\perp$ follows from Proposition 3.2, because the timelike straight lines are evidently non-degenerate. Since the signature of the Minkowski metric is (1,3) in $\mathcal{M}$ and $-(1,0)$ in $L^\perp$, it must be (0,3) in $L^\perp$, which completes the proof.

We shall now study the geometric meaning of inner products. The indeterminacy of the Minkowski metric leads to remarkable differences from the Euclidean situation, which can be of important physical significance. The most striking facts stem from the fact that the Cauchy-Bunyakovskii-Schwarz inequality for timelike vectors is reversed.

12.5. Proposition. *Let $(l_1, l_1) > 0$, $(l_2, l_2) > 0$, $l_i \in \mathcal{M}$. Then*

$$(l_1, l_2)^2 \geq (l_1, l_1)(l_2, l_2).$$

The equality holds if and only if l_1, l_2 are linearly independent.

Proof. We first verify that the quadratic trinomial $(tl_1 + l_2, tl_1 + l_2)$ always has a real root t_0. In the matrix realization of $\mathcal{M}$ the condition $(l_2, l_2) > 0$ means that $\det l_2 > 0$, that is, that l_2 has real characteristic roots with the same sign, say $\epsilon_2(+1$ or $-1)$. Analogoulsy, let ϵ_1 be the sign of the eigenvalues of l_1. Then as $t \to -(\epsilon_1\epsilon_2)\infty$ the matrix $tl_1 + l_2$ has eigenvalues which are approximately proportional to the eigenvalues of l_1 (because $l_1 + t^{-1}l_2$ approaches l_1), and their sign will be $-\epsilon_2$, while at $t = 0$ the matrix $0l_1 + l_2 = l_2$ has eigenvalues with sign ϵ_2. Therefore as t varies from 0 to $-(\epsilon_1\epsilon_2)\infty$ the eigenvalues of $tl_1 + l_2$ pass through zero, and $\det(tl_1 + l_2)$ vanishes. This means that the discriminant of this trinomial is non-negative, so that

$$(l_1, l_2)^2 \geq (l_1, l_1)(l_2, l_2).$$

If it equals zero, then some value of $t_0 \in \mathbf{R}$ is a double root, and the matrix $t_0 l_1 + l_2$, having two zero eigenvalues and being diagonalizable (it is Hermitian !), equals zero. Therefore, l_1 and l_2 are linearly independent.

12.6. Corollary ("the inverse triangle inequality"). *If l_1, l_2 are timelike and $(l_1, l_2) \geq 0$, then $l_1 + l_2$ is timelike and*

$$|l_1 + l_2| \geq |l_1| + |l_2|$$

(where $|l| = (l, l)^{1/2}$) and the equality holds if and only if l_1 and l_2 are linearly independent.

Proof.

$$|l_1 + l_2|^2 = |l_1|^2 + 2(l_1, l_2) + |l_2|^2 \geq$$

$$\geq |l_1|^2 + 2|l_1|\,|l_2| + |l_2|^2 = (|l_1| + |l_2|)^2.$$

The equality holds only when $(l_1, l_2) = |l_1|\,|l_2|$.

We now give the physical interpretations of these facts.

12.7. "The twin paradox". We shall call timelike vectors l_1, l_2 with $(l_1, l_2) \geq 0$ *identically time-oriented vectors.* It is evident from Proposition 12.5 that for them $(l_1, l_2) > 0$. Imagine two twin observers: one is inertial and moves along his world line from the point zero to the point $l_1 + l_2$, and the other reaches the same point from the origin, moving first inertially from zero to l_1, and then from l_1 to $l_1 + l_2$: near zero and near l_1 he turns on the engines of his space ship, so as to first fly away from his brother and then again in order to return to him. According to Corollary 12.6, the elapsed proper time for the travelling twin will be strictly shorter than the elapsed time for the twin who stayed at home.

12.8. The Lorentz factor. If l_1 and l_2 are timelike and identically time-oriented, then Proposition 12.5 implies that $\frac{(l_1, l_2)}{|l_1|\,|l_2|} \geq 1$ and we cannot interpret this quantity as the cosine of an angle. In order to understand what it does represent, we resort once again to the physical interpretation.

Let $|l_1| = 1$, $|l_2| = 1$; in particular, an inertial observer l_1 has lived a unit of proper time from the moment at which the measurement began. At the point l_1, the physical space of simultaneous events for him is $l_1 + (\mathbf{R}l_1)^{\perp}$. The world line of the observer $\mathbf{R}l_2$ intersects this point at the point xl_2, where x is obtained from the condition

$$(xl_2 - l_1, l_1) = 0,$$

that is, $x = (l_1, l_2)^{-1}$. The distance from l_1 to xl_2 is spacelike for observer $\mathbf{R}l_1$; this is the distance over which $\mathbf{R}l_2$ has moved away from him within a unit time, that is, the relative velocity of $\mathbf{R}l_2$. It equals (we must take into account the fact that the sign in the metric in $(\mathbf{R}l_1)^\perp$ must be changed !)

$$v = [-(xl_2 - l_1, xl_2 - l_1)]^{1/2} = [-(xl_2 - l_1, xl_2)]^{1/2} =$$

$$= [-x^2(l_2, l_2) + x(l_1, l_2)]^{1/2} = [-(l_1, l_2)^{-2} + 1]^{1/2},$$

whence

$$(l_1, l_2) = \frac{1}{\sqrt{1 - v^2}}.$$

This is the famous *Lorentz factor*; it is often written in the form $\frac{1}{\sqrt{1-v^2/c^2}}$, indicating explicitly that the velocities are measured with respect to the velocity of light. In particular,

$$x = \frac{1}{(l_1, l_2)} = \sqrt{1 - v^2},$$

that is, at the moment that the proper time equals one for the first observer, the second observer is located in his physical space, when the clocks of the first observer show $\sqrt{1 - v^2}$. This is the quantitative expression of the "time contraction" effect for a moving observer, qualitatively described in the preceding section.

12.9. Euclidean angles. In the space $(\mathbf{R}l_0)^\perp$, where l_0 is a timelike vector, the geometry is Euclidean, and there the inner product has the usual meaning. Let l_1 and l_2 be two timelike vectors with the same orientation. We can project them onto $(\mathbf{R}l_0)^\perp$ and calculate the cosine of the angle between the projections. We leave it to the reader to verify that for the observer $\mathbf{R}l_0$, this is the angle between the directions at which the observers $\mathbf{R}l_1$ and $\mathbf{R}l_2$ move away from him in his physical space. This angle does not have an absolute value; another observer $\mathbf{R}l_0'$ will see a different angle.

12.10. Four orientations of Minkowski space. Let $\{e_i\}$, $\{e_i'\}$, $i = 0, \ldots, 3$, be two orthonormal bases in $\mathcal{M}$: $(e_0, e_0) = (e_0', e_0') = 1$, $(e_i, e_i) = (e_i', e_i') = -1$ with $i = 1, \ldots, 3$. By analogy with the previous definitions we shall say that they are *identically oriented* if one is transformed into the other by a continuous system of *isometries* $f_t : \mathcal{M} \to \mathcal{M}$, $0 \le t \le 1$, $f_0 = \mathrm{id}$, $f_1(e_i) = e_i'$. There are evidently two necessary conditions for identical orientability:

a) $(e_0, e_0') > 0$. Indeed, Proposition 12.5 implies that $(e_0, f_t(e_0))^2 \ge 1$ so that the sign of $(e_0, f_t(e_0))$ cannot change when t changes, and $(e_0, f_0(e_0)) = 1$. Above, we called e_0 and e_0' with this property identically time-oriented.

b) The determinant of the mapping of the orthogonal projection $\sum_{i=1}^3 \mathbf{R}e_i \to \sum_{i=1}^3 \mathbf{R}e_i'$, written in the bases $\{e_i\}$ or $\{e_i'\}$, is positive.

Indeed, the projection $\sum_{i=1}^{3} \mathbf{R}e_i \to \sum_{i=1}^{3} \mathbf{R}f_t(e_i)$ is non-degenerate for all t: otherwise a spacelike vector from $\sum_{i=1}^{3} \mathrm{R}e_i$ would be orthogonal to

$$\sum_{i=1}^{3} \mathbf{R}f_t(e_i') = (f_t(e_0'))^{\perp},$$

that is, proportional to $f_t(e_0')$, which is a timelike vector, and this is impossible. Therefore, the determinants of these projections have the same sign for all t, while at $t = 0$ the determinant equals zero.

We can say that pairs of bases with the property b) are *identically space-oriented.*

Conversely, if two orthonormal bases in $\mathcal{M}$ have the same spatial and temporal orientation, then they are identically oriented, that is, they are transformed into one another by a continuous system of isometries f_t. In order to construct it we first set $f_t(e_0) = \frac{te_0'+(1-t)e_0}{|te_0'+(1-t)e_0|}$. From the condition $(e_0, e_0') \geq 1$ it follows that $f_t(e_0)$ is timelike and the square of the length equals one for all $0 \leq t \leq 1$. Next we choose for $f_t(e_1, e_2, e_3)$ an orthonormal basis of $f_t(e_0)^{\perp}$, obtained from a projection of $\{e_1, e_2, e_3\}$ onto $f_t(e_0)^{\perp}$ by the Gram–Schmidt orthogonalization process; it obviously is a continuous function of t. It is clear that $f_1(e_0) = e_0'$, and $\{f_1(e_1), f_1(e_2), f_1(e_3)\}$ and $\{e_1', e_2', e_3'\}$ are identically oriented orthonormal bases of $(e_0')^{\perp}$. They can be transformed into one another by a continuous family of purely Euclidean rotations $(e_0')^{\perp}$, leaving e_0' unchanged. This completes the proof.

We denote by Λ the *Lorentz group*, that is the group of isometries of the space $\mathcal{M}$, or $O(1,3)$. We denote further by $\Lambda_+^{\uparrow}$ the subgroup of Λ that preserves the orientation of some orthonormal basis; by $\Lambda_-^{\uparrow}$ the subset of Λ that changes its spatial, but not temporal orientation; by $\Lambda_+^{\downarrow}$ the subset of Λ that changes its temporal, but not its spatial orientation: and, by $\Lambda_-^{\downarrow}$ the subset of Λ that changes its temporal and spatial orientation. It is easy to verify that these subsets do not depend on the choice of the starting basis. We have proved the following result:

12.11. Theorem. *The Lorentz group Λ consists of four connected components:* $\Lambda = \Lambda_+^{\uparrow} \cup \Lambda_-^{\uparrow} \cup \Lambda_+^{\downarrow} \cup \Lambda_-^{\downarrow}$.

The identity mapping obviously lies in $\Lambda_+^{\uparrow}$. The following result is the analogue of Theorem 11.12.

12.12. Theorem. *We realize $\mathcal{M}$ as the space of Gram matrices of the Hermitian metrics in $\mathcal{H}$ in the basis $\{h_1, h_2\}$. For any matrix $V \in SL(2, \mathbf{C})$ we associate with the matrix $l \in \mathcal{M}$ a new matrix*

$$s(V)l = V^t l \overline{V}.$$

The mapping s defines a surjective homomorphism of $SL(2, \mathbf{C})$ onto $\Lambda_+^{\uparrow}$ with kernel $\{\pm E_2\}$.

Proof. It is evident that $s(V)l$ is linear with respect to l and preserves the squares of lengths: $\det(V^t l \overline{V}) = \det l$. Therefore $s(V) \in \Lambda$. Since the group $SL(2,\mathbf{C})$ is connected, any element in it can be continuously deformed into the unit element, while remaining inside $SL(2,\mathbf{C})$ — the Lorentz transformation $s(V)$ can be continuously deformed into an identity, so that $s(V) \in \Lambda_+^\uparrow$. Since $s(\mathrm{id}) = \mathrm{id}$ and $s(V_1V_2) = s(V_1)s(V_2)$, s is a homomorphism of groups. If $V^t l \overline{V} = l$ for all $l \in \mathcal{M}$, then, in particular, $V^t \sigma_i \overline{V} = \sigma_i$, where $\sigma_0 = E_2$ and $\sigma_1, \sigma_2, \sigma_3$ are the Pauli matrices. The condition $V^t\overline{V} = E_2$ means that V is unitary; then the condition $V^t \sigma_i \overline{V} = V^t \sigma_i (V^t)^{-1} = \sigma_i$ indicates that $V = \pm E_2$: this was proved in §12.11. Thus $\ker s = \{\pm E_2\}$.

It remains to establish that s is surjective. Let $f : \mathcal{M} \to \mathcal{M}$ be a Lorentz transformation from $\Lambda_+^\uparrow$, transforming the orthonormal basis $\{e_i\}$ into $\{e_i'\}$. Metrics in $\mathcal{H}$, corresponding to e_0 and e_0' are definite, because the eigenvalues of both e_0 and e_0' have the same sign, so that $\det e_0 = \det e_0' = 1$. It follows from $(e_0, e_0') > 0$ that these metrics are simultaneously positive- or negative-definite. Indeed, we verified above that the segment $te_0' + (1 - te_0)$, $0 \le t \le 1$ connecting them consists entirely of timelike vectors. This already implies the existence of a matrix $V \in SL(2,\mathbf{C})$ such that $s(V)$ transforms e_0 into e_0', that is, $e_0' = V^t e_0 V$, where e_0 and e_0' are identified with their Gram matrices. Indeed, V is the matrix of the isometry of $(\mathcal{H}, e_0)$ to $(\mathcal{H}, e_0')$; a priori, its determinant can equal -1, but this would contradict the possibility of connecting V with E_2 in $SL(2,\mathbf{C})$ by deforming V_q, where $e_0' = (V_q)^t f_q(e_0) V_q$, and f_q is the corresponding deformation in $\Lambda_+^\uparrow$.

So, $s(V)$ transforms e_0 into e_0'. It now remains to be shown that a Euclidean rotation of $\{s(V)e_1,\ s(V)e_2,\ s(V)e_3\}$ into $\{e_1', e_2', e_3'\}$ can be realized with the help of $s(U)$, where $U \in SL(2,\mathbf{C})$ and $s(U)$ leaves e_0 alone. It may be assumed that e_0' is represented by the matrix σ_0 in the basis $\{h_1, h_2\}$. Then we must choose U to be unitary with the condition $U(s(V)e_i)U^{-1} = e_i'$ for $i = 1,2,3$. According to Theorem 12.11 this can be done, because the bases $\{s(V)e_i\}$ and $\{e_i'\}$, $i = 1,2,3$, in $(e_0')^\perp$ are orthonormal and have the same orientation. The proof is complete.

12.13. Euclidean rotations and boosts. Let e_0, e_0' be two identically time-oriented timelike vectors of unit length and let L_0, L_0' be their orthogonal complements. There exists a standard Lorentz transformation from $\Lambda_+^\uparrow$ that transforms e_0 into e_0', which in the physics literature is called a *boost*. When $e_0 = e_0'$ it is the identity transformation. When $e_0 \ne e_0'$, it is defined as follows. We consider the plane $(L_0 \cap L_0')^\perp$. It contains e_0 and e_0'. The signature of the Minkowski metric in it equals (1,1). Therefore, there exists a pair of unit spacelike vectors $e_1, e_1' \in (L_0 \cap L_0')^\perp$ that are orthogonal to e_0 and e_0' respectively. The boost leaves alone all vectors in $L_0 \cap L_0'$ and transforms e_0 into e_0' and e_1 into e_1' respectively. To calculate the elements of the transformation matrix $\{e_0, e_1\} \begin{pmatrix} a & c \\ b & d \end{pmatrix} = \{e_0', e_1'\}$, we note first of all that $a = (e_0, e_0') = \frac{1}{\sqrt{1-v^2}}$, where v is the relative velocity of inertial observers, corresponding to e_0 and e_0'. Then the Gram matrices of $\{e_0, e_1\}$ and $\{e_0', e_1'\}$ are $\begin{pmatrix} 1 & 0 \\ 0 & -1 \end{pmatrix}$, so that

$$a^2 - b^2 = 1,\ ac - bd = 0,\ c^2 - d^2 = -1.$$

From the first equation, knowing a, we find $b = \frac{v}{\sqrt{1-v^2}}$. Adding here the condition that the determinant of the boost $ad - bc$ equals 1, we obtain $d = a$ and $c = b$. Finally, the matrix of the boost in the basis $\{e_0, e_1, e_2, e_3\}$, where $\{e_2, e_3\}$ is an orthonormal basis of $(L_0 \cap L_0')^\perp$, has the form

$$\begin{pmatrix} \frac{1}{\sqrt{1-v^2}} & \frac{v}{\sqrt{1-v^2}} & 0 & 0 \\ \frac{v}{\sqrt{1-v^2}} & \frac{1}{\sqrt{1-v^2}} & 0 & 0 \\ 0 & 0 & 1 & 0 \\ 0 & 0 & 0 & 1 \end{pmatrix}$$

or, in terms of space-time coordinates,

$$x_0 = \frac{x_0' + vx_1'}{\sqrt{1-v^2}},\ x_1 = \frac{vx_0' + x_1'}{\sqrt{1-v^2}},\ x_2 = x_2',\ x_3 = x_3'.$$

The matrix in the upper left hand corner can be written in the same form as the matrix of "hyperbolic rotation":

$$\begin{pmatrix} \cosh\theta & \sinh\theta \\ \sinh\theta & \cosh\theta \end{pmatrix},$$

where θ is found from the conditions

$$\cosh\theta = \frac{e^\theta + e^{-\theta}}{2} = \frac{1}{\sqrt{1-v^2}},\quad \sinh\theta = \frac{e^\theta - e^{-\theta}}{2} = \frac{v}{\sqrt{1-v^2}}.$$

If we start from two identically oriented orthonormal bases $\{e_0, e_1, e_2, e_3\}$ and $\{e_0', e_1', e_2', e_3'\}$, then the Lorentz transformation which transforms one into the other can be represented as the product of a boost, trnasforming e_0 into e_0', followed by a Euclidean rotation in $(e_0')^\perp$, which transforms the image of the basis $\{e_1, e_2, e_3\}$ after the boost into the basis $\{e_1', e_2', e_3'\}$ leaving $e_0^\perp$ alone.

12.14. Spatial and temporal reflections. Any three-dimensional subspace $L \subset \mathcal{M}$, in which the Minkowski metric is (anti) Euclidean (that is, the straight line $L^\perp$ is timelike), defines a Lorentz transformation which is the identity in L and changes the sign in $L^\perp$. All such operators are called *time reversals*.

Any three-dimensional subspace $L \subset \mathcal{M}$, in which the Minkowski metric has the signature (1,2) (that is, the straight line $L^\perp$ is spacelike) also defines a Lorentz transformation, which is an identity in L and changes the sign in $L^\perp$. All such operators are called *spatial reflections*.

If a time reversal T and a spatial reflection P are fixed, then all elements of $\Lambda_+^\downarrow, \Lambda_-^\uparrow, \Lambda_-^\downarrow$ will be obtained from all elements of $\Lambda_+^\uparrow$ by multiplication by T, P, PT respectively.

§13. Symplectic Spaces

13.1. In this chapter we shall study finite-dimensional linear spaces L over a field $\mathcal{K}$ with characteristic $\neq 2$, equipped with a *non-degenerate* skew-symmetric inner product $[\ ,\ ]: L \times L \to \mathcal{K}$; we call them *symplectic spaces*. We recall the properties of symplectic spaces, which have already been established in §3.

The dimension of a symplectic space is always even. If it equals $2r$, then there exists in the space a symplectic basis $\{e_1, \ldots, e_r; e_{r+1}, \ldots, e_{2r}\}$, that is, a basis with a Gram matrix of the form

$$\begin{pmatrix} 0 & E_r \\ -E_r & 0 \end{pmatrix}.$$

In particular, all symplectic spaces with the same dimension over a common field of scalars are isometric.

A subspace $L_1 \subset L$ is called *isotropic* if the restriction of the inner product $[\ ,\ ]$ to it identically equals zero. All one-dimensional subspaces are isotropic.

13.2. Proposition. *Let L be a symplectic space with dimension $2r$ and $L_1 \subset L$ an isotropic subspace with dimension r_1. Then $r_1 \leq r$, and if $r_1 < r$, then L_1 is contained in an isotropic subspace with the maximum possible dimension r.*

Proof. Since the form $[\ ,\ ]$ is non-degenerate, it defines an isomorphism $L \to L^*$, which associates the vector $l \in L$ with the linear functional $l' \mapsto [l, l']$. From here it follows that for any subspace $L_1 \subset L$ we have $\dim L_1^\perp = \dim L - \dim L_1$ (cf. §7 of Chapter I). If in addition L_1 is isotropic, then $L_1 \subset L_1^\perp$, whence $r_1 = \dim L_1 \leq$ $\leq \dim L_1^\perp = \dim L - \dim L_1 = 2r - r_1$, so that $r_1 \leq r$.

We now examine the restriction of the form $[\ ,\ ]$ to $L_1^\perp$. In the entire space L, the orthogonal complement to $L_1^\perp$ has the dimension $\dim L - \dim L_1^\perp = \dim L_1$ according to the previous discussion. On the other hand, L_1 lies in this orthogonal complement and therefore coincides with it. This means that L_1 is precisely the kernel of the restriction of $[\ ,\ ]$ to $L_1^\perp$. But, $L_1^\perp$ has a symplectic basis in the variant studied in §3, where degenerate spaces were permitted:

$$\{e_1, \ldots, e_{r-r_1}; e_{r-r_1+1}, \ldots, e_{2(r-r_1)}; e_{2(r-r_1)+1}, \ldots, e_{2r-r_1}\},$$

with the Gram matrix

$$\left(\begin{array}{c|c|c} 0 & E_{r-r_1} & 0 \\ \hline -E_{r-r_1} & 0 & 0 \\ \hline 0 & 0 & 0 \end{array}\right).$$

The size of a block is $\frac{1}{2}(\dim L_1^\perp - \dim L_1) = r - r_1$. The vectors $e_{2(r-r_1)+1}, \ldots, e_{2r-r_1}$ generate the kernel of the form on $L_1^\perp$, that is, L_1; adding to them, for example, $e_1, \ldots, e_{r-r_1}$, we obtain an r-dimensional isotropic subspace containing L_1.

13.3. Proposition. *Let L be a symplectic space with dimension $2r$, and let $L_1 \subset L$ be an isotropic subspace with dimension r. Then there exists another isotropic subspace $L_2 \subset L$ with dimension r, such that $L = L_1 \oplus L_2$, and the inner product induces an isomorphism $L_2 \to L_1^*$.*

Proof. We shall prove a somewhat stronger result, which is useful in applications, namely, we shall establish the existence of the subspace L_2 from among the finite number of isotropic subspaces, associated with the fixed symplectic basis $\{e_1, \ldots, e_r; e_{r+1}, \ldots, e_{2r}\}$ of L.

Namely, let a decomposition $\{1, \ldots, r\} = I \cup J$ into two non-intersecting subsets be given. Then the vectors $\{e_i, e_{r+j} | i \in I,\ j \in J\}$ generate an r-dimensional isotropic subspace in L, called a coordinate subspace (with respect to the chosen basis). Obviously, there are 2^r of them. We shall show that L_2 is a coordinate subspace.

Let M be spanned by $\{e_1, \ldots, e_r\}$ and $\dim(L_1 \cap M) = s$, $0 \leq s \leq r$. There exists a subset $I \subset \{1, \ldots, r\}$ consisting of $r - s$ elements, such that $L_1 \cap M$ is transverse to N, spanned by $\{e_i | i \in I\}$, that is, $L_1 \cap M \cap N = \{0\}$. Indeed, the set $\{\text{basis of } L_1 \cap M\} \cup \{e_1, \ldots, e_r\}$ generates M, so that Proposition 2.10 of Chapter I implies that the basis of $L_1 \cap M$ can be extended to a basis of M with the help of $r - s$ vectors from $\{e_1, \ldots, e_r\}$. The numbers of these vectors form the set I sought, because $L_1 \cap M + N = M$, so that $L_1 \cap M \cap N = \{0\}$.

We now set $J = \{1, \ldots, r\} \setminus I$ and show that the isotropic subspace L_2 spanned by $\{e_i, e_{r+j} | i \in I,\ j \in J\}$, is the direct complement of L_1. It is sufficient to verify that $L_1 \cap L_2 = \{0\}$. Indeed, from the proof of Proposition 13.2 it follows that $L_1^\perp = L_1$, $L_2^\perp = L_2$. But $L_1 \cap M$ is contained in L_1 and N is contained in L_2, so that the sum $M = L_1 \cap M + N$ is orthogonal to $L_1 \cap L_2$. But M is isotropic and r-dimensional, so that $M^\perp = M$ and $L_1 \cap L_2 \subset M$. Therefore finally

$$L_1 \cap L_2 = (L_1 \cap M) \cap (L_2 \cap M) = (L_1 \cap M) \cap N = \{0\}.$$

The linear mapping $L_2 \to L_1^*$ associates with a vector $l \in L_2$ the linear form $m \mapsto [l, m]$ on L_1. It is an isomorphism because $\dim L_2 = \dim L_1^* = r$, and its kernel is contained in the kernel of the form $[\ ,\ ]$ which, by definition, is non-degenerate. This completes the proof.

13.4. Corollary. *All pairs of mutually complementary isotropic subspaces of L are identically arranged: if $L = L_1 \oplus L_2 = L'_1 \oplus L'_2$, then there exists an isometry $f : L \to L$ such that $f(L_1) = L'_1$, $f(L_2) = L'_2$.*

Proof. We select a basis $\{e_1, \ldots, e_r\}$ of L_1 and its dual basis $\{e_{r+1}, \ldots, e_{2r}\}$ in L_2 relative to the identity $L_2 \to L_1^*$ described above. Obviously, $\{e_1, \ldots, e_{2r}\}$ is a symplectic basis of L. Analogously, we construct a symplectic basis $\{e'_1, \ldots, e'_{2r}\}$ with respect to the decomposition $L'_1 \oplus L'_2$. The linear mapping $f : e_i \mapsto e'_i$, $i = 1, \ldots, 2r$ is clearly the required isometry.

It follows from this corollary and Propositions 13.2 and 13.3 that any isotropic subspaces of equal dimension in L are transformed into one another by an appropriate isometry.

13.5. Symplectic group. The set of all isometries $f : L \to L$ of a symplectic space forms a group. The set of matrices, representing this group in a symplectic basis $\{e_1, \ldots, e_{2r}\}$, is called a *symplectic group* and is denoted by $Sp(2r, \mathcal{K})$, if $\dim L = 2r$. The condition $A \in Sp(2r, \mathcal{K})$ is equivalent to the fact that the Gram matrix of the basis $\{e_1, \ldots, e_{2r}\}A$ equals $I_{2r} = \begin{pmatrix} 0 & E_r \\ -E_r & 0 \end{pmatrix}$, that is, $A^t I_{2r} A = I_{2r}$ so that $\det A = \pm 1$; we shall prove below that $\det A = 1$ (see §13.11). Since $I_{2r}^2 = -E_{2r}$, this condition can also be written in the form $A = -I_{2r}(A^t)^{-1} I_{2r}$. Hence we have the following proposition.

13.6. Proposition. *The characteristic polynomial of $P(t) = \det(tE_{2r} - A)$ of a symplectic matrix A is reciprocal, that is, $P(t) = t^{2r} P(t^{-1})$.*

Proof. We have, using the fact that $\det A = 1$,

$$\det(tE_{2r} - A) = \det(tE_{2r} + I_{2r}(A^t)^{-1} I_{2r}) = \det(tE_{2r} - (A^t)^{-1}) =$$

$$= \det(tA^t - E_{2r}) = t^{2r} \det(t^{-1}E_{2r} - A^t) = t^{2r} \det(t^{-1}E_{2r} - A).$$

13.7. Corollary. *If $\mathcal{K} = \mathbf{R}$ and A is a symplectic matrix, then together with every eigenvalue λ of A there exist eigenvalues λ^{-1}, $\bar{\lambda}$ and $\bar{\lambda}^{-1}$.*

Proof. Since A is non-singular, $\lambda \neq 0$ and $P(\lambda^{-1}) = \lambda^{-2r} P(\lambda) = 0$. Since the coefficients of P are real, $P(\bar{\lambda}) = \overline{P(\lambda)} = 0$.

Complex conjugation is a symmetry relative to the real axis, and the mapping $\lambda \mapsto \bar{\lambda}^{-1}$ is a symmetry relative to the unit circle. Therefore, the complex eigenvalues of A are quadruples, symmetric simultaneously relative to the real axis and the unit circle, whereas the real eigenvalues are pairs.

13.8. Pfaffian. Let $\mathcal{K}^{2r}$ be a coordinate space and A a non-singular skew-symmetric matrix of order $2r$ over $\mathcal{K}$. The inner product $[\vec{x}, \vec{y}] = \vec{x}^t A \vec{y}$ in $\mathcal{K}^{2r}$ is non-degenerate and skew-symmetric. Transforming from the starting basis to the symplectic basis, we find that for any matrix A there exists a non-singular matrix B such that

$$A = B^t \begin{pmatrix} 0 & E_r \\ -E_r & 0 \end{pmatrix} B,$$

whence $\det A = (\det B)^2$. Thus the determinant of every skew-symmetric matrix is an exact square. This suggests that we attempt to extract the square root of the determinant, which would be a *universal polynomial* of the elements of A. This is indeed possible.

13.9. Theorem. *There exists a unique polynomial with integer coefficients PfA of the elements of a skew-symmetric matrix A such that $\det A = (PfA)^2$ and $Pf\begin{pmatrix} 0 & E_r \\ -E_r & 0 \end{pmatrix} = 1$. This polynomial is called the Pfaffian and has the following property:*

$$Pf(B^tAB) = \det B \cdot PfA$$

for any matrix B. (If $\mathcal{K} \neq 0$ the coefficients of Pf are "integers" in the sense that they belong to a simple subfield of $\mathcal{K}$, that is, they are sums of ones.)

Proof. Consider $r(2r-1)$ independent variables over the field $\mathcal{K}$: $\{a_{ij} | 1 \leq i \leq \leq j \leq 2r\}$. Denote by $\mathcal{K}$ the field of rational functions (ratios of polynomials) of a_{ij} with coefficients from a simple subfield of $\mathcal{K}$. We set $A = (a_{ij})$, where $a_{ij} = -a_{ji}$ for $i > j$, $a_{ii} = 0$, and we introduce on the coordinate space K^{2r} a non-degenerate skew symmetric inner product $\vec{x}^t A\vec{y}$. Transforming to the symplectic basis with the help of some matrix B, we find, as above, that $\det A = (\det B)^2$. A priori, $\det B$ is only a rational function of a_{ij} with coefficients from $\mathbf{Q}$ or a simple field with a finite characteristic. But since $\det A$ is a polynomial with integer coefficients, its square root also must have integer coefficients (here we use the theorem on the unique factorization of polynomials in the ring $\mathbf{Z}[a_{ij}]$ or $\mathbf{F}_p[a_{ij}]$). The sign of $\sqrt{\det A}$ is, evidently, uniquely fixed by the requirement that the value of $\sqrt{\det I_{2r}}$ must equal one.

The last equality is established as follows. First, B^tAB and A are skew-symmetric, so that

$$\mathrm{Pf}^2(B^tAB) = \det(B^tAB) = (\det B)^2 \det A = (\det B)^2 \mathrm{Pf}^2 A.$$

Therefore

$$\mathrm{Pf}(B^tAB) = \pm \det B \,\mathrm{Pf}\, A.$$

To establish the sign it is sufficient to determine it in the case $A = I_{2r}$, $B = E_{2r}$, where it is obviously positive.

13.10. Examples.

$$\mathrm{Pf}\begin{pmatrix} 0 & a_{12} \\ -a_{12} & 0 \end{pmatrix} = a_{12};$$

$$\mathrm{Pf}\begin{pmatrix} 0 & a_{12} & a_{13} & a_{14} \\ -a_{12} & 0 & a_{23} & a_{24} \\ -a_{13} & -a_{23} & 0 & a_{34} \\ -a_{14} & -a_{24} & -a_{34} & 0 \end{pmatrix} = -a_{12}a_{34} + a_{13}a_{24} - a_{14}a_{23}.$$

13.11. Corollary. *The determinant of any symplectic matrix equals one.*

Proof. From the condition $A^t I_{2r} A = I_{2r}$ and Theorem 13.9 it follows that

$$1 = \mathrm{Pf}\, I_{2r} = \mathrm{Pf}\,(A^t I_{2r} A) = \det A\, \mathrm{Pf}\, I_{2r},$$

which proves the required result.

We used this fact in proving Proposition 13.6.

13.12. Relationship between the orthogonal, unitary and symplectic groups. Let $\mathbf{R}^{2r}$ be a coordinate space with two inner products: a Euclidean product (,) and a symplectic product [,]:

$$(\vec{x}, \vec{y}) = \vec{x}^{\,t}\vec{y};$$

$$[\vec{x}, \vec{y}] = \vec{x}^{\,t} I_{2r} \vec{y} = (\vec{x}, I_{2r}\vec{y}).$$

Since $I_{2r}^2 = -E_{2r}$, the operator I_{2r} determines on $\mathbf{R}^{2r}$ a complex structure (see §12 of Chapter I) with the complex basis $\{e_j + ie_{r+j} \mid j = 1, \ldots, r\}$, relative to which there exists a Hermitian inner product

$$\langle \vec{x}, \vec{y} \rangle = (\vec{x}, \vec{y}) - i[\vec{x}, \vec{y}]$$

(see Proposition 6.2).

In terms of these structures, we have

$$U(r) = O(2r) \cap Sp(2r) = GL(r, \mathbf{C}) \cap Sp(2r) = GL(r, \mathbf{C}) \cap O(2r).$$

The verification is left to the reader as an exercise.

§14. Witt's Theorem and Witt's Group

1. In this section we shall present the results obtained by Witt in the theory of finite-dimensional orthogonal spaces over arbitrary fields. They refine the classification theorem of §3, and they can be regarded as far-reaching generalizations of the inertia theorem and the concept of signature. We shall start with some definitions. As usual, we assume that the characteristic of the field of scalars does not equal 2.

A *hyperbolic plane* is a two-dimensional space L with a non-degenerate symmetric inner product $(\, , \,)$ containing a non-zero isotropic vector.

A *hyperbolic space* is a space that decomposes into a direct sum of pairwise orthogonal hyperbolic planes.

An *anisotropic space* is a space that does not have (non-zero) isotropic vectors.

Anisotropic spaces L over a real field have the signature $(n,0)$ or $(0,n)$, where $n = \dim L$. We shall now show that hyperbolic spaces are a generalization of spaces with signature (m,m).

14.2. Lemma. *A hyperbolic plane L always has bases $\{e_1', e_2'\}$ with the Gram matrix $\begin{pmatrix} 1 & 0 \\ 0 & -1 \end{pmatrix}$ and $\{e_1, e_2\}$ with the Gram matrix $\begin{pmatrix} 0 & 1 \\ 1 & 0 \end{pmatrix}$.*

Proof. Let $l \in L, (l,l) = 0$. If $l_1 \in L$ is not proportional to l, then $(l_1, l) \neq 0$, because L is non-degenerate. It may be assumed that $(l_1, l) = 1$. We set $e_1 = l$, $e_2 = l_1 - \frac{(l_1,l_1)}{2} l$. Then $(e_1, e_1) = (e_2, e_2) = 0$, $(e_1, e_2) = 1$. We set $e_1' = \frac{e_1+e_2}{2}$, $e_2' = \frac{e_1-e_2}{2}$. Then $(e_1', e_1') = 1$, $(e_2', e_2') = -1$, $(e_1', e_2') = 0$. The lemma is proved.

We shall call the basis $\{e_1, e_2\}$ *hyperbolic*. Analogously, in a general hyperbolic space we shall call a basis whose Gram matrix consists of diagonal blocks $\begin{pmatrix} 0 & 1 \\ 1 & 0 \end{pmatrix}$ hyperbolic.

14.3. Lemma. *Let $L_0 \subset L$ be an isotropic subspace in a non-degenerate orthogonal space L and let $\{e_1, \ldots, e_m\}$ be a basis of L_0. Then there exist vectors $e_1', \ldots, e_m' \in L$ such that $\{e_1, e_1', \ldots, e_m, e_m'\}$ form a hyperbolic basis of their linear span.*

Proof. We set $L_1 =$ (linear span of $\{e_2, \ldots, e_m\}$). Since L_1 is strictly smaller than L_0, $L_1^\perp$ is strictly larger than $L_0^\perp$ because of the non-degeneracy of L. Let $e_1'' \in L_1^\perp \backslash L_0^\perp$. Then $(e_1'', e_i) = 0$ for $i \geq 2$ but $(e_1'', e_1) \neq 0$. It may be assumed that $(e_1'', e_1) = 1$ so that e_1'' is not proportional to e_1. As in the proof of Lemma 14.2, we set $e_1' = e_1'' - \frac{(e_1'', e_1'')}{2} e_1$. Then $\{e_1, e_1'\}$ form a hyperbolic basis of their linear span. Its orthogonal complement is non-degenerate and contains an isotropic subspace spanned by $\{e_2, \ldots, e_m\}$. An analogous argument can be applied to this pair, and induction on m gives the required result.

14.4. Witt's theorem. *Let L be a non-degenerate finite-dimensional orthogonal space and let $L', L'' \subset L$ be two isometric subspaces of it. Then any isometry $f' : L' \to L''$ can be extended to an isometry $f : L \to L$, coinciding with f' on L'.*

Proof. We analyse several cases successively.

a) *$L' = L''$ and both spaces are non-degenerate.* Then $L = L' \oplus (L')^\perp$ and we can set $f = f' \oplus \mathrm{id}_{(L')^\perp}$.

b) *$L' \neq L''$, $\dim L' = \dim L'' = 1$, and both spaces are non-degenerate.* The isometry $f' : L' \to L''$ can be extended to the isometry $f'' : L' + L'' \to L' + L''$, setting $f''(l) = f'(l)$ for $l \in L'$, and $f''(l) = (f')^{-1}(l)$ for $l \in L''$. If $L' + L''$ is non-degenerate, then f'' is extended to f according to the preceding case. If $L' + L''$ is degenerate, then the kernel of the inner product on $L' + L''$ is one-dimensional. Let e_1 generate this kernel and let e_2 generate L'. The orthogonal complement of e_2 (with respect to L) contains a vector e_1' such that the basis $\{e_1, e_1'\}$ generated by these vectors in the plane is hyperbolic. This is possible according to Lemma 14.3. We shall show that the subspace L_0, spanned by $\{e_1, e_1', e_2\}$, is non-degenerate, and the isometry $f' : L' \to L''$ is extended to the isometry $f : L_0 \to L_0$. After this, case a) can be applied.

Non-degeneracy follows from the fact that $(e_2, e_2) \neq 0$, and the Gram matrix of the vectors $\{e_1, e_1', e_2\}$ has the form

$$\begin{pmatrix} 0 & 1 & 0 \\ 1 & 0 & 0 \\ 0 & 0 & (e_2, e_2) \end{pmatrix}.$$

To extend f' we first note that the orthogonal complement to $f'(e_2)$ in L_0 is two-dimensional, non-degenerate, and contains the isotropic vector e_1. Hence it is a hyperbolic plane, like the orthogonal complement to e_2 in L_0. Lemma 14.2 implies that there exists an isometry of the second plane to the first one. Its direct sum with f' is the extension sought.

c) $\dim L' = \dim L'' > 1$ and L', L'' are *non-degenerate*. We perform induction on $\dim L'$. Since L' has an orthogonal basis, the decomposition $L' = L_1' \oplus L_2'$ into an orthogonal direct sum of subspaces with non-zero dimension exists. Since f' is an isometry, $L'' = L_1'' \oplus L_2''$, where $L_i'' = f'(L_i')$ and this sum is orthogonal. By induction, the restriction of f' to L_1' is extended to the isometry $f_1' : L \to L$. It transforms $(L_1')^\perp \supset L_2'$ into $(L_1'')^\perp \supset L_2''$. Again by induction, there exists an isometry of $(L_1')^\perp$ to $(L_1'')^\perp$, which coincides on L_2' with the restriction of f'. Extending it by the restriction of f' to L_1' we obtain the required result.

d) *L' is degenerate.* We shall reduce this last case to the case already analysed. Let $L_0' \subset L'$ be the kernel of the restriction of the metric to L'. Selecting an orthonormal basis of L', we can construct a direct decomposition $L' = L_1' \oplus L_0'$ where L_1' is non-degenerate. The orthogonal complement to L_1' contains in L a subspace $\bar{L}_0'$ such that the sum $\bar{L}_0' \oplus L_1' \oplus L_0$ is a direct sum and the space $\bar{L}_0' \oplus L_0'$ is hyperbolic, as in Lemma 14.3; in particular, $\bar{L}_0' \oplus L_1' \oplus L_0'$ is non-degenerate. Analogously, we construct $\bar{L}_0'' \oplus L_1'' \oplus L_0''$, starting from the space L''. Evidently, the isometry $f' : L' \to L''$ can be extended to the isometries of these first sums, because all hyperbolic spaces with the same dimension are isometric. The possibility of further extension of this isometry now follows from case c). The theorem is proved.

14.5. Corollary. *Let L_1, L_2 be isometric non-degenerate spaces and let L_1', L_2' be isometric subspaces of them. Then their orthogonal complements $(L_1')^\perp, (L_2')^\perp$ are isometric.*

14.6. Corollary. *Let L be a non-degenerate orthogonal space. Then any isotropic subspace of L is contained in a maximal isotropic subspace, and for two maximal isotropic subspaces L' and L'' there exists an isometry $f : L \to L$ transforming L' into L''.*

Proof. The first assertion is trivial. To prove the second assertion, we assume that $\dim L' \leq \dim L''$. Any linear injection $f' : L' \to L''$ is an isometry of L' to $\operatorname{im} f'$. Therefore it can be extended to the isometry $f : L \to L$. Then $L' \subset f^{-1}(L'')$ and $f^{-1}(L'')$ is isotropic. Since L' is maximal, $\dim L' = \dim f^{-1}(L'') = \dim L''$.

14.7. Corollary. *For any orthogonal space L there exists an orthogonal direct decomposition $L_0 \oplus L_h \oplus L_d$, where L_0 is isotropic, L_h is hyperbolic, and L_d is anisotropic. For any two such decompositions there exists an isometry $f : L \to L$, transforming one of them into the other.*

Proof. We take for L_0 the kernel of the inner product. We decompose L into the direct sum $L_0 \oplus L_1$. In L_1 we take a maximal isotropic subspace and embed it into the hyperbolic subspace with the doubled dimension of L_h, as in Lemma 14.3. For L_d we take the orthogonal complement of L_h in L_1. It does not contain isotropic vectors, because otherwise such a vector could be added to the starting isotropic subspace, which would not be maximal. This proves the existence of the decomposition of the required form.

Conversely, in any such decomposition $L_0 \oplus L_h \oplus L_d$ the space L_0 is a kernel. Further, a maximal isotropic subspace in L_h is simultaneously maximally isotropic in $L_h \oplus L_d$, so that the dimension of L_h is determined uniquely. Therefore, for two decompositions $L_0 \oplus L_h \oplus L_d$ and $L_0 \oplus L_h' \oplus L_d'$ there exists an isometry transforming L_0 into L_0 and L_h into L_h'. It is extended by the isometry of L_d into L_d' according to Witt's theorem, which completes the proof. We call L_d the ***anisotropic part*** of the space L; it is determined up to isometry.

This corollary represents an extension of the principle of inertia to arbitrary fields of scalars, reducing the classification of orthogonal spaces to the classification of anisotropic spaces or, in the language of quadratic forms, to the classification of forms, *not representing zero*, for which $q(l) = 0$ implies that $l = 0$.

14.8. Witt's group. Let $\mathcal{K}$ be a field of scalars. We denote by $W(\mathcal{K})$ the set of classes of anisotropic orthogonal spaces over $\mathcal{K}$ (up to isometry), supplemented by the class of the null space. We introduce on $W(\mathcal{K})$ the following addition operation: if L_1, L_2 are two anisotropic spaces, and $[L_1], [L_2]$ are the classes in $W(\mathcal{K})$, then $[L_1] + [L_2]$ is the class of anisotropic parts of $L_1 \oplus L_2$ (the orthogonal external direct sum stands on the right).

It is not difficult to verify that the definition is correct. Further, this addition operation is associative, and the class of the null space serves as the zero in $W(\mathcal{K})$. Moreover:

14.9. Theorem. a) *$W(\mathcal{K})$ with the addition operation introduced above is an abelian group, called the Witt group of the field $\mathcal{K}$.*

b) *Let L_a be a one-dimensional coordinate space over $\mathcal{K}$ with the inner product axy, $a \in \mathcal{K}\backslash\{0\}$. Then $[L_a]$ depends only on the coset $a(\mathcal{K}^*)^2$, and the elements of $[L_a]$ form a system of generators of the group $W(\mathcal{K})$.*

Proof. We have only to verify that for every element of $W(\mathcal{K})$ there exists an inverse element. Indeed, let L be an anisotropic space with a metric, which in the orthogonal basis $\{e_1, \dots, e_n\}$ is given by the form $\sum_{i=1}^n a_i x_i^2$. We denote by $\bar{L}$ the space L with the metric $-\sum_{i=1}^n a_i x_i^2$ and we show that $L \oplus \bar{L}$ is hyperbolic, so that $[L] + [\bar{L}] = [0]$ in $W(\mathcal{K})$. Indeed the metric in $L \oplus \bar{L}$ is given by the form $\sum_{i=1}^n a_i(x_i^2 - y_i^2)$. But the plane with the metric $a(x^2 - y^2)$ is evidently hyperbolic, because the form is non-degenerate, and the vector $(1, 1)$ is isotropic. The fact that $[L_a]$ depends only on $a(\mathcal{K}^*)^2$ was verified in §2.7. In addition, every n-dimensional orthogonal space can be decomposed into a direct orthogonal sum of one-dimensional spaces of the form L_a. This completes the proof.

§15. Clifford Algebras

1. We shall call an associative ring with identity A, containing the field $\mathcal{K}$, such that $\mathcal{K}$ lies at the centre of A, that is, it commutes with all elements of A an *algebra*. In particular, A is a $\mathcal{K}$-linear space.

Consider a finite-dimensional orthogonal space L with the metric g. In this section we shall construct an algebra $C(L)$ and a $\mathcal{K}$-linear imbedding $\rho : L \to C(L)$ such that for any element $l \in L$ the relation

$$\rho(l)^2 = g(l, l) \cdot 1,$$

holds, that is, the *inner square* of any vector in L will be realized as its *square* in the sense of multiplication in $C(L)$. In addition, the elements of $\rho(L)$ will be *multiplicative generators* of $C(L)$, that is, any element in $C(L)$ can be represented as a linear combination of "non-commutative" monomials of elements of $\rho(L)$. The algebra $C(L)$ (together with the mapping ρ) with these properties will be called the *Clifford algebra* of the space L.

15.2. Theorem. a) *For any finite-dimensional orthogonal space L, the Clifford algebra $C(L)$ exists and has the dimension 2^n over $\mathcal{K}$, where $n = \dim L$.*

b) *Let $\sigma : L \to D$ be any $\mathcal{K}$-linear mapping of L into the $\mathcal{K}$-algebra D, for which $\sigma(l)^2 = g(l,l) \cdot 1$ for all $l \in L$. Then there exists a unique homomorphism of $\mathcal{K}$-algebras $\tau : C(L) \to D$ such that $\sigma = \tau \circ \rho$. In particular, $C(L)$ is defined uniquely up to isomorphism.*

Proof. a) Choose an orthogonal basis $\{e_1, \ldots, e_n\}, (e_i, e_i) = a_i$ of L. By definition, the relations

$$\rho(e_i)^2 = a_i, \ \rho(e_i)\rho(e_j) = -\rho(e_j)\rho(e_i), \ i \neq j$$

must be satisfied in $C(L)$. The second of these relations follows from the fact that $[\rho(e_i + e_j)]^2 = \rho(e_i)^2 + \rho(e_i)\rho(e_j) + \rho(e_j)\rho(e_i) + \rho(e_j)^2 = \rho(e_i)^2 + \rho(e_j)^2$. Decomposing the elements $l_1, \ldots, l_m \in L$ with respect to the basis $\{e_i\}$ and using the fact that multiplication in L is $\mathcal{K}$-linear with respect to each of the cofactors (this follows from the fact that $\mathcal{K}$ lies at the centre), we can represent any product $\rho(l_1) \ldots \rho(l_m)$ as a linear combination of monomials relative to $\rho(e_i)$. Replacing $\rho(e_i)^3$ by a_i and $\rho(e_i)\rho(e_j)$ for $i > j$ by $-\rho(e_j)\rho(e_i)$, we can put any monomial into the form $a\rho(e_{i_1}) \ldots \rho(e_{i_m})$, where $a \in \mathcal{K}$, $i_1 < i_2 < \ldots < i_m$. Further relations between such expressions are not evident; there are 2^m monomials $\rho(e_{i_1}) \ldots \rho(e_{i_m})$ (including the trivial monomial 1 for $m = 0$).

The plan of the proof is to make these introductory considerations more rigorous, working more formally.

To this end, for each subset $S \subset \{1, \ldots, n\}$ we introduce the symbol e_S (which will subsequently turn out to be equal to $\rho(e_{i_1}) \ldots \rho(e_{i_m})$, if $S = \{i_1, \ldots i_m\}$, $i_1 < \ldots < i_m$); we also set $e_\emptyset = 1$ ($\emptyset$ is the empty subset). We denote by $C(L)$ the $\mathcal{K}$-linear space with the basis $\{e_S\}$. We define multiplication in $C(L)$ as follows. If $1 \leq s, \ t \leq n$, we set

$$(s,t) = \begin{cases} 1 & \text{for } s \leq t, \\ -1 & \text{for } s > t \end{cases}.$$

For two subsets $S, T \subset \{1, \ldots, n\}$ we set

$$a(S,T) = \prod_{s \in S, t \in T} (s,t) \prod_{i \in S \cap T} a_i,$$

where, we recall, $a_i = g(e_i, e_i)$. Empty products are assumed to be equal to unity. Finally we define the product of linear combinations

$$\sum a_S e_S, \sum b_T e_T \in C(L);\ a_S, b_T \in \mathcal{K},$$

by the formula

$$(\sum a_S e_S)(\sum b_T e_T) = \sum a_S b_T a(S,T) e_{S\nabla T},$$

where $S\nabla T = (S\cup T)\backslash(S\cap T)$ is the symmetric difference of the sets S and T. All axioms of the $\mathcal{K}$-algebra are verified trivially, with the exception of associativity. It is sufficient to prove associativity for the elements of the basis, that is, to establish the identity

$$(e_S e_T)e_R = e_S(e_T e_R).$$

Since $e_S e_T = a(S,T)e_S\nabla_r$, we have

$$(e_S e_T)e_R = a(S,T)a(S\nabla T, R)e_{(S\nabla T)\nabla R},$$
$$e_S(e_T e_R) = a(S, T\nabla R)a(T,R)e_{S\nabla(T\nabla R)}.$$

It is not difficult to verify that

$$(S\nabla T)\nabla R = S\nabla(T\nabla R) =$$

$$= \{(S\cup T\cup R)\backslash[(S\cap T)\cup(S\cap R)\cup(T\cap R)]\}\cup(S\cap T\cap R).$$

Therefore it remains to verify only that the scalar coefficients are equal. The part $a(S,T)a(S\nabla T, R)$ referring to the signs has the form

$$\prod_{s\in S,\, t\in T}(s,t)\prod_{u\in S\nabla T,\, r\in R}(u,r).$$

Letting u in the second product run first through all elements of S and then through all elements of T (with fixed r), we introduce the cofactors $(u,r)^2, u\in S\cap T$, equal to one, so that this "sign" can be written in a symmetric form with respect to S, T and R

$$\prod_{s\in S,\, t\in T}(s,t)\prod_{u\in S,\, r\in R}(u,r)\prod_{u\in T,\, r\in R}(u,r).$$

The sign corresponding to $a(S, T\nabla R)a(T,R)$ is transformed analogously, leading to the same result. It remains to analyse the factors that contain the inner squares of a_i. For $a(S,T)a(S\nabla T, R)$ they have the form

$$\prod_{i\in S\cap T} a_i \prod_{j\in(S\nabla T)\cap R} a_j.$$

But $(S\nabla T)\cap R=(S\cap R)\nabla(T\cap R)$ and the intersection of $S\cap T$ with this set is empty and $(S\cap T)\cup[(S\cap R)\nabla(T\cap R)]$ consists of the elements of $S\cup T\cup R$ that are contained in more than one of these three sets. Therefore, our factor depends symmetrically on S,T and R. The part of the coefficient $a(S,T\nabla R)a(T,R)$ that we need is calculated analogously, leading to the same result. This completes the proof of the associativity of the algebra $C(L)$.

Finally we define the $\mathcal{K}$-linear mapping $\rho: L\to C(L)$ by the condition $\rho(e_i)=$ $=e_{\{i\}}$. According to the multiplication formulas, $e_\emptyset$ is unity in $C(L)$ and

$$\rho(e_i)\rho(e_j)=e_{\{i\}}e_{\{j\}}=\begin{cases} a_ie_\emptyset & \text{for } i=j,\\ -e_{\{j\}}e_{\{i\}} & \text{for } i\neq j\end{cases}.$$

Therefore, $(\rho(C(L))$ is a Clifford algebra for L.

b) The last assertion is proved formally. Let $\sigma: L\to D$ be a $\mathcal{K}$-linear mapping with $\sigma(l)^2=g(l,l)\cdot 1$. There exists a unique $\mathcal{K}$-linear mapping $\tau: C(L)\to D$, which in the elements of the basis e_S is defined by the formula

$$\tau(e_{\{i_1\dots i_m\}})=\sigma(e_{i_1})\dots\sigma(e_{i_m}),$$

$$\tau(e_\emptyset)=1_D.$$

Here $\tau\circ\rho=\sigma$, because this is so on the elements of the basis of L. Finally, τ is a homomorphism of algebras. Indeed,

$$\tau(e_Se_T)=\tau(a(S,T)e_{S\nabla T})=a(S,T)\tau(e_{S\nabla T}),$$

and it is easy to verify that $\tau(e_S)\tau(e_T)$ can be put into the same form, using the relations

$$\sigma(e_i)^2=a_i,\quad \sigma(e_i)\sigma(e_j)=-\sigma(e_j)\sigma(e_i)\quad\text{for}\quad i\neq j.$$

This completes the proof.

15.3. Examples. a) Let L be the two-dimensional real plane with the matrix $-(x^2+y^2)$. The Clifford algebra $C(L)$ has the basis $(1,e_1,e_2,e_1e_2)$ with the multiplication relations

$$e_1^2=e_2^2=-1,\ e_1e_2=-e_2e_1.$$

It is not difficult to verify that the mapping $C(L)\to H: 1\mapsto 1, e_1\mapsto \mathbf{i}$, $e_2\mapsto \mathbf{j}$, $e_1e_2\mapsto \mathbf{k}$ defines an isomorphism of $C(L)$ to the algebra of quaternions H.

b) Let L be a linear space with a *zero* metric. The algebra $C(L)$ is generated by the generators $\{e_1,\dots,e_n\}$ with the relations

$$e_i^2=0,\ e_ie_j=-e_je_i\quad\text{for}\quad i\neq j.$$

It is called the *exterior algebra*, or a *Grassmann algebra*, of the linear space L. We shall return to this algebra in Chapter 4.

c) Let $L = \mathcal{M}^{\mathbf{C}}$ be complexified Minkowski space with the metric $x_0^2 - \sum_{i=1}^3 x_i^2$ relative to the orthonormal basis $\{e_i\}$ of $\mathcal{M}$, which is simultaneously a basis of $\mathcal{M}^{\mathbf{C}}$. We shall show that the Clifford algebra $C(\mathcal{M}^{\mathbf{C}})$ is isomorphic to the algebra of complex 4×4 matrices. To this end we examine the Dirac matrices, written in the block form

$$\gamma_0 = \begin{pmatrix} \sigma_0 & 0 \\ 0 & -\sigma_0 \end{pmatrix}, \quad \gamma_j = \begin{pmatrix} 0 & \sigma_j \\ -\sigma_j & 0 \end{pmatrix}, \quad j = 1, 2, 3.$$

Using the properties of the Pauli matrices σ_j it is not difficult to verify that the γ_j satisfy the same relations in the algebras of the matrices $M_4(\mathbf{C})$ as do the $\rho(e_j)$ in the algebra $C(\mathcal{M}^{\mathbf{C}})$:

$$\gamma_0^2 = -\gamma_1^2 = -\gamma_2^2 = -\gamma_3^2 = 1$$

(that is, E_4); $\gamma_i\gamma_j + \gamma_j\gamma_i = 0$ for $i \neq j$. Therefore the $\mathbf{C}$-linear mapping $\sigma : \mathcal{M}^{\mathbf{C}} \to M_4(\mathbf{C})$ induces a homomorphism of algebras $\tau : C(\mathcal{M}^{\mathbf{C}}) \to M_4(\mathbf{C})$ for which $\tau \cdot \rho(e_i) = \gamma_i$. By direct calculation it can be shown that the mapping τ is surjective, and since both $\mathbf{C}$-algebras $C(\mathcal{M}^{\mathbf{C}})$ and $M_4(\mathbf{C})$ are 16-dimensional algebras, τ is an isomorphism.

CHAPTER 3

Affine and Projective Geometry

§1. Affine Spaces, Affine Mappings, and Affine Coordinates

1.1. Definition. *An affine space over a field $\mathcal{K}$ is a triple $(A, L, +)$, consisting of a linear space L over a field $\mathcal{K}$, a set A whose elements are called points, and an external binary operation $A \times L \to A : (a, l) \mapsto a + l$, satisfying the following axioms:*

a) $(a + l) + m = a + (l + m)$ *for all* $a \in A$; $l, m \in L$;

b) $a + 0 = a$ *for all* $a \in A$;

c) *for any two points* $a, b \in A$ *there exists a unique vector* $l \in L$ *with the property* $b = a + l$.

1.2. Example. The triple $(L, L, +)$, where L is a linear space and $+$ coincides with addition in L, is an affine space. It is convenient to say that it defines the affine structure of the linear space L. This example is typical; later we shall see that any affine space is isomorphic to this space.

1.3. Terminology. We shall often call the pair (A, L) or even simply A an affine space, omitting the operation $+$. The linear space L is said to be associated with the affine space A. The mapping $A \to A : a \mapsto a + l$ is called a translation by the vector l; it is convenient to denote it by the special notation t_l. We shall write $a - l$ instead of $t_{-l}(a)$ or $a + (-l)$.

1.4. Proposition. *The mapping $l \mapsto t_l$ defines an injective homomorphism of the additive group of the space L into the group of permutations of the points of the affine space A, that is, the effective action of L on A. This action is transitive, that is, for any pair of points $a, b \in A$ there exists an $l \in L$ such that $t_l(a) = b$.*

Conversely, the specification of a transitive effective action of the additive group of L on the set A determines on A the structure of an affine space with associated space L.

Proof. It follows from axioms a) and b) that for any $l \in L$ and $a \in A$ the equation $t_l(x) = a$ has the solution $x = a + (-l)$, so that all t_l are surjective. If $t_l(a) = t_l(b)$, then having found by axiom c) a vector $m \in L$ such that $b = a + m$, we obtain

$a + l = (a + m) + l = (a + l) + m$. But $a + l = (a + l) + 0$, and therefore from the uniqueness condition in axiom c) it follows that $m = 0$, so that $a = b$. Therefore all the t_l are injective.

Axiom a) means that $t_m \circ t_l = t_{l+m}$ and axiom b) means that $t_0 = \mathrm{id}_A$. Therefore the mapping $l \mapsto t_l$ is a homomorphism of the additive group L into the group of bijections of A with itself. Axioms b) and c) imply that its kernel equals zero.

Conversely, let $L \times A \to A : (l, a) \mapsto a + l$ be the effective transitive action of L on A. Axioms a) and b) are then obtained directly from the definition of the action and axiom c) is obtained by combining the properties of effectiveness and transitivity.

1.5. Remark. With regard to the action of groups (not necessarily abelian) on sets, see §7.2 of Introduction to Algebra. The set on which the group acts transitively and effectively is called the *principal homogeneous space* over this group.

We note that the axioms of an affine space do not explicitly include the structure of multiplication by scalars in L. It appears only in the definition of affine mappings and then in barycentric combinations of the points of A. But, we shall first say a few words about the formalism.

1.6. Computational rules. It is convenient to denote the unit vector $l \in L$ for which $b = a + l$ by $b - a$. This operation of "exterior subtraction" $A \times A \to L : (b, a) \mapsto b - a$ has the following properties.

a) $(c - b) + (b - a) = c - a$ for all $a, b, c \in A$; the addition on the left is addition in L.

Indeed, let $c = b + l$, $b = a + m$; then $c = a + (l + m)$, so that $c - a = l + m = (c - b) + (b - a)$.

b) $a - a = 0$ for all $a \in A$.

c) $(a + l) - (b + m) = (a - b) + (l - m)$ for all $a, b \in A$, $l, m \in L$.

Indeed, it is sufficient to verify that $(b + m) + (a - b) + (l - m) = a + l$, or $b + (a - b) = a$, and this is the definition of $a - b$.

Generally the use of the signs $\pm$ for different operations $L \times L \to L$, $A \times L \to L$, $A \times A \to L$ obeys the following formal rules. The expression $\pm a_1 \pm a_2 \pm \ldots \pm a_m + l_1 + \ldots + l_n$ for $a_j \in A$, $l_k \in L$ makes sense only if m is either even and all $\pm a_j$ can be combined into pairs of the form $a_i - a_j$, or it is odd and all points can be combined into such pairs except for one, which enters with the $+$ sign. In the first case, the entire sum lies in L and in the second case it lies in A. In addition, it is commutative and associative. For example, $a_1 - a_2 + l$ can be calculated as $(a_1 - a_2) + l$ or $(a_1 + l) - a_2$ or $a_1 - (a_2 - l)$; we shall write $a + l$ as well as $l + a$.

We shall not prove this rule in general form. Whenever it is used, the reader will be able to separate without difficulty the required calculation into a series of elementary steps, each of which will reduce to the application of one of the axioms, or the formulas a)–c) at the beginning of this subsection.

We note that the *sum* $a + b$, where $a, b \in A$, like the *expression* xa, where $x \in \mathcal{K}$ is, generally speaking, *meaningless* (exception: $A = L$). Nevertheless, in what follows we shall give a unique meaning, for example, to the expression $\frac{1}{2}a + \frac{1}{2}b$ (but not to its terms !).

Intuitively, the affine space $(A, L, +)$ must be imagined to be a linear space L whose origin of coordinates 0 is "forgotten". Only the operation of translation by vectors in L, summation of translations, and multiplication of the translation vector by a scalar are retained.

1.7. Affine mappings. Let $(A_1, L_1), (A_2, L_2)$ be two affine spaces over the same field $\mathcal{K}$. The pair (f, Df), where $f : A_1 \to A_2$, $Df : L_1 \to L_2$ satisfying the following conditions:

a) Df is a linear mapping and

b) for any $a_1, a_2 \in A$

$$f(a_1) - f(a_2) = Df(a_1 - a_2).$$

(both expressions lie in L_2), is called an *affine linear* or simply an *affine mapping* of the first space into the second space.

Df (or $D(f)$) is the linear part of the affine mapping f. Since $a_1 - a_2$ runs through all vectors in L_1, when $a_1, a_2 \in A_1$, the linear part of Df is defined with respect to f uniquely. This makes it possible to denote affine mappings simply as $f : A_1 \to A_2$.

1.8. Examples. a) Any linear mapping $f : L_1 \to L_2$ induces an affine mapping of the spaces $(L_1, L_1, +) \to (L_2, L_2, +)$. For it, $Df = f$.

b) Any translation $t_l : A \to A$ is affine, and $D(t_1) = \mathrm{id}_L$. Indeed,

$$t_l(a_1) - t_l(a_2) = (a_1 + l) - (a_2 + l) = a_1 - a_2.$$

c) If $f : A_1 \to A_2$ is an affine mapping and $l \in L_2$ then the mapping $t_l \circ f : A_1 \to$ $\to A_2$ is affine and $D(t_l \circ f) = D(f)$. Indeed,

$$t_l \circ f(a_1) - t_l \circ f(a_2) = (f(a_1) + l) - (f(a_2) + l) = f(a_1) - f(a_2) = Df(a_1 - a_2).$$

d) *An affine function* $f : A \to \mathcal{K}$ is defined as an affine mapping of A into $(\mathcal{K}^1, \mathcal{K}^1, +)$, where $\mathcal{K}^1$ is a one-dimensional coordinate space. Thus f assumes values in $\mathcal{K}$, while Df is a linear functional on L. Any constant function f is affine: $Df = 0$.

1.9. Theorem. a) *Affine spaces together with affine mappings form a category.*

b) *A mapping that associates with an affine* (A, L) *a linear space* L *and with an affine mapping* $f : (A_1, L_1) \to (A_2, L_2)$ *the linear mapping* $Df : L_1 \to L_2$, *is a functor from the category of affine spaces into the category of linear spaces.*

Proof. The validity of the general category axioms (see §13 of Chapter I) follows from the following facts.

The identity mapping $\mathrm{id}: A \to A$ is an affine mapping. Indeed, $a_1 - a_2 = \mathrm{id}_L(a_1 - a_2)$. In particular, $D(\mathrm{id}_A) = \mathrm{id}_L$.

A composition of affine mappings $A \xrightarrow{f_1} A \xrightarrow{f_2} A$ is an affine mapping. Indeed, let $a, b \in A$. Then $f_1(a) - f_1(b) = Df_1(a - b)$ and, further

$$f_2 f_1(a) - f_2 f_1(b) = Df_2[f_1(a) - f_1(b)] = Df_2 \circ Df_1(a - b).$$

We have proved the required result as well as the fact that $D(f_2 f_1) = Df_2 \circ Df_1$. Together with the formula $D(\mathrm{id}_A) = \mathrm{id}_L$ this proves the assertion b) of the theorem.

The following important result characterizes isomorphisms in our category.

1.10. Proposition. *The following three properties of affine mappings* $f : A_1 \to A_2$ *are equivalent:*

a) f *is an isomorphism,*

b) Df *is an isomorphism,*

c) f *is a bijection in the set-theoretic sense.*

Proof. According to the general categorical definition, $f : A_1 \to A_2$ is an isomorphism if and only if there exists an affine mapping $g : A_2 \to A_1$ such that $gf = \mathrm{id}_{A_1}$, $fg = \mathrm{id}_{A_2}$. If it does exist, then $D(fg) = \mathrm{id}_{L_2} = D(f)D(g)$ and $D(gf) = \mathrm{id}_{L_1} = D(g)D(f)$, whence it follows that $D(f)$ is an isomorphism.

We shall now show that Df is an isomorphism if and only if f is a bijection. We fix a point $a_1 \in A_1$ and set $a_2 = f(a_1)$. Any element A_i can be uniquely represented in the form $a_i + l_i$, $l_i \in L_i$, $i = 1, 2$. From the basic identity

$$f(a_1 + l_1) - f(a_1) = Df[(a_1 + l_1) - a_1] = Df(l_1)$$

it follows that $f(a_1 + l_1) = a_2 + Df(l_1)$. Therefore, f is a bijection if and only if $Df(l_1)$ as $l_1 \in L_1$ runs through all elements of L_2 once, that is, Df is a bijection. But a linear mapping is a bijection if and only if it is invertible, that is, it is an isomorphism.

Finally we shall show that a bijective affine mapping is an affine isomorphism. For this, we must verify that the set-theoretic mapping inverse to f is affine. But in the notation of the preceding item, this mapping is defined by the formula

$$f^{-1}(a_2 + l_2) = a_1 + (Df)^{-1}(l_2), \quad l_2 \in L_2.$$

Therefore

$$f^{-1}(a_2+l_2)-f^{-1}(a_2+l_2')=(Df)^{-1}(l_2)-(Df)^{-1}(l_2')=(Df)^{-1}(l_2-l_2')$$

because of the linearity of $(Df)^{-1}$. Thus f^{-1} is affine and $D(f^{-1})=D(f)^{-1}$.

Finally, we have established the implications a) $\Rightarrow$ b) $\Leftrightarrow$ c) $\Rightarrow$ a), whence follows the proposition.

The construction of specific affine mappings is often based on the following result.

1.11. Proposition. *Let (A_1,L_1), (A_2,L_2) be two affine spaces. For any pair of points $a_1\in A_1$, $a_2\in A_2$ and any linear mapping $g:L_1\to L_2$ there exists a unique affine mapping $f:A_1\to A_2$ such that $f(a_1)=a_2$ and $Df=g$.*

Indeed, we set

$$f(a_1+l_1)=a_2+g(l_1)$$

for $l_1\in L_1$. Since any point in A_1 can be uniquely represented in the form a_1+l_1, this formula defines a set-theoretic mapping $f:A_1\to A_2$. It is affine, $f(a_1)=a_2$ and $Df=g$ because

$$f(a_1+l_1)-f(a_1+l_1')=g(l_1)-g(l_1')=g(l_1-l_1')=$$

$$=g[(a_1+l_1)-(a_1+l_1')].$$

This proves the existence of f. Conversely, if f is a mapping with the required properties, then

$$f(a_1+l)-f(a_1)=g(l),$$

whence $f(a_1+l)=a_2+g(l)$ for all $l\in L$.

1.12. An important particular case of Proposition 1.11 is obtained by applying it to (A,L), (L,L), $a\in A$, $0\in L$ and $g=\mathrm{id}_L$. We find that *for any point $a\in A$ there exists a unique affine isomorphism $f:A\to L$ that transforms this point into the origin of coordinates and has the same linear part.* This is the precise meaning of the statement that an affine space is a "linear space whose origin of coordinates is forgotten".

In particular, affine spaces are isomorphic if and only if the associated linear spaces are isomorphic. The latter are classified by their dimension, and we can call the dimension of an affine space the dimension of the corresponding linear space.

1.13. Corollary. *Let $f_1,f_2:A_1\to A_2$ be two affine mappings. Then their linear parts are equal if and only if f_2 is the composition of f_1 with a translation by some unique vector from L_2.*

Proof. The sufficiency of the condition was checked in Example 1.8c. To prove necessity we select any point $a \in A_1$ and set $f_2' = t_{f_2(a)-f_1(a)} \circ f_1$. Evidently, $f_2'(a) = f_2(a)$ and $D(f_2') = D(f_2)$. By Proposition 1.11, $f_2' = f_2$. Conversely, if $f_2 = t_l \circ f_1$, then $l = f_2(a) - f_1(a)$; this vector is independent of $a \in A$ because f_1 and f_2 have the same linear parts.

1.14. Affine coordinates. a) *A system of affine coordinates* in an affine space (A, L) is a pair consisting of points $a_0 \in A$ (origin of coordinates) and a basis $\{e_1, \ldots, e_n\}$ of the associated linear space L. The *coordinates* of the point $a \in A$ in this system form the vector $(x_1, \ldots, x_n) \in \mathcal{K}^n$, uniquely defined by the condition $a = a_0 + \sum_{i=1}^n x_i e_i$.

In other words, we identify A with L with the help of a mapping which has the same linear part and transforms a_0 into 0, and takes the coordinates of the image of the point a in the basis $\{e_1, \ldots, e_n\}$: this will be $x_1, \ldots, x_n$.

Choose a system of coordinates in the spaces A_1 and A_2 such that they identify the spaces with $\mathcal{K}^m$, $\mathcal{K}^n$ respectively. Then any affine mapping $f : A_1 \to A_2$ can be written in the form

$$f(\vec{x}) = B\vec{x} + \vec{y},$$

where B is the matrix of the mapping Df in the corresponding bases of L_1 and L_2, while $\vec{y}$ are the coordinates of the vector $f(a_0') - a_0''$ in the basis L_2; a_0' is the origin of coordinates in A_1 and a_0'' is the origin of coordinates in A_2. Indeed, the mapping $\vec{x} \mapsto B\vec{x} + \vec{y}$ is an affine mapping, transforms a_0' into $f(a_0')$ and has the same linear part as f.

b) Another variant of this definition of a coordinate system consists in replacing the vectors $\{e_1, \ldots, e_n\}$ by the points $\{a_0 + e_1, \ldots, a_0 + e_n\}$ in A. We set $a_i = a_0 + e_i$, $i = 1, \ldots, n$. The coordinates of the point $a \in A$ are then found from the representation $a = a_0 + \sum_{i=1}^n x_i(a_i - a_0)$. One is tempted to "reduce similar terms" and to write the expression on the right in the form $(1 - \sum_{i=1}^n x_i) a_0 + \sum_{i=1}^n x_i a_i$. The individual terms in this sum are meaningless ! It turns out that sums of this form can nonetheless be studied, and they are very useful.

1.15. Proposition. *Let $a_0, \ldots, a_s$ be any points in an affine space A. For any $y_0, \ldots, y_s \in \mathcal{K}$ with the condition $\sum_{i=0}^s y_i = 1$, we define the formal sum $\sum_{i=0}^s y_i a_i$ by an expression of the form*

$$\sum_{i=0}^{s} y_i a_i = a + \sum_{i=0}^{s} y_i(a_i - a),$$

where a is any point in A. It is asserted that the expression on the right does not depend on a. Therefore the point $\sum_{i=0}^s y_i a_i$ is correctly defined. It is called the barycentric combination of the points $a_0, \ldots, a_s$ with coefficients $y_0, \ldots, y_s$.

Proof. Replace the point a by the point $a+l$, $l \in L$. We obtain

$$a+l+\sum_{i=0}^{s} y_i(a_i-a-l)=a+\sum_{i=0}^{s} y_i(a_i-a),$$

because $(1-\sum_{i=0}^{s} y_i)l=0$. We used here the rules formulated in §1.6. It will be instructive for the reader to carry out this calculation in detail.

1.16. Corollary. *The system $\{a_0; a_1-a_0,\dots,a_n-a_0\}$, consisting of the points $a_0 \in A$ and the vectors a_i-a_0 in L, forms a system of affine coordinates in A if and only if any point in A can be uniquely represented in the form of a barycentric combination $\sum_{i=0}^{n} x_i a_i$, $x_i \in \mathcal{K}$, $\sum_{i=0}^{n} x_i=1$.*

When this condition is satisfied, the system of points $\{a_0,\dots,a_n\}$ is called *a barycentric system of coordinates* in A, and the numbers $x_0,\dots,x_n$ are the *barycentric coordinates* of the point $\sum_{i=0}^{n} x_i a_i$.

Proof. Everything follows directly from the definitions, if $\sum_{i=0}^{n} x_i a_i$ is calculated from the formula $a_0+\sum_{i=1}^{n} x_i(a_i-a_0)$. Indeed, since any point in A can be uniquely represented in the form a_0+l, $l \in L$, the system $\{a_0, a_1-a_0,\dots,a_n-a_0\}$ is an affine coordinate system in A if and only if any vector $l \in L$ can be uniquely represented as a linear combination $\sum_{i=1}^{n} x_i(a_i-a_0)$, that is, if $\{a_1-a_0,\dots,a_n-a_0\}$ form a basis of L. The barycentric coordinates of the point a_0+1 are reconstructed uniquely from the coordinates $x_1,\dots,x_n$ of the vector l in the form $1-\sum_{i=1}^{n} x_i = x_0, x_1,\dots,x_n$.

1.17. Barycentric combinations can in many ways be treated like ordinary linear combinations in a linear space. For example, terms with zero coefficients can be dropped. A more useful remark is that a barycentric combination of several barycentric combinations of points $a_0,\dots,a_s$ is in its turn a barycentric combination of these points, whose coefficients can be calculated from the expected formal rule:

$$x_1\sum_{i=0}^{s} y_{i1}a_i+x_2\sum_{i=0}^{s} y_{i2}a_i+\dots+x_m\sum_{i=0}^{s} y_{im}a_i=\sum_{i=0}^{s}\left(\sum_{k=1}^{m} x_k y_{ik}\right)a_i.$$

Indeed,

$$\sum_{i=0}^{s}\sum_{k=1}^{m} x_k y_{ik}=\sum_{k=1}^{m} x_k\sum_{i=0}^{s} y_{ik}=\sum_{k=1}^{m} x_k=1,$$

so that the latter combination is barycentric. Calculating the left and right sides of this equality according to the rule formulated in Proposition 1.15, with the help of the same point $a \in A$ and applying the formalism in §1.6, we easily find that they are equal.

Finally, under affine mappings barycentric combinations behave like linear combinations.

1.18. Proposition. a) *Let $f : A_1 \to A_2$ be an affine mapping, $a_0, \ldots, a_s \in A_1$. Then*

$$f\left(\sum_{i=0}^{s} x_i a_i\right) = \sum_{i=0}^{s} x_i f(a_i),$$

if $\sum_{i=0}^{s} x_i = 1$.

b) *Let $a_0, \ldots, a_n$ define a barycentric coordinate system in A_1. Then for any points $b_0, \ldots, b_n \in A_2$ there exists a unique affine mapping f transforming a_i into b_i, $i = 1, \ldots, n$.*

Proof. Choosing $a \in A_1$, we obtain

$$f\left(\sum_{i=0}^{s} x_i a_i\right) = f\left(a + \sum_{i=0}^{s} x_i(a_i - a)\right) = f(a) + Df\left(\sum_{i=0}^{s} x_i(a_i - a)\right) =$$

$$= f(a) + \sum_{i=0}^{s} x_i Df(a_i - a) = f(a) + \sum_{i=0}^{s} x_i(f(a_i) - f(a)) = \sum_{i=0}^{s} x_i f(a_i)$$

according to Proposition 1.15, which proves assertion a).

If $a_0, \ldots, a_n$ form a barycentric coordinate system in A_1, then by Corollary 1.16 any point in A can be represented by a unique barycentric combination $\sum_{i=0}^{n} x_i a_i$. We then define a set-theoretic mapping $f : A_1 \to A_2$ by the formula $f\left(\sum_{i=0}^{n} x_i a_i\right) = \sum_{i=0}^{n} x_i b_i$. Part a) implies that this is the only possible definition, and we have only to check that f is an affine mapping. Indeed, calculating as in Proposition 1.15, we obtain

$$f\left(\sum_{i=0}^{n} x_i a_i\right) - f\left(\sum_{i=0}^{n} y_i a_i\right) = \sum_{i=0}^{n} x_i b_i - \sum_{i=0}^{n} y_i b_i = b_0 + \sum_{i=0}^{n} x_i(b_i - b_0) -$$

$$- \left[b_0 + \sum_{i=0}^{n} y_i(b_i - b_0)\right] = \sum_{i=0}^{n}(x_i - y_i)(b_i - b_0) = Df\left(\sum_{i=0}^{n} x_i a_i - \sum_{i=0}^{n} y_i a_i\right),$$

where $Df : L_1 \to L_2$ is a linear mapping, transforming $a_i - a_0$ into $b_i - b_0$ for all $i = 1, \ldots, n$. It exists, because $a_1 - a_0, \ldots, a_n - a_0$ by definition form a basis of L_1.

1.19. Remark. In the affine space $\mathbf{R}^n$ the barycentric combination $\sum_{i=1}^{m} \frac{1}{m} a_i$ represents the position of the "centre of gravity" of a system of individual masses positioned at the points a_i. This explains the terminology. If $a_i = (0, \ldots, 1, \ldots, 0)$ (the one is in the ith place), then the set of points with barycentric coordinates $x_1, \ldots, x_n$, $0 \le x_i \le 1$ comprises the intersection of the linear manifold $\sum_{i=1}^{n} x_i = 1$ with the positive octant (more precisely, a "2^n-tant"). In topology this set is called the *standard* $(n-1)$*-dimensional simplex*. A one-dimensional simplex is a segment of a straight line; a two-dimensional simplex is a triangle; a three-dimensional simplex is a tetrahedron. In general, the set

$$\left\{\sum_{i=1}^{n} x_i a_i \middle| \sum_{i=1}^{n} x_i = 1,\ 0 \le x_i \le 1\right\}$$

is a *closed simplex* with vertices $a_1, \ldots, a_n$ in a real affine space. They are said to be degenerate if the vectors $a_2 - a_1, \ldots, a_n - a_1$ are linearly dependent.

§2. Affine Groups

2.1. Let A be an affine space over the field $\mathcal{K}$. Proposition 1.10 implies that the set of affine bijective mappings $f : A \to A$ forms a group, which we shall call the affine group and denote it by $\mathrm{Aff}\, A$.

Its mapping $D : \mathrm{Aff}\, A \to GL(L)$, where $GL(L)$ is the group of linear automorphisms of the associated vector space, is a homormorphism. Definition 1.11 implies that it is surjective, and according to Corollary 1.13 its kernel is the group of translations $\{t_l | l \in L\}$. Proposition 1.4 implies that this group of translations is isomorphic to the additive group of the space L. Thus $\mathrm{Aff}\, A$ is the extension of the group $GL(L)$ with the help of the additive group L, which is a normal subgroup in $\mathrm{Aff}\, A$.

This extension is a semidirect product of $GL(L)$ and L. To verify this we choose any point $a \in A$ and examine the subgroup $G_a \subset \mathrm{Aff}\, A$, consisting of mappings that leave a unchanged. By Definition 1.11 every element $f \in G_a$ is uniquely determined by its linear part Df, and Df can be selected arbitrarily. Therefore, D induces an isomorphism of G_a to $GL(L)$. For any mapping $f \in \mathrm{Aff}\, A$ there exists a unique mapping $f_a \in G_a$ with the same linear part, and $f = t_l \circ f_a$ for an appropriate $l \in L$ by Corollary 1.13. Having fixed a, we shall write $t_l \circ f_a$ as a pair $[g; l]$, where $g = Df = Df_a \in GL(L)$. The rules of multiplication in the group $\mathrm{Aff}\, A$ in terms of such pairs assume the following form.

2.2. Proposition. *We have*

$$[g_1; l_1][g_2; l_2] = [g_1 g_2; g_1(l_2) + l_1],$$

$$[g; l]^{-1} = [g^{-1}; -g^{-1}(l)].$$

Proof. According to the definitions, $[g; l]$ transforms the point $a + m \in A$ into $a + g(m) + l$, whence

$$[g_1; l_1][g_2; l_2](a + m) = [g_1; l_1](a + g_1(m) + l_1) =$$

$$= a + g_1(g_2(m) + l_2) + l_1 = a + g_1g_2(m) + g_1(l_2) + l_1 =$$
$$= [g_1g_2; g_1(l_2) + l_1](a + m),$$

which proves the first formula. By means of this,the product $[g; l][g^{-1}; -g^{-1}(l)]$ can be calculated. We obtain $[\mathrm{id}_L; 0]$, and this pair represents the identity element of $\mathrm{Aff}\,A$. This completes the proof of the proposition and shows that $\mathrm{Aff}\,A$ is a semidirect product.

2.3. Now let $G \subset GL(L)$ be some subgroup. The set of all elements $f \in \mathrm{Aff}\,A$ whose linear parts belong to G, obviously forms a subgroup of $\mathrm{Aff}\,A$ — the inverse image of G with respect to the canonical homomorphism $\mathrm{Aff}\,A \to GL(L)$. We shall call it the *affine extension of the group G.*

The case when the linear space associated with A is equipped with an additional structure, an inner product, and G represents the corresponding group of isometries is especially important. Two groups of importance in applications are constructed in this manner: the *group of motions* of the affine Euclidean space $G = O(n)$) and the *Poincaré group* (Minkowski space L and the Lorentz group G). We shall study this group of motions in greater detail.

2.4. Definition. a) *An affine Euclidean space is a pair, consisting of an affine finite-dimensional space A over the field of real numbers and a metric d (in the sense of Definition 10.1 of Chapter I) with the following property: for any points $a, b \in A$ the distance $d(a, b)$ depends only on $a - b \in L$ and equals the length of the vector $a - b$ in an appropriate Euclidean metric of the space L (independent of a and b).*

b) *A motion of an affine Euclidean space A is an arbitrary distance-preserving mapping $f : A \to A$: $d(f(a), f(b)) = d(a, b)$ for all $a, b \in A$.*

2.5. Theorem. *The motions of an affine Euclidean space A form a group, which coincides with the affine extension of the group of orthogonal isometries $O(L)$ of the Euclidean space L associated with A.*

Proof. We first verify that any affine mapping $f : A \to A$ with $Df \in O(L)$ is a motion. Indeed, by definition

$$d(f(a), f(b)) = |f(a) - f(b)| = |Df(a - b)| = |a - b| = d(a, b);$$

in the third equality we made use of the fact that $Df \in O(L)$.

The main problem is to prove the converse.

First of all, it is obvious that a composition of motions is a motion. Further, we have already established that translations are motions. Let $a \in A$ be an arbitrary fixed point and let f be a motion. We set $g = t_{a-f(a)} \circ f$. This is a motion that does not displace the point a. It is sufficient to prove that it is affine and that $Dg \in O(L)$. We identify A with L, as in §1.12, with the help of a mapping with the same linear part and transforming a into $0 \in L$. Then g will transform into the mapping $g : L \to L$ with the properties $g(0) = 0$ and $|g(l) - g(m)| = |l - m|$ for all $l, m \in L$, and it is sufficient to establish that such a mapping lies in $O(L)$.

We first verify that g preserves inner products. Indeed, for any $l, m \in L$

$$|l|^2 - 2(l,m) + |m|^2 = |l-m|^2 = |g(l) - g(m)|^2 =$$

$$= |g(l)|^2 - 2(g(l), g(m)) + |g(m)|^2,$$

whence follows the required result, because $|g(l)|^2 = |l|^2$, $|g(m)|^2 = |m|^2$. We now show that g is additive: $g(l+m) = g(l) + g(m)$. Setting $l + m = n$ and making use of the preceding property, we have

$$0 = |n - l - m|^2 = |n|^2 + |l|^2 + |m|^2 - 2(n,l) - 2(n,m) + 2(l,m) =$$

$$= |g(n)|^2 + |g(l)|^2 + |g(m)|^2 - 2(g(n), g(l)) - 2(g(n), g(m)) +$$

$$+ 2(g(l), g(m)) = |g(n) - g(l) - g(m)|^2,$$

whence $g(n) = g(l) + g(m)$.

Finally, we show that $g(xl) = xg(l)$ for all $x \in \mathbf{R}$, $l \in L$. Setting $m = xl$, we have

$$0 = |m - xl|^2 = |m|^2 - 2x(m,l) + x^2|l|^2 =$$

$$= |g(m)|^2 - 2x(g(m), g(l)) + x^2|g(l)|^2 = |g(m) - xg(l)|^2.$$

Thus g is a linear mapping that preserves inner products, that is, $g \in O(L)$. The theorem is proved.

2.6. Theorem. *Let $f : A \to A$ be a motion in a Euclidean affine space with a linear part Df. Then there exists a vector $l \in L$ such that $Df(l) = l$ and $f = t_l \circ g$, where $g : A \to A$ is a motion with a fixed point $a \in A$.*

Proof. First we shall clarify the geometric meaning of this assertion. Identifying A with L by means of an affine mapping with an identical linear part and which maps a into zero, we find that f is a composition of the orthogonal transformation g and a translation by a vector l, which is fixed with respect to g (because $Df = Dg$). In other words, this is a "screw motion" if $\det g = 1$ or a screw motion combined with a reflection if $\det g = -1$. Indeed, g is completely determined by its restriction g_0 to $l^\perp$: $g = g_0 \oplus \mathrm{id}_{\mathbf{R}l}$, so that g is a rotation around the axis $\mathbf{R}l$ (possibly with a reflection).

We now proceed with the proof. Let $L_2 = \ker(Df - \mathrm{id}_L)$, $L_1 = L_2^\perp$. We have $L = L_1 \oplus L_2$; L_2 consists of Df-invariant vectors, the space L_1 is invariant with respect to $Df - \mathrm{id}_L$ (because Df is orthogonal), and the restriction of $Df - \mathrm{id}_L$ to L_1 is invertible.

We first select an arbitrary point $a' \in A$ and set $g' = t_{a'-f(a')} \circ f$. Evidently, $g'(a') = a'$. We set $f(a') - a' = l_1 + l_2$ where $l_1 \in L_1$, $l_2 \in L_2$. Then $f = t_{l_2} \circ t_{l_1} \circ g'$ and $Df(l_2) = l_2$ by definition. We shall show that $g = t_{l_1} \circ g'$ has a fixed point $a = a' + m$ for some $m \in L_1$. We have

$$t_{l_1} \circ g'(a' + m) = g'(a' + m) + l_1 = a' + Df(m) + l_1.$$

The right side equals $a' + m$ if and only if $(Df - \mathrm{id}_L)m + l_1 = 0$. But, as we have already noted, in L_1 the operator $Df - \mathrm{id}_L$ is invertible and $l_1 \in L_1$. Therefore, m exists. We have obtained the required decomposition $f = t_{l_2} \circ g$ and we have completed the proof.

Motions f with the property $\det Df = 1$ are sometimes called *proper motions*, and other motions (with $\det Df = -1$) are called improper motions. We shall present more graphic information about the motions of affine Euclidean spaces with dimension $n \leq 3$, contained in Theorem 2.6. In the next section the notation of this theorem is retained.

2.7. Examples. a) $n = 1$. Since $O(1) = \{\pm 1\}$, the proper motions consist only of translations. If f is improper, then $Df = -1$, and from $Df(l) = l$ it follows that $l = 0$. Therefore, any improper motion of a straight line has a fixed point and consequently is a reflection relative to this point.

b) $n = 2$. The proper motion f with $Df = \mathrm{id}$ is a translation. If $Df \neq \mathrm{id}$ and $\det Df = 1$, then Df, being a rotation, does not have fixed vectors, so that once again $l = 0$ and f has a fixed point, relative to which f is a rotation.

If f is an improper motion, then Df is a reflection of the plane relative to a straight line, and f is a combination of such a reflection and a translation along this straight line. This means that if the improper motion of a plane has a fixed point, then it has an entire straight line of fixed points and represents a reflection relative to the straight line.

c) $n = 3$. If $\det Df = 1$, then Df always has an eigenvalue equal to one and a fixed vector. Therefore, all proper motions of a three-dimensional Euclidean space are screw motions along some axis (including translations, that is, degenerate screw motions with zero rotation). This is the so-called Chasles' theorem.

If the motion $f = t_l g$ is improper and $l \neq 0$, then the restriction of g to the plane orthogonal to l and passing through the fixed point a is an improper motion of this plane. Therefore, it is a reflection with respect to a straight line in this plane. We denote by P the plane spanned by l and this straight line. Then $t_l g$ is a combination of a reflection with respect to the plane P and a translation by a vector l lying in P.

Finally, if $l = 0$, that is, f is an improper motion and has a fixed point, then, identifying it with zero in L and f with Df and making use of the fact that f has a characteristic straight line L_0 with an eigenvalue equal to -1, we obtain the geometric description of f as a composition of a rotation in $L_0^\perp$ and a reflection with respect to $L_0^\perp$.

Using the polar decomposition of linear operators, we can also analyse the geometric structure of any invertible affine transformation of a Euclidean affine space.

2.8. Theorem. *Any affine transformation of an n-dimensional Euclidean space f can be represented in the form of a composition of three mappings: a) n dilations (with positive coefficients) along n pairwise orthogonal axes, passing through some point $a_0 \in A$; b) motions leaving the point a_0 fixed; and c) translations.*

Proof. Replacing f by its composition with an appropriate translation, as in the proof of Theorem 2.5, we can assume that f already has a fixed point a_0. Identifying A with L and a_0 with 0, we can decompose $f = Df$ into a composition of a positive-definite symmetric operator and an orthogonal operator. Reducing the first transformation to principal axes and transferring these axes into A, we obtain the required result.

2.9. In conclusion, we note that in this section we made extensive use of linear varieties in A (straight lines, planes), defining them constructively as the inverse images of linear spaces in L with different identifications of A with L, depending on the choice of the origin of coordinates. In the next section these concepts are studied more systematically.

§3. Affine Subspaces

3.1. Definition. *Let (A, L) be some affine space. The subset $B \subset A$ is called an affine subspace in A, if it is empty or if the set*

$$M = \{b_1 - b_2 \in L | b_1, b_2 \in B\} \subset L$$

is a linear subspace in L and $t_m(B) \subset B$ for all $m \in M$.

3.2. Remarks. a) If the requirements of the definition are satisfied and B is not empty, then the pair (B, M) forms an affine space, which justifies the terminology (it is presumed that the translation of B by a vector from M is obtained by restricting the same translation to B in all of A). Indeed, an examination of the conditions of Definition 1.1 immediately shows that they are satisfied for (B, M). In particular, choosing any point $b \in B$, we obtain $B = \{b + m | m \in M\}$.

b) We shall call the linear subspace $M = \{b_1 - b_2 | b_1, b_2 \in B\}$ the *orienting* subspace for the affine subspace B. The dimension of B equals the dimension of M. Evidently, it follows from $B_1 \subset B_2$ that $M_1 \subset M_2$ and therefore, $\dim B_1 \leq \dim B_2$. Two affine subspaces with the same dimension with a common orienting space are said to be *parallel*.

3.3. Proposition. *Affine subspaces with the same dimension $B_1, B_2 \subset A$ are parallel if and only if there exists a vector $l \in L$, such that $B_2 = t_l(B_1)$. Any two vectors with this property differ by a vector from the orienting space for B_1 and B_2.*

Proof. If $B_2 = t_l(B_1)$ and M_2, M_1 are the orienting subspaces of B_2 and B_1 respectively, then

$$M_2 = \{a - b | a, b \in B_2\} = \{(a' + l) - (b' + l) | a', b' \in B_1\} = M_1,$$

so that B_1 and B_2 are parallel.

Conversely, let M be the common orienting subspace for B_1 and B_2. Choose the points $b_1 \in B$ and $b_2 \in B_2$. We have $B_1 = \{b_1 + l | l \in M\}$, $B_2 = \{b_2 + l | l \in M\}$, whence $B_2 = t_{b_2 - b_1}(B_1)$. Finally, it is easy to see that $t_{l_1}(B_1) = t_{l_2}(B_2)$ if and only if $l_1 - l_2 \in M$.

3.4. Corollary. *Affine spaces in L (with an affine structure) are linear subvarieties of L in the sense of Definition 6.1 of Chapter 1, that is, translations of linear subspaces.*

3.5. Corollary. *Parallel affine subspaces with the same dimension either do not intersect or they coincide.*

Proof. If $b \in B_1 \cap B_2$, then by the above, $B_1 = \{b + m | m \in M\} = B_2$ where M is the common orienting subspace of B_1 and B_2.

3.6. Two affine subspaces B_1 and B_2 with not necessarily the same dimensions are said to be *parallel* if one of their orienting subspaces is contained in the other. Slightly modifying the preceding proofs, it is easy to prove the following facts. Let B_1 and B_2 be parallel and let $\dim B_1 \leq \dim B_2$. Then there exists a vector $l \in L$ such that $t_l(B_1) \subset B_2$ and two vectors with this property differ by an element from M_1. In addition, either B_1 and B_2 do not intersect or B_1 is contained in B_2.

3.7. Proposition. *Let (B_1, M_1), (B_2, M_2) be two affine subspaces in A. Then $B_1 \cap B_2$ is either empty or an affine subspace with the orienting subspace $M_1 \cap M_2$.*

Proof. Let $B_1 \cap B_2$ be non-empty and let $b \in B_1 \cap B_2$. Then $B_1 = \{b + l_1 | l \in M_1\}$, $B_2 = \{b + l_2 | l \in M_2\}$, whence $B_1 \cap B_2 = \{b + l | l \in M_1 \cap M_2\}$ which proves the required result. (Corollary 3.5 evidently follows from here.)

3.8. Affine spans. Let $S \subset A$ be a set of points in an affine space A. The smallest affine subspace containing S is called the *affine span* of S. It exists and coincides with the intersection of all affine subspaces containing S. We can describe an affine span in terms of barycentric linear combinations (Proposition 1.11).

3.9. Proposition. *The affine span of a set S equals the set of barycentric combinations of elements from S:*

$$\tilde{S} = \left\{ \sum_{i-1}^{n} x_i s_i \middle| \sum_{i=1}^{n} x_i = 1 \right\},$$

where $\{s_1, \ldots, s_n\} \subset S$ runs through all possible finite subsets of S.

Proof. We shall first show that the barycentric combinations form an affine subspace in A. Indeed, we denote by $M \subset L$ the linear subspace spanned by all possible vectors $s - t$; $s, t \in S$. Any two barycentric combinations of the points S can be represented in the form $\sum_{i=1}^{n} x_i s_i$, $\sum_{i=1}^{n} y_i s_i$ with the same set $\{s_1, \ldots, s_n\}$, by taking the union of the two starting sets and setting the extra coefficients equal to zero. Since $\sum_{i=1}^{n} x_i - \sum_{i=1}^{n} y_i = 0$, the difference of these combinations can be represented in the form

$$\sum_{i=1}^{n} (x_i - y_i)(s_i - s_1)$$

and therefore it lies in M. Conversely, any element from M of the form $\sum_{i=1}^{n} x_i(s_i - t_i)$ is the difference of the points

$$\sum_{i=1}^{n} x_i s_i + \left(1 - \sum_{i=1}^{n} x_i\right) s_1 \text{ and } \sum_{i=1}^{n} x_i t_i + \left(1 - \sum_{i=1}^{n} x_i\right) s_1$$

from $\tilde{S}$. Therefore $M = \{b_1 - b_2 | b_1, b_2 \in S\}$. The same argument shows that $t_m(\tilde{S}) \subset \tilde{S}$ for all $m \in M$. Therefore, $\tilde{S}$ is an affine subspace with orienting space M. It is clear that $S \subset \tilde{S}$.

Conversely, let $B \supset S$ be any affine subspace, $\{s_1, \ldots, s_n\} \subset S$. Then for any $x_1, \ldots, x_n \in \mathcal{K}$, $\sum_{i=1}^{n} x_i = 1$, we have

$$\sum_{i=1}^{n} x_i s_i = s_1 + \sum_{i=1}^{n} x_i (s_i - s_1).$$

Since $s_1, \ldots, s_n \in B$, the vector $\sum_{i=1}^{n} x_i(s_i - s_1)$ lies in the orienting space B and therefore the translation s_1 on it lies in B. Therefore, $\tilde{S} \subset B$ and $\tilde{S}$ are indeed the smallest affine subspaces containing S.

3.10 Proposition. *Let $f : A_1 \to A_2$ be an affine mapping of two affine spaces and let $B_1 \subset A_1$ and $B_2 \subset A_2$ be affine subspaces. Then $f(B_1) \subset A_2$ and $f^{-1}(B_2) \subset A_1$ are affine subspaces.*

Proof. Let $B_1 = \{b + l | l \in M_1\}$, where M_1 is the orienting space for B_1. Then $f(B_1) = \{f(b) + Df(l) | l \in M_1\} = \{f(b) + l' | l' \in \operatorname{im} Df\}$. Therefore $f(B_1)$ is the affine subvariety with the orienting space $\operatorname{im} Df$.

In particular, $f(A_1)$ is an affine subvariety in A_2, $B_2 \cap f(A_1)$ is an affine subvariety, and $f^{-1}(B_2) = f^{-1}(B_2 \cap f(A_1))$ by virtue of the general set-theoretic definitions. Replacing A_2 by $f(A_1)$ and B_2 by $B_2 \cap f(A_1)$, we can confine our attention to the case when f is surjective. Let M_2 be the orienting space for B_2. Then $B_2 = \{b + m | m \in M_2\}$ and $f^{-1}(B_2) = \{b' + m' | f(b') = b,\ Df(m') \in M_2\}$. On the right, we need study only one value $b' \in f^{-1}(b)$: the remaining values are obtained from it by translations on $\ker Df$. It follows that $f^{-1}(B_2)$ has the form $\{b' + m | m \in Df^{-1}(M_2)\}$ and is therefore an affine subspace with orienting space $(Df)^{-1}(M_2)$.

3.11. Corollary. *The level set of any affine function is an affine subspace.*

Proof. Indeed, the level set of the affine function $f : A \to \mathcal{K}^1$ consists of the inverse images of points in $\mathcal{K}^1$. But any point in an affine space is an affine subspace (with orienting space $\{0\}$).

3.12. Proposition. *Let $f_1, \ldots, f_n$ be affine functions on an affine space A. Then the set $\{a \in A | f_1(a_1) = \ldots = f_n(a_n) = 0\}$ is an affine subspace of A. If A is finite-dimensional, then any of its affine subspaces have this form.*

Proof. The indicated set is a finite intersection of level sets of affine functions. Therefore, it is affine by virtue of Corollary 3.11 and Proposition 3.7. Conversely, let $B \subset A$ be an affine subspace in the finite-dimensional affine space A, and let $M \subset L$ be the corresponding linear spaces. If B is empty, then it can be defined by the equation $f = 0$, where f is a constant non-zero function on A (any such function is obviously affine, $Df = 0$). Otherwise, let $g_1 = \ldots = g_n = 0$ be a system of linear equations on L, defining M; for $g_1, \ldots, g_n$ we can take, for example, the basis of the subspace $M^\perp \subset L^*$. We select a point $b \in B$ and construct the affine functions $f_i : A \to \mathcal{K}^1$ with the conditions $f_i(b) = 0$, $Df_i = g_i$, $i = 1, \ldots, n$. Evidently, $f_i(b + l) = g_i(l)$. Therefore, all functions f_i vanish at the points $b + l \in A$ if and only if $l \in M$, that is, if and only if $b + l \in B$. This completes the proof.

3.13. By a *configuration* in an affine space A we mean a finite ordered system of affine subspaces $\{B_1, \ldots, B_n\}$. We call two configurations $\{B_1, \ldots, B_n\}$ and $\{B'_1, \ldots, B'_n\}$ *affine-congruent* if there exists an affine automorphism $f \in \operatorname{Aff} A$, such that $f(B_i) = B'_i$, $i = 1, \ldots, n$. Variants of this concept are possible when f can be chosen only from some subgroup of $\operatorname{Aff} A$, for example, the group of motions, when A is Euclidean. In the last case, we shall call the configurations *metrically congruent*. Important concepts and results of affine geometry are associated with the search for the invariants of configurations with respect to the congruence relation. We note that it is the affine variant of the concept "the same arrangement", which we studied in §5 of Chapter I.

We shall prove several basic results about congruence.

Let A be an affine space with dimension n. In accordance with the results of §§1.9–11 we call a configuration of $(n+1)$ points $\{a_0, \dots, a_n\}$ in A a *coordinate configuration* if its affine span coincides with A.

3.14. Proposition. a) *Any two coordinate configurations are congruent and are transformed into one another by a unique mapping $f \in \mathrm{Aff}\, A$.*

b) *Two coordinate configurations $\{a_0, \dots, a_n\}$ and $\{a'_0, \dots, a'_n\}$ in a Euclidean space A are metrically congruent if and only if $d(a_i, a_j) = d(a'_i, a'_j)$ for any $i, j \in$ $\in 1, \dots, n$.*

Proof. a) We set $e_i = a_i - a_0$, $e'_i = a'_i - a'_0$. The systems $\{e_i\}$ and $\{e'_i\}$ form a basis in L. Let $g : L \to L$ be a linear mapping transforming e_i into e'_i. We construct an affine mapping $f : A \to A$ with the property $Df = g$ and $f(a_0) = a'_0$. It exists by virtue of Proposition 1.11 and lies in $\mathrm{Aff}\, A$, because g is invertible. In addition,

$$f(a_i) = f(a_0) + g(a_i - a_0) = a'_0 + e'_i = a'_0 + (a'_i - a'_0) = a'_i$$

for all $i = 1, \dots, n$. The same formula shows that f is unique, because Df must transform e_i into e'_i and $f(a_0) = a'_0$.

b) By virtue of what was proved above, it is sufficient to verify that f is a motion if and only if $d(a_i, a_j) = d(a'_i, a'_j)$ for all i, j. Indeed, $d(a_i, a_j) = |a_i - a_j| =$ $= |e_i - e_j|$, where $e_0 = a_0 - a_0 = 0$, and analogously $d(a'_i, a'_j) = |e'_i - e'_j|$. If f is a motion, then Df is orthogonal and preserves the lengths of vectors, so that the condition is necessary. Conversely, assume that it is satisfied. Then $|e_i| = |e'_i|$ for all $i = 1, \dots, n$ and we find from the equalities $|e_i - e_j|^2 = |e'_i - e'_j|^2$ that $(e_i, e_j) = (e'_i, e'_j)$ for all i and j. Therefore the Gram matrices of the bases $\{e_i\}$ and $\{e'_i\}$ are equal. But then, the mapping g, transforming $\{e_i\}$ into $\{e'_i\}$, is an isometry, so that f is a motion. This completes the proof.

We shall now examine the configurations (b, B), consisting of points and an affine subspace. In the Euclidean case we call the distance

$$d(b, B) = \inf\{|l| \,|\, b + l \in B\}$$

the distance from b to B.

3.15. Proposition. a) *The configurations (b, B) and (b', B') are affinely congruent if and only if $\dim B = \dim B'$ and either $b \notin B, b' \notin B'$ or $b \in B, b' \in B'$ at the same time.*

b) *The configurations* (b, B) *and* (b', B') *are metrically congruent if and only if* $\dim B = \dim B'$ *and* $d(b, B) = d(b', B')$.

Proof. a) The stated conditions are evidently necessary. Assume that they are satisfied. We denote by M and M' the orienting subspaces of B and B' respectively and we select a linear automorphism $g : L \to L$ for which $g(M) = M'$. If $b \in B$ and $b' \in B'$, then we construct an affine mapping $f : A \to A$ with the conditions $Df = g$ and $f(b) = b'$. Evidently, $f(b + l) = b' + g(l)$, so that $f(B) = B'$.

If $b \notin B$ and $b' \notin B'$, we impose additional conditions on g. We select points $a \in B$, $a' \in B'$ and require that g transform the vector $b - a$ into the vector $b' - a'$. Both vectors are non-zero and lie outside M and M' respectively, so that the standard construction, starting from the bases of L of the form {basis of $M, b-a$, complement} and {basis of $M', b' - a'$, complement}, shows that g exists. Next we once again construct the affine mapping $f : A \to A$ with $Df = g$ and $f(b) = b'$. We verify that $f(B) = B$. Indeed, first of all, $f(a) = a'$, because

$$f(a) = f(b - (b - a)) = f(b) - g(b - a) = b' - (b' - a') = a'.$$

Furthermore, $f(a + l) = f(a) + g(l)$ and the condition $l \in M$ is equivalent to the condition $g(l) \in M'$ so that $f(B) = B'$.

b) The necessity of the condition is once again obvious. To prove sufficiency we impose additional requirements on the selections made in the preceding discussion. First of all, we identify A with L, and we select the origin of coordinates in B. Then B is identified with M and b becomes a vector in L. Let a be the orthogonal projection of b on M. In the linear version we already know that $d(b, B) = |b - a|$. Analogously, we define a point a' in M', or, in our identification, in B'. For g we choose an isometry of L transforming M into M' and b into b'. It exists: we extend the orthonormal bases in M and M' to orthonormal bases in L, containing $(b - a)/|b - a|$ and $(b' - a')/|b' - a'|$, respectively, and define g as an isometry transforming the first basis into the second. Then the affine mapping $f : A \to A$ with $Df = g$ and $f(b) = b'$ will be a motion, transforming (b, B) into (b', B').

3.16. Finally, we shall examine the configurations consisting of two subspaces B_1, B_2. The complete classification of these configurations, up to affine congruence, can be made with the help of the corresponding result for linear subspaces, proved in §5.5 of Chapter 1. The complete metric classification is quite cumbersome: it requires an examination of the distance between B_1 and B_2 and a series of angles. We shall confine ourselves to a discussion of the only metric invariant – a distance which, as usual, we shall define by the formula

$$d(B_1, B_2) = \inf\{|b_1 - b_2|\, |b_1 \in B_1,\ b_2 \in B_2\}.$$

We shall call a pair of points $b_1 \in B_1$, $b_2 \in B_2$ such that the vector $b_1 - b_2$ is orthogonal to the orienting spaces B_1 and B_2 the *common perpendicular* to B_1

and B_2. (It would be more accurate to call the common perpendicular the segment $\{tb_1 + (1-t)b_2 | 0 \le t \le 1\}$.)

3.17. Proposition. a) *A common perpendicular to B_1 and B_2 always exists. The set of common perpendiculars is bijective to the orienting spaces of B_1 and B_2.*

b) *The distance between B_1 and B_2 equals the length of any common perpendicular to the $|b_1 - b_2|$.*

Proof. a) Let M_1, M_2 be the orienting subspaces of B_1 and B_2 and let $b_1' \in B_1$, $b_2' \in \in B_2$. We project the vector $b_1' - b_2'$ orthogonally onto $M_1 + M_2$ and represent the projection in the form $m_1 + m_2$, $m_i \in M_i$. We set $b_1 = b_1' - m_1, b_2 = b_2' + m_2$. Evidently, $b_i \in B_i$ and

$$b_1 - b_2 = b_1' - b_2' - (m_1 + m_2) \in (M_1 + M_2)^\perp.$$

Hence, $\{b_1, b_2\}$ is a common perpendicular to B_1, B_2.

Let $\{b_1, b_2\}$ and $\{b_1', b_2'\}$ be two common perpendiculars. Then $b_1 - b_1' \in M_1$, $b_2 - b_2' \in M_2$ and in addition,

$$b_1 - b_2 \in (M_1 + M_2)^\perp, \; b_1' - b_2' \in (M_1 + M_2)^\perp.$$

Hence the difference $(b_1 - b_1') - (b_2 - b_2')$ lies simultaneously in $M_1 + M_2$ and $(M_1 + M_2)^\perp$. It therefore equals zero. Hence $b_1 - b_1' = b_2 - b_2' \in M_1 \cap M_2$. Conversely, if $\{b_1, b_2\}$ is a fixed common perpendicular and $m \in M_1 \cap M_2$, then $\{b_1 + m, \; b_2 + m\}$ is also a common perpendicular. This completes the proof of the first part of the assertion.

b) Let $\{b_1, b_2\}$ be a common perpendicular to B_1 and B_2 and let $b_1' \in B_1$, $b_2' \in \in B_2$ be any other pair of points. It is sufficient to prove that $|b_1 - b_2| \le |b_1' - b_2'|$. Indeed,

$$b_1' - b_2' = (b_1 - b_2) + (b_1' - b_1) + (b_2 - b_2').$$

But $(b_1' - b_1) + (b_2 - b_2') \in M_1 + M_2$, and the vector $b_1 - b_2$ is orthogonal to $M_1 + M_2$. Therefore by Pythagoras's theorem

$$|b_1' - b_2'|^2 = |b_1 - b_2|^2 + |b_1' - b_1 + b_2 - b_2'|^2 \ge |b_1 - b_2|^2,$$

which completes the proof.

In conclusion we shall establish a useful result characterizing affine subspaces.

3.18. Proposition. *A subset $S \subset A$ is an affine subspace if and only if together with any two points $s, t \in S$ it contains the entire straight line passing through these points, that is, their affine span.*

Proof. The straight line passing through the points $s, t \in S$ is the set $\{xs + +(1-x)t | x \in \mathcal{K}\}$. Therefore the necessity of the condition follows from Proposition

3.9. Conversely, assume that the condition holds. Since by virtue of the same proposition the affine span S consists of all possible barycentric combinations of points in S, we must verify that such combinations $\sum_{i=1}^{n} x_i s_i$ lie in S. We perform induction on n. For $n = 1$ and 2, the result is obvious. Let $n > 2$ and assume that the result is proved for the smallest values of n. We represent $\sum_{i=1}^{n} x_i s_i$ in the form

$$y_1 \sum_{i=1}^{n-2} \frac{x_i}{y_1} s_i + y_2 \sum_{i=n-1}^{n} \frac{x_i}{y_2} s_i,$$

where $y_1 = \sum_{i=1}^{n-2} x_i$, $y_2 = x_{n-1} + x_n$ (we can assume that both of these sums differ from zero, otherwise $\sum_{i=1}^{n} x_i s_i \in S$ by the induction assumption). Evidently,

$$\sum_{i=1}^{n-2} \frac{x_i}{y_1} = \sum_{i=n-1}^{n} \frac{x_i}{y_2} = y_1 + y_2 = 1.$$

Therefore, $\sum_{i=1}^{n-2} \frac{x_i}{y_1} s_i$ and $\sum_{i=n-1}^{n} \frac{x_i}{y_2} s_i$ lie in S, and their barycentric combination with coefficients y_1 and y_2 lies in S. This completes the proof.

EXERCISES

1. We call the ith median of the system of points $a_1, \ldots, a_n \in A$ a segment connecting the point a_i with the centre of gravity of the remaining points $\{a_j / j \neq i\}$. Prove that all medians intersect at one point – the centre of gravity of $a_1, \ldots, a_n$.

2. The angle between two straight lines in a Euclidean affine space A is the angle between their orienting spaces. Prove that two configurations of two straight lines in A are metrically congruent if and only if the angles and distances between the straight lines are the same in both configurations.

3. The angle between a straight line and an affine subspace with dimension ≥ 1 is defined as the angle between the orienting straight line and its projection on the orienting subspace. Using this definition, extend the result of Exercise 2 to configurations consisting of a straight line and a subspace.

§4. Convex Polyhedra and Linear Programming

4.1. Formulation of the problem. The basic problem of linear programming is stated as follows. A finite-dimensional affine space A over the field of real numbers $\mathbf{R}$ and $m+1$ affine functions $f_1, \dots, f_m; f : A \to \mathbf{R}$ are given. The problem is to find a point (or points) $a \in A$ satisfying the conditions $f_1(a) \geq 0, \dots, f_m(a) \geq 0$, for which the function f assumes the greatest possible value under these restrictions.

A variant in which some of the inequalities are reversed, $f_i(a) \leq 0$, and/or the problem is to find points at which f assumes the smallest possible value, can be reduced to the preceding case by changing the sign of the corresponding functions. The condition $f_i(a) = 0$ is equivalent to the combined conditions $f_i(a) \geq 0$ and $-f_i(a) \geq 0$. All functions f_i may be assumed not to be constant.

4.2. Motivation. Consider the following mathematical model of production. Consider an enterprise that consumes m types of different resources and produces n types of different products. The resources and products are measured in their own units by non-negative real numbers (we shall not consider the case when these numbers are integers, for example, the number of automobiles; for large volumes of production and consumption of resources the continuous model is a good approximation).

The production plan is the vector $(x_1, \dots, x_n) \in \mathbf{R}^n$, indicating the amount x_j of the jth product that must be produced. The following linear model of the consumption of resources is used: if a_{ij} units of the ith resource are consumed in the production of a unit of the jth product, then $\sum_{j=1}^n a_{ij}x_j$ units of the ith resource are required in order to fulfil the plan $(x_1, \dots, x_n)$. The resources supplied to the enterprise are determined by the vector $(b_1, \dots, b_m)$: b_i units of resource i are supplied. Therefore the plan $(x_1, \dots, x_n)$ can be fulfilled only if the following system of restrictions is satisfied:

$$f_i(x_1, \dots, x_n) = b_i - \sum_{j=1}^{n} a_{ij}x_j \geq 0, \quad i = 1, \dots, m.$$

We shall always assume that these inequalities are compatible.

Assume that the enterprise produces the ith product at a price c_i per unit. Then the profit from production will be given by

$$f(x_1, \dots, x_n) = \sum_{i=1}^{n} c_i x_i.$$

The production plan $(x_1, \dots, x_n)$ is said to be optimal with respect to the profit if $f(x)$ attains the greatest possible value under the restrictions $f_j \geq 0$, $x_i \geq 0$ $(i = 1, \dots, n)$ (the last condition means that the enterprise does not obtain the products it itself produces elsewhere either for sale or for spare parts).

We can see that the problem of formulating the optimal plan is a particular case of the problem formulated in §4.1.

Of course, the practical applications of linear programming involve the development of specific algorithms for searching for the optimal plan, which can be implemented manually or with the help of computer. In this section we shall study only the geometric aspects of the problem, which, of course, are the basis for all the algorithms.

4.3. Basic geometric concepts. We fix a finite-dimensional affine space A over a field $\mathbf{R}$. The letters f with indices will indicate affine functions on A.

A *half-space* is a set of points of the form $\{a \in A | f(a) \geq 0\}$, where f is a non-constant affine function. A *polyhedron* is an intersection of a finite number of half-spaces.

We recall that a subset $S \subset A$ is convex if $a_1, a_2 \in S$ and $0 \leq x \leq 1$ imply that $xa_1 + (1-x)a_2 \in S$. Since $f(xa_1 + (1-x)a_2) = xf(a_1) + (1-x)f(a_2)$, all half-spaces are convex. Since the intersection of any family of convex sets is convex, all polyhedra are convex. We shall call any point $xa_1 + (1-x)a_2$, $0 < x < 1$, an *interior point* of the segment with ends a_1 and a_2.

Let S be a convex set. A convex subset $T \subset S$ is called a *face* of S if any segment whose ends are contained in S, any interior point of which is contained in T, is wholly contained in T. The entire set S is its own face. A face of S consisting of one point is called a *vertex* of S. The reader should imagine a cube, an octahedron, and a polyhedral angle in three-dimensional space, in order to have a clear picture of the basic situation that is important for linear programming. The faces of these figures, in the sense of our definition, are the faces, edges, and vertices of high-school geometry together with the figure itself. All points on the surface of a sphere are vertices of the sphere.

The most important result of this section is that the maximum of an affine function on a bounded polyhedron (this is the case most often encountered in practice) is reached on one of the vertices of the polyhedron; there are a finite number of vertices. But first we shall have to study the structure of polyhedra and their faces in greater detail.

4.4. Lemma. *The intersection of a family of faces and the face of a face of a convex set S is a face of S.*

Proof. a) Let $T = \bigcap T_i$, where the T_i are faces of S. Any segment whose ends are contained in S, an interior point of which is contained in T_i, is wholly contained in T_i. This means that if its interior point is contained in T, then it is also contained in T.

b) Let $T_1 \subset T \subset S$, T being a face of S. Any segment with ends in S, an interior point of which is contained in T_1, is wholly contained in T, because T is a face of S; hence its ends are contained in T, and therefore it is contained in T_1, because T_1 is a face of T.

4.5. Lemma. *Let S be a polyhedron, defined by the inequalities $f_i \geq 0$, $i = 1,\ldots,m$. Then for any i the polyhedron $S_i = S \cap \{a|f_i(a) = 0\}$ is either empty or is a face of S.*

Proof. Let S_i be non-empty, let $a_1, a_2 \in S$, and let the interior point of the segment $xa_1 + (1-x)a_2$ be contained in S_i. The function $f_i(xa_1 + (1-x)a_2)$, $0 \leq x \leq 1$, linear with respect to x, vanishes for some $0 < x_0 < 1$ and, in addition, is non-negative at $x = 0$ and $x = 1$. Therefore, it identically equals zero, so that the entire segment is contained in S_i.

4.6. Lemma. *The non-constant affine function f on a polyhedron $S = \{a|f_i(a) \geq \geq 0\}$ cannot assume a maximum value at a point $a \in S$ for which all $f_i(a) > 0$.*

Proof. Since f is not constant, $Df \not\equiv 0$. We select in the vector space L, associated with A , a vector $l \in L$ for which $Df(l) \neq 0$. It may be assumed that $Df(l) > 0$, changing the sign of l if necessary. If the number $\epsilon > 0$ is sufficiently small and $a \in S$,then $f_i(a+\epsilon l) > 0$ for all $i = 1,\ldots,m$: it is sufficient to take $\epsilon < \min_i \frac{f_i(a)}{|Df_i(l)|}$. Therefore, $a+\epsilon l \in S$ for such ϵ. But $f(a+\epsilon l) = f(a)+\epsilon Df(l) > f(a)$, so that $f(a)$ is not the maximum value of f.

We can now prove our main result.

4.7. Theorem. *Assume that an affine function f is bounded above on the polyhedron S. Then, it assumes its maximum value at all points of some face of S that is also a polyhedron. If S is bounded, then f assumes its maximum value at some vertex of S.*

Proof. We perform induction on the dimension of A. The case $\dim A = 0$ is obvious. Let $\dim A = n$ and assume that for low dimensions the theorem is proved. Let S be given by the system of inequalities $f_1 \geq 0,\ldots,f_m \geq 0$. Since the set S is closed , the function f which is bounded above assumes a maximum value on S at some point a. If $f_1(a) > 0,\ldots,f_m(a) > 0$, then Lemma 4.6 implies that f can only be a constant; in particular, it assumes its only value on all of S. Otherwise, $f_i(a) = 0$ for some i. This means that f assumes a maximum value at a point of the non-empty polygon S_i which is a face of S and lies in the affine space $\{a|f_i(a) = 0\}$ with dimension $n-1$ because f_i is not constant. By the induction hypothesis, the restriction of f to S_i assumes its maximum value at all points of some polyhedral face of S_i. Lemma 4.4 implies that it will be a face of S. It will be a polyhedron, because to the inequalities defining it in S_i, whose left sides are continued onto all of A, we must add the equality $f_i = 0$.

Now, by induction on the dimension of the affine span of S we show that any bounded polyhedron necessarily has a vertex. Indeed, in zero dimension this is obvious. Let the dimension be greater than zero. We may assume that the affine span of S is all of A. We take any variable affine function on A. It must assume its maximum value on S, because S is bounded and closed. Consequently, S contains a non-empty face at all points of which this value is assumed. It is a bounded polyhedron, whose affine span has a strictly lower dimension. By the induction hypothesis it has a vertex which, according to Lemma 4.6, is also a vertex of S.

Finally, let S be bounded and let T be a polyhedral face of S, on which the starting function f assumes its maximum value. Then any vertex of T, whose existence has been proved, is the vertex of S sought.

§5. Affine Quadratic Functions and Quadrics

5.1. Definition. *A quadratic function Q on an affine space (A, L) over a field $\mathcal{K}$ is a mapping $Q : A \to \mathcal{K}$, for which there exists a point $a_0 \in A$, a quadratic form $q : L \to \mathcal{K}$, a linear form $l : L \to \mathcal{K}$ and a constant $c \in \mathcal{K}$, such that*

$$Q(a) = q(a - a_0) + l(a - a_0) + c$$

for all $a \in A$.

The form q is called the quadratic part of Q, while l is the linear part of Q relative to the point a_0. Obviously, $c = Q(a_0)$. We shall first show that the quadratic nature of Q is independent of the choice of a_0. More precisely, let g be the symmetric bilinear form on L that is the polarization of q. As usual, we shall assume that the characteristic of $\mathcal{K}$ does not equal 2.

5.2. Proposition. *If $Q(a) = q(a - a_0) + l(a - a_0) + c$, then for any point $a_0' \in A$*

$$Q(a) = q(a - a_0') + l'(a - a_0') + c',$$

where

$$l'(m) = l(m) + 2g(m, a_0' - a_0), \; c' = Q(a_0').$$

Thus the transition to a different point changes the linear part of Q and the constant.

Proof. Indeed,

$$q(a - a_0) = q((a - a_0') + (a_0' - a_0)) =$$

$$= q(a - a_0') + 2g(a - a_0', a_0' - a_0) + q(a_0' - a_0),$$

$$l(a - a_0) = l((a - a_0') + (a_0' - a_0)) = l(a - a_0') + l(a_0' - a_0),$$

which proves the required result.

5.3. We shall call the point a_0 a *central* point for the quadratic function Q if the linear part of Q relative to a_0 equals zero. This terminology is explained by the remark that the point a_0 is central if and only if $Q(a) = Q(a_0 - (a - a_0))$ for all a; indeed, the difference between the left and right sides in the general case equals $2l(a - a_0)$ because $q(a - a_0) = q(-(a - a_0))$. Geometrically, this means that after the identification of A with L such that a_0 transforms into the origin of coordinates, the function Q becomes symmetric relative to the reflection $m \mapsto -m$.

We shall call the set of central points of the function Q the *centre* of Q.

5.4. Theorem. a) *If the quadratic part q of the function Q is non-degenerate, then the centre of Q consists of one point.*

b) *If q is degenerate, then the centre of Q is either empty or it is an affine subspace of A with dimension* $\dim A - \operatorname{rk} q$ (rk q *is the rank of* q) *, whose orienting subspace coincides with the kernel of* q.

Proof. We begin with any point $a_0 \in A$ and represent Q in the form $q(a - a_0)+$ $+l(a - a_0) + c$. According to Proposition 5.2, the point $a_0' \in A$ will be central for Q if and only if the condition

$$l(m) = -2g(m, a_0' - a_0)$$

holds for all $m \in L$. When a_0' runs through all points of A, the vector $a_0' - a_0$ runs through all elements of L and the linear function of $m \in L$ of the form $-2g(m, a_0' - a_0)$ runs through all elements of L^*, contained in the image of the canonical mapping $\tilde{g} : L \to L^*$, associated with the form g.

If q is non-degenerate, then $\tilde{g}$ is an isomorphism. In particular, for the functional $-l/2 \in L^*$ there exists a unique vector $a_0' - a_0 \in L$ with the property $g(\cdot, a_0' - a_0) = -\frac{1}{2}l(\cdot)$. The point a_0' is, in this case, the only central point of Q.

If q is degenerate, then there are two possible cases. Either $-l/2$ is not contained in the image of $\tilde{g}$, and then there are no central points, or $-l/2$ is contained in the image of $\tilde{g}$. Then for any two points a_0', a_0'' with the condition

$$g(\cdot, a_0' - a_0) = g(\cdot, a_0'' - a_0) = -\frac{1}{2}l(\cdot)$$

we have $a_0' - a_0'' \in \ker \tilde{g}$ and, conversely, if $g(\cdot, a_0' - a_0) = -\frac{1}{2}l(\cdot)$ and $a_0'' \in a_0' + \ker \tilde{g}$, then

$$g(\cdot, a_0'' - a_0) = -\frac{1}{2}l(\cdot).$$

Thus the centre is an affine subspace and $\ker \tilde{g}$, that is, the kernel of q, is the orienting subspace. This completes the proof.

We can now prove the theorem on the reduction of a quadratic form Q to canonical form in an appropriate affine coordinate system $\{a_0, e_1, \dots, e_n\}$,where

$\{e_i\}$ is a basis of L and $a_0 \in A$. We recall that the point $a \in A$ in it is represented by the vector $(x_1, \ldots, x_n)$, if $a = a_0 + \sum_{i=1}^n x_i e_i$.

5.5. Theorem. *Let Q be a quadratic function on the affine space A. Then there exists an affine coordinate system in A in which Q assumes one of the following forms.*

a) *If q is non-degenerate, then $Q(x_1, \ldots x_n) = \sum_{i=1}^n \lambda_i x_i^2 + c$; $\lambda_i, c \in \mathcal{K}$.*

b) *If q is degenerate and has rank r and the centre of Q is non-empty, then*

$$Q(x_1, \ldots, x_n) = \sum_{i=1}^r \lambda_i x_i^2 + c; \ \lambda_i, c \in \mathcal{K}.$$

c) *If q is degenerate and has rank r and the centre of Q is empty then*

$$Q(x_1, \ldots, x_n) = \sum_{i=1}^r \lambda_i x_i^2 + x_{r+1}.$$

Proof. If q is non-degenerate, we choose the central point of Q as a_0. Then $Q(a) = q(a - a_0) + c$. For $e_1, \ldots, e_n$ we select a basis of L in which q is reduced to the sum of squares with coefficients. The same technique also leads to the goal if the centre is non-empty.

If the centre of Q is empty, then we start with an arbitrary point a_0 and the basis $\{e_1, \ldots, e_n\}$ in which the quadratic part Q has the form $\sum_{i=1}^r \lambda_i x_i^2$. Let the linear part have the form $l = \sum_{i=1}^n \mu_i x_i$. We assert that $\mu \neq 0$ for some $j > r$. Indeed, otherwise $l = \sum_{i=1}^r \mu_i x_i$, and then Q can be represented in the form

$$\sum_{i=1}^r \lambda_i x_i^2 + \sum_{i=1}^r \mu_i x_i + c = \sum_{i=1}^r \lambda_i \left(x_i + \frac{\mu_i}{2\lambda_i} \right)^2 + c'.$$

Therefore the point $a_0 - \sum_{i=1}^r \frac{\mu_i}{2\lambda_i} e_i$ will be a central point for Q which contradicts the assumption that the centre is empty.

But if $\mu_j > 0$ for some $j > r$, then the system of functionals $\{e^1, \ldots, e^r, l\}$ in L^* is linearly independent. We can extend it up to a basis of L^* and in the dual basis of L we can obtain for Q an expression of the form $\sum_{i=1}^r \lambda_i x_i^2 + x_{r+1} + c$, where x_{r+1}, as a function on L, is simply l. It is now clear that there exists a point at which Q vanishes, for example, $x_1 = \ldots = x_r = 0$, $x_{r+1} = -c$, $x_{r+2} = \ldots = x_n = 0$ in this system of coordinates. Having begun the construction at this point, we obtain the representation of Q in the form $\sum_{i=1}^r \lambda_i x_i^2 + x_{r+1}$.

5.6. Supplement. a) The question of the uniqueness of a canonical form reduces to a previously solved problem about quadratic forms. If q is non-degenerate and has the form $\sum_{i=1}^n \lambda_i x_i^2 + c$ in some coordinate system, then the point $(0, \ldots, 0)$

is the centre and is therefore determined uniquely, the constant c is determined uniquely as the value of Q at the centre, and the arbitrariness in the selection of axes and coefficients is the same as for quadratic forms. In particular, over **R** it may be assumed that $\lambda_i = \pm 1$, and the signature is a complete invariant. Over **C** it may be assumed that all $\lambda_i = 1$.

In the singular case with a non-empty centre, the origin of coordinates can be placed at the centre arbitrarily, but the constant c is still determined uniquely, because the value of Q at all points of the centre is constant: if a and a_0 are contained in the centre, then $l(a - a_0) = 0$ and $q(a - a_0) = 0$, because $a - a_0$ is contained in the kernel of q. The previous remarks are applicable to the quadratic part.

Finally, in the degenerate case with an empty centre, any point at which Q vanishes can be chosen for the origin of coordinates; the previous remarks are applicable to the quadratic part.

b) If A is an affine Euclidean space, then Q is reduced to canonical form in an orthonormal basis. The numbers $\lambda_1, \ldots, \lambda_n$ are determined uniquely. The arbitrariness in the selection of the centre is the same as in the affine case, and the arbitrariness in the selection of axes is the same as for quadratic forms in a linear Euclidean space.

5.7. Affine quadrics. An affine quadric is a set $\{a \in A | Q(a) = 0\}$, where Q is some quadratic function on A. A glance at the canonical forms Q shows that all of the results of §10 of Chapter 2 are applicable to the study of the types of quadrics.

Consider the problem of the uniqueness of a function Q defining a given affine quadric over the field **R**. First of all, the quadric can be an affine space in A (possibly empty): the equation $\sum_{i=1}^{r} x_i^2 = 0$ is equivalent to the system of equations $x_1 = \ldots = x_r = 0$. For $r > 1$ there exist many quadratic functions that are not proportional to one another but which give the same quadric, for example, $\sum_{i=1}^{r} \lambda_i x_i^2 = 0$ for any $\lambda_i > 0$. We shall show that for other quadrics the answer is simpler.

5.8. Proposition. *Let the affine quadric X, which is not an affine subspace, be given by the equations $Q_1 = 0$ and $Q_2 = 0$ where Q_1, Q_2 are quadratic functions. Then $Q_1 = \lambda Q_2$ for an appropriate scalar $\lambda \in$ **R**.*

Proof. First, X does not reduce to one point. Proposition 3.18 implies that there exist two points $a_1, a_2 \in X$ whose affine span (straight line) is not wholly contained in X.

Let $a_1, a_2 \in X$ and assume that the straight line passing through the points a_1, a_2 is not wholly contained in X. We introduce in A a coordinate system $\{a_1; e_1, \ldots, e_n\}$, where $e_n = a_2 - a_1$. We write the function Q_1 in this coordinate system as a quadratic trinomial in x_n:

$$Q_1(x_1,\dots,x_n) = \lambda x_n^2 + l_1'(x_1,\dots,x_{n-1})x_n + l_1''(x_1,\dots,x_{n-1}),$$

where l_1', l_1'' are affine functions, that is polynomials of degree ≤ 1 in $x_1,\dots,x_{n-1}$. Since the straight line through the points $a_1 = (0,\dots,0)$ and $a_2 = (0,\dots,0,1)$ is not wholly contained in X, $\lambda \neq 0$ and $l_1'(0)^2 - 4\lambda l_1''(0) > 0$. Dividing Q_1 by λ, it may be assumed that $\lambda = 1$. Analogously, it may be assumed that

$$Q_2(x_1,\dots,x_n) = x_n^2 + l_2'(x_1,\dots,x_{n-1})x_n + l_2''(x_1,\dots,x_{n-1})$$

and $l_2'(0)^2 - 4l_2''(0) > 0$. We now know that Q_1 and Q_2 have the same set of real zeros and we want to prove that $Q_1 = Q_2$.

We fix the vectors $(c_1,\dots,c_{n-1}) \in \mathbf{R}^{n-1}$ and examine the vectors

$$(tc_1,\dots,tc_{n-1}), t \in \mathbf{R}.$$

For small absolute values of t, the discriminants of the trinomials $Q_1(tc_1,\dots$ $\dots,tc_{n-1},x_n)$ and $Q_2(tc_1,\dots,tc_{n-1},x_n)$ with respect to x_n remain positive, and their real roots, corresponding to the points of intersection of the same straight line with X, are the same. Hence $l_1' = l_2'$ and $l_1'' = l_2''$ at the points $(tc_1,\dots,tc_{n-1})$. Therefore, $l_1' \equiv l_2'$ and $l_1'' \equiv l_2''$, because affine functions which are equal on an open set are equal. Indeed, their difference vanishes in a neighbourhood of the origin of coordinates and therefore the set of its zeros cannot be a proper linear subspace. This completes the proof.

§6. Projective Spaces

6.1. Affine spaces are obtained from linear spaces by eliminating the origin of coordinates. Projective spaces can be constructed from linear spaces by at least two methods:

a) By adding points at infinity to the affine space.

b) By realizing a projective space as a set of straight lines in a linear space.

We shall choose b) as the basic definition: it shows more clearly the homogeneity of the projective space.

6.2. Definition. *Let L be a linear space over a field $\mathcal{K}$. The set $P(L)$ of straight lines (that is, one-dimensional linear subspaces) in L is called the projective space associated with L, and the straight lines in L are themselves called points of $P(L)$.*

The number $\dim L - 1$ is the dimension of $P(L)$, and it is denoted by $\dim P(L)$. One- and two-dimensional projective spaces are called respectively a projective straight line or a projective plane. An n-dimensional projective space over a field $\mathcal{K}$ is also denoted by $\mathcal{K}P^n$ or $P^n(\mathcal{K})$ or simply P^n. The meaning of the identity $\dim P(L) = \dim L - 1$ will now become clear.

6.3. Homogeneous coordinates. We select a basis $\{e_0, \dots, e_n\}$ in the space L. Every point $p \in P(L)$ is uniquely determined by any non-zero vector on the corresponding straight line in L. The coordinates $x_0, \dots, x_n$ of this vector are called *homogeneous coordinates of the point* p. They are defined up to a non-zero scalar factor: the point $(\lambda x_0, \dots, \lambda x_n)$ lies on the same straight line p and all points of the straight line are obtained in this manner. Therefore, the vector of homogeneous coordinates of the point p is traditionally denoted by $(x_0 : x_1 : \dots : x_n)$.

Thus the *n-dimensional projective coordinate space* $P(\mathcal{K}^{n+1})$ is the set of orbits of the multiplicative group $\mathcal{K}^* = \mathcal{K}\backslash\{0\}$, acting on the set of non-zero vectors $\mathcal{K}^{n+1}\backslash\{0\}$ according to the rule $\lambda(x_0, \dots, x_n) = (\lambda x_0, \dots, \lambda x_n)$; $(x_0 : x_1 : \dots : x_n)$ is the symbol for the corresponding orbit.

In terms of homogeneous coordinates one can easily imagine the structure of P_n as a set by several different methods.

a) *Affine covering of* P^n. Let

$$U_i = \{(x_0 : \dots : x_n) | x_i \neq 0\},\ i = 0, \dots, n.$$

Obviously, $P^n = \bigcup_{i=0}^n U_i$. The collection of vectors of the projective coordinates of any point $p \in U_i$ contains a unique vector with the ith coordinate equal to 1: $(x_0 : \dots : x_i : \dots : x_n) = (x_0/x_i : \dots : 1 : \dots : x_n/x_i)$. Dropping this number one, we find that U_i is bijective to the set $\mathcal{K}^n$, which we can interpret as an n-dimensional linear or affine coordinate space. We note, however, that we do not yet have any grounds for assuming that U_i has some natural linear or affine structure, independent of the choice of coordinates. Later we shall show that it is possible to introduce on U_i in an invariant manner only an entire class of affine structures, which, however, are associated with canonical isomorphisms, so that the geometry of affine configurations will be the same in all of them.

We shall call the set $U_i \cong \mathcal{K}^n$ the *ith affine map of* P^n (in a given coordinate system). The points $(y_1^{(i)}, \dots, y_n^{(i)}) \in U_i$ and $(y_1^{(j)}, \dots, y_n^{(j)}) \in U_j$ with $i \neq j$ correspond to the same point in P^n, contained in the intersection $U_i \cap U_j$, if and only if by inserting the number one in the ith place in the vector $(y_1^{(i)}, \dots, y_n^{(i)})$ and in the jth place in $(y_1^{(j)}, \dots, y_n^{(j)})$, we obtain proportional vectors.

In particular, $P^1 = U_0 \cup U_1$, $U_0 \cong U_1 \cong \mathcal{K}$; the point $y \in U_0$ corresponds to the point $1/y \in U_1$ with $y \neq 0$; the point $y = 0$ in U_0 is not contained in U_1, while the point $1/y = 0$ in U_1 is not contained in U_0. It is natural to assume that P^1 is obtained from $U_0 \cong \mathcal{K}$ by adding one point with the coordinate $y = \infty$. Generalizing this construction, we obtain the following.

b) *Cellular partition of* P^n. Let

$$V_i = \{(x_0 : \ldots : x_n) | x_j = 0 \quad \text{for} \quad j < i, \ x_i \neq 0\}.$$

Obviously, $V_0 = U_0$ and $P^n = \bigcup_{i=0}^n V_i$, but this time all the V_i are pairwise disjoint. The collection of projective coordinates of any point $p \in V_i$ contains a unique representative with the number one in the ith place; dropping the number one and the preceding zeros, we obtain a bijection of V_i with $\mathcal{K}^{n-i}$. Finally

$$P^n \cong \mathcal{K}^n \cup \mathcal{K}^{n-1} \cup \mathcal{K}^{n-2} \cup \ldots \cup \mathcal{K}^0 \cong \mathcal{K}^n \cup P^{n-1}.$$

In other words, P^n is obtained by adding to $U_0 \cong \mathcal{K}^n$ an $(n-1)$-dimensional projective space, infinitely far away and consisting of the points $(0 : x_1 : \ldots : x_n)$; in its turn, it is obtained from the affine subspace V_1 by adding a projective space P^{n-2}, infinitely far away (with respect to V_1), and so on.

c) *Projective spaces and spheres.* In the case $\mathcal{K} = \mathbf{R}$ or $\mathbf{C}$, there is a convenient method for normalizing the homogeneous coordinates in P^n, which does not require the selection of a non-zero coordinate and division by it. Namely, any point in P^n can be represented by the coordinates $(x_0 : \ldots : x_n)$ with the condition $\sum_{i=0}^n |x_i|^2 = 1$, that is, by a point on the n-dimensional (with $\mathcal{K} = \mathbf{R}$) or $(2n+1)$-dimensional (with $\mathcal{K} = \mathbf{C}$) Euclidean sphere. The degree of the remaining non-uniqueness is as follows: the point $(\lambda x_0 : \ldots : \lambda x_n)$ as before lies on the unit sphere if and only if $|\lambda| = 1$, that is, $\lambda = \pm 1$ for $\mathcal{K} = \mathbf{R}$, and $\lambda = e^{i\phi}$, $0 \leq \phi < 2\pi$ for $\mathcal{K} = \mathbf{C}$.

In other words, an n-dimensional real projective space $\mathbf{R}P^n$ is obtained from an n-dimensional sphere S^n by identifying antipodal pairs. In particular, $\mathbf{R}P^1$ is arranged like a circle and $\mathbf{R}P^2$ is arranged like a Möbius strip, to whose boundary a circle is sewn.

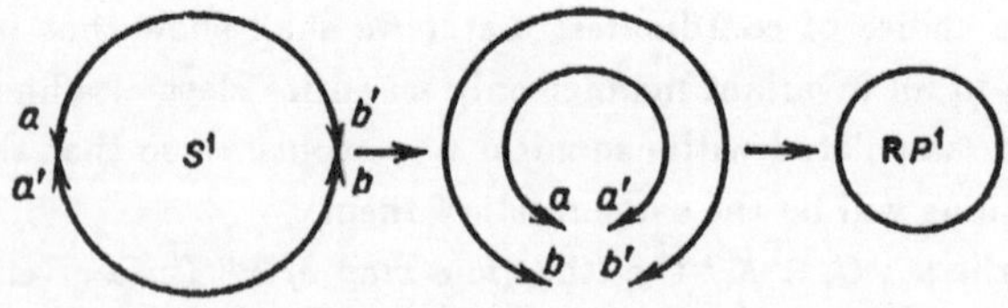

Fig. 1

It is more difficult to visualise $\mathbf{C}P^n$: an entire great circle of the sphere S^{2n+1} consisting of the points $(x_0 e^{i\phi}, \ldots, x_n e^{i\phi})$ with variable ϕ, is sewn to one point of $\mathbf{C}P^n$. From the description of $\mathbf{C}P^1$ in item c) above as $\mathbf{C} \cup \{\infty\}$, it is clear that $\mathbf{C}P^1$ can be represented by a two-dimensional Riemann sphere, in which ∞ is represented by the North Pole, like in a stereographic projection (Fig. 3). Therefore, our new

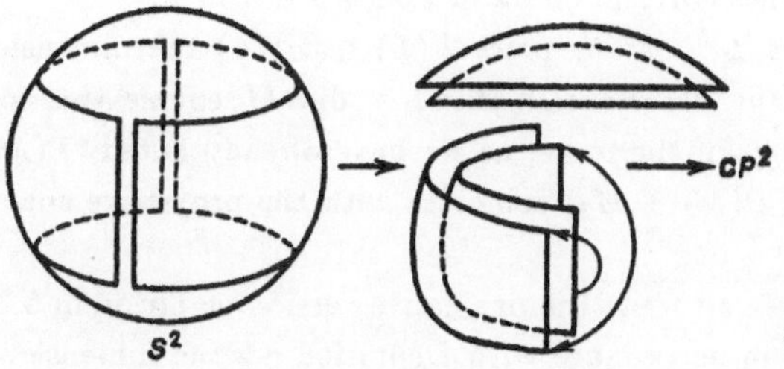

Fig. 2

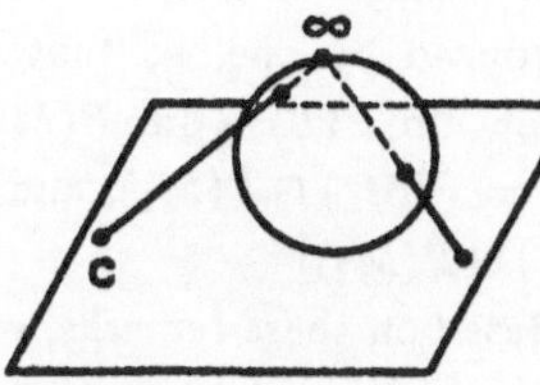

Fig. 3

representation of $\mathbf{C}P^1$ in the form of a quotient space of S^3 gives the remarkable mapping $S^3 \to S^2$, whose fibres are circles in S^1. It is called the *Hopf mapping*.

In the exposition of this subsection we have completely ignored the linear structure, which is the basis for $\mathbf{R}P^n$ and $\mathbf{C}P^n$, though we were able to see clearly the topological properties of the spaces, primarily their compactness. (Strictly speaking, no topology entered into the definition of P^n; it is more convenient to introduce it precisely with the help of mappings of spheres, agreeing that open sets in $\mathbf{R}P^n$ and $\mathbf{C}P^n$ are sets whose inverse images in S^n and S^{2n+1} are open.) Henceforth we shall not use topology, and we shall return to the study of the linear geometry of projective spaces. It is not, however, an exaggeration to say that the importance of $\mathbf{R}P^n$ and $\mathbf{C}P^n$ is to a large extent explained by the fact that these are the only compactifications of $\mathbf{R}^n$ and $\mathbf{C}^n$ that make it possible to extend the basic features of a linear structure to infinity. Even over an abstract field $\mathcal{K}$, not carrying any topology, this "compactness" of projective spaces appears in many algebraic variants. The following is a typical example: two different straight lines in an affine space, generally speaking, intersect at one point, but they can also be parallel. This means that their point of intersection has receded to infinity, and it is successfully observed by transferring to a projective plane: any two projective straight lines in a plane intersect.

We now return to the systematic study of the geometry of P^n.

6.4. Projective subspaces. Let $M \subset L$ be any linear subspace of L. Then $P(M) \subset P(L)$ because every straight line in M is at the same time a straight line in L.

Sets of the type $P(M)$ are called *projective subspaces of* $P(L)$. Evidently, $P(M_1 \cap M_2) = P(M_1) \cap P(M_2)$ and the same is true for the intersection of any family. Therefore, a family of projective subspaces is closed with respect to intersections. For this reason, the set of projective subspaces $P(L)$, containing a given set $S \subset$ $\subset P(L)$ contains a smallest set – the intersection of all such subspaces. It is called the *projective span of* S and is denoted by $\bar{S}$; it coincides with $P(M)$, where M is the linear span of all straight lines corresponding to points $s \in S$ in L.

In transforming from pairs $L \subset M$ to pairs $P(L) \subset P(M)$ the dimensions are reduced by one, so that the *codimension* $\dim L - \dim M$ equals the codimension $\dim P(L) - \dim P(M)$. Furthermore, as we have already noted, $P(M_1 \cap \cap M_2) = P(M_1) \cap P(M_2)$, and $P(M_1 + M_2)$ coincides with the projective span of $P(M_1) \cup P(M_2)$.

Based on these remarks, we can write the projective version of Theorem 5.3 of Chapter 1. We note only that in accordance with Definition 6.2 the dimension of an empty projective space must be set equal to -1: this case is entirely realistic, because the intersection of non-empty subspaces can be empty.

6.5. Theorem. *Let P_1 and P_2 be two finite-dimensional projective subspaces of the projective space P. Then*

$$\dim P_1 \cap P_2 + \dim \overline{P_1 \cup P_2} = \dim P_1 + \dim P_2.$$

6.6. Examples. a) Let P_1 and P_2 be two different points. Then $\dim P_1 \cap P_2 = = -1$, $\dim P_1 = \dim P_2 = 0$, whence $\dim \overline{P_1 \cup P_2} = 1$, that is, the projective span of two points is a straight line. According to the definition of a projective span, it is the only projective straight line passing through these points.

b) Let $\dim P_1 + \dim P_2 \geq \dim P$. Then, since $\overline{P_1 \cup P_2} \leq \dim P$, we have $P_1 \cap P_2 \leq 0$. In other words, two projective subspaces, the sum of whose dimensions is greater than or equal to the dimension of the enveloping space, have a non-empty intersection. In particular, there are no "parallel" straight lines in the projective plane: any two straight lines either intersect at one point or at two points and then (according to item a) above) they coincide. Analogously, two projective planes in a three-dimensional projective space necessarily intersect along a straight line or they coincide. A projective plane and a straight line in the three-dimensional space intersect at a point, or the straight line lies in the plane.

c) The condition $P_1 \cap P_2 = \emptyset$ in the case $P_i = P(M_i)$ means that $M_1 \cap M_2 = \{0\}$, that is, the sum $M_1 + M_2$ is a *direct* sum.

6.7. Representation of projective subspaces by equations. The linear function $f : L \to \mathcal{K}$ on a linear space L *does not define* any function on $P(L)$ (except the case $f \equiv 0$), because there is always a straight line in L on which this function is not constant, and it is impossible to fix its value at the corresponding point of $P(L)$. But the *equation* $f = 0$ defines a linear subspace of L and therefore a projective subspace of $P(L)$. If L is finite-dimensional, then any subspace of L and therefore any subspace of $P(L)$ can be defined by the system of equations

$$f_1 = \ldots = f_m = 0.$$

This effect is manifested as follows in the homogeneous coordinates P^n: the system of linear homogeneous equations

$$\sum_{j=0}^{n} a_{ij}x_j = 0, \ i = 1, \dots, m,$$

defines a projective subspace of P^n, consisting of points whose homogeneous coordinates $(x_0 : \dots : x_n)$ satisfy this system. Multiplication of all coordinates by λ does not destroy the fact that the left sides vanish.

6.8 Affine subspaces and hyperplanes. Let $M \subset L$ be a projective subspace of codimension one. Then $P(M) \subset P(L)$ has codimension one, and we shall call such subspaces *hyperplanes*.

We shall now show how to introduce on the complement A_M of the hyperplane $P(M)$ the structure of an affine space $(A_M, M, +)$. We choose in L a linear variety $M' = m' + M$, not passing through the origin of coordinates. It has a unique affine structure: a translation by $m \in M$ into M' is induced by a translation by m in L, that is, it consists of adding m.

On the other hand, there is a bijective correspondence between A_M and M': a point in A_M is a straight line not contained in M and it intersects M' at one point, which we associate with the starting point of A_M. In this manner all points are obtained once each. With the help of this bijective correspondence the affine structure on M' can be transferred to A_M. However, the choice of M' is not unique, and therefore the affine structure of A_M is not unique either. In order to compare two such structures, we shall show that the set-theoretic identity mapping of A_M into itself is an affine isomorphism of these two structures.

6.9. Proposition. *Let $(A_M, M, +')$ and $(A_M, M, +'')$ be two affine structures on A_M, constructed with the help of the procedure described above. Then the identity mapping of A_M into itself is an affine isomorphism, whose linear part is some homothety of M.*

Proof. Let the two structures correspond to the subvarieties $m' + M$ and $m'' + M$. The sets $m' + M$ and $m'' + M$ in the one-dimensional quotient space L/M are proportional. Therefore, it may be assumed that $m'' = am', a = \mathcal{K}$. Multiplication by a in L transforms $m' + M$ into $m'' + M$ and induces the identity mapping of $P(L)$ into itself and therefore of A_M into itself. On the other hand, a translation by a vector $m \in M$ in $m' + M$ under a homothety transforms into a translation by a vector $am \in M$ in $m'' + M$. This completes the proof.

6.10. Corollary. *The set of affine subspaces of A_M together with their identity relations as well as sets of affine mappings of A_M into other affine spaces are independent of the arbitrariness in the choice of affine structure of A_M.*

This justifies the possibility of regarding the complement of any hyperplane in a projective space simply as an affine space without further refinements.

We shall now examine the appearance of the projective space $P(M)$ "from the viewpoint" of an affine space A_M.

6.11. Proposition. *There is a bijective correspondence between the points of $P(M)$ and the sets of parallel straight lines in A_M. In other words, every point of $P(M)$ is "a path to infinity" in A_M.*

Proof. We identify A_M with $m' + M$. The set of parallel straight lines in $m' + M$ is determined by its orienting subspace in M, that is, a point in $P(M)$, and this correspondence is bijective.

6.12. In fact, we can say more: every straight line l in A_M uniquely determines a straight line in $P(L)$ containing it, namely, its projective span $\bar{l}$. The projective span is obtained by adding to l one point that is contained precisely in $P(M)$ and is the "point at infinity" on this straight line. The entire set of parallel straight lines in A_M has a common point at infinity in $P(M)$. Under the identification of A_M with $m' + M$ the span of $\bar{l}$ corresponds to all straight lines of a plane in L passing through l and its orienting subspace, and the point at infinity in $\bar{l}$ is the orienting subspace itself.

More generally, let $A \subset A_M$ be any affine subspace. Then its projective span $\bar{A}$ in $P(L)$ has the following properties:

a) $\bar{A} \backslash A \subset P(M)$: only the points at infinity are added.

b) $\dim A = \dim \bar{A}$.

c) $\bar{A} \backslash A$ is a projective space in $P(M)$ with dimension $\dim A - 1$. (Therefore $\bar{A}$ is also called the projective closure of A.)

The identification of A_M with $m' + M$ reduces the verification of these properties to the direct application of the definitions. Indeed, $\bar{A}$ consists of straight lines contained in the linear span $A \subset m' + M$. This linear span is spanned by the orienting subspace L_0 of A and any vector from A. Therefore, its dimension equals $\dim L_0 + 1 = \dim A + 1$, and hence $\dim A = \dim \bar{A}$. All straight lines in this linear span intersect $m' + M$, that is, they correspond to points in A, with the exception of straight lines contained in the orienting subspace L_0. The latter lie in $P(M)$ and form a projective space with dimension $\dim L_0 - 1 = \dim A - 1$.

§7. Projective Duality and Projective Quadrics

7.1. Let L be a linear space over the field $\mathcal{K}$ and L^* its dual space of linear functionals on L. *The projective space $P(L^*)$ is called the dual space of the projective space $P(L)$.*

Every point in $P(L^*)$ is a straight line $\{\lambda f\}$ in the space of linear functionals on L. The hyperplane $f = 0$ in $P(L)$ does not depend on the choice of functional f on this straight line and uniquely determines the entire line. Therefore, we can say that *the points of the dual projective space are hyperplanes of the starting projective space.*

If dual bases are chosen for L and L^* and a corresponding system of homogeneous coordinates in $P(L)$ and $P(L^*)$ is also chosen, then this correspondence acquires a simple form: a hyperplane, represented by the equation

$$\sum_{i=0}^{n} a_i x_i = 0$$

in $P(L)$ corresponds to a point with the homogeneous coordinates $(a_0 : \ldots : a_n)$ in $P(L^*)$. The canonical isomorphism $L \to L^{**}$ shows that the duality relationship between two projective spaces is symmetric. More generally, translating the results of §7 of Chapter 1 into projective language, we obtain the following duality relationship between systems of projective subspaces in $P(L)$ and $P(L^*)$ (we assume below that L is finite-dimensional).

a) The subspace $P(M) \subset P(L)$ corresponds to its dual subspace $P(M^\perp) \subset$ $\subset P(L^*)$. In this case,

$$\dim P(M) + \dim P(M^\perp) = \dim P(L) - 1.$$

b) The intersection of projective subspaces corresponds to the projective span of their dual spaces, while the projective span corresponds to the intersection. In particular, the incidence relation between two subspaces (that is, the inclusion of one in the other) transfers into an incidence relation.

This makes it possible to formulate the following *principle of projective duality*, which is, strictly speaking, *a metamathematical principle*, because it is an assertion about the language of projective geometry.

7.2. Principle of projective duality. *Suppose that we have proved a theorem about the configurations of projective subspaces of projective spaces, whose formulation involves only the properties of dimensionality, incidence, intersection, and the selection of a projective span. Then the duality assertion, in which all terms are replaced by their duals according to the rules of the preceding item, is also a theorem about projective geometry.*

Here is a simple example: the theorem "two different planes in a 3-dimensional projective space intersect along one straight line" is the dual of the theorem: "one straight line passes through any two points in a 3-dimensional projective space". (Much more interesting theorems about projective configurations will be presented in §9.)

7.3. Projective duality and quadrics. If a linear space L is equipped with the isomorphism $L \to L^*$, then $P(L^*)$ can be identified with $P(L)$, and the duality mapping between the projective subspaces $P(L)$ and $P(L^*)$ will become a *duality mapping between subspaces of* $P(L)$.

The specification of an isomorphism $L \to L^*$ is equivalent to the specification of a non-degenerate inner product $g : L \times L \to \mathcal{K}$. We shall examine in greater detail the geometry of projective duality for the case when the inner product g is *symmetric*. As usual, we shall assume that the characteristic of the field $\mathcal{K}$ is not two. Then g is uniquely reconstructed from the quadratic form $q(l) = g(l, l)$.

The equation $q(l) = 0$ defines a quadric Q_0 in L. We shall also call its image in $P(L)$ a quadric and, in application to the duality theory, a *polar quadric*. We note that Q_0 is a cone centred at the origin: if $l \in Q_0$, then all straight lines $\mathcal{K}l$ are contained in Q_0. Identifying $P(L)$ with the points of L at infinity, we can identify Q with the base of the cone Q_0.

According to the general theory, g and q define a duality mapping of the set of projective subspaces $P(L)$ into itself; the hyperplane in $P(L)$, dual to the point $p \in P(L)$, is said to be *polar to p* (relative to q or Q). To explain the geometric structure of this mapping we first introduce an equation for a polar hyperplane in homogeneous coordinates. We can at first work in L. Let the equation of Q_0 have the form

$$q(x_0, \ldots, x_n) \equiv \sum_{i,j=0}^{n} a_{ij} x_i x_j = 0, \quad a_{ij} = a_{ji}.$$

The point $(x_0^0, \ldots, x_n^0)$ in L with the isomorphism $L \to L^*$, associated with q, corresponds to the linear function $\sum_{i,j=0}^{n} a_{ij} x_i^0 x_j$ of $(x_0, \ldots, x_n) \in L$. Therefore, the equation of the polar hyperplane has the form

$$\sum_{i,j=0}^{n} a_{ij} x_i^0 x_j = 0.$$

In particular, *if* $(x_0^0 : \ldots : x_n^0) \in Q$, *then the polar hyperplane to this point contains the point.* Moreover, in this case its equation can be rewritten in the form

$$\sum_{j=1}^{n} \frac{\partial q}{\partial x_j}(x_0^0, \ldots, x_n^0)(x_j - x_j^0) = \sum_{i,j=0}^{n} a_{ij} x_i^0 (x_j - x_j^0) = 0.$$

In elementary analytic geometry (over $\mathbf{R}$), such an equation defines the *tangential hyperplane* to Q_0 at its point $(x_0^0, \ldots, x_n^0)$. This motivates the following general definition.

7.4. Definition. *By the* tangential hyperplane *to a non-degenerate quadric* $Q \subset P(L)$ *at the point* $p \in Q$, *one means the hyperplane polar to* p *relative to the quadratic form* q, *specifying* Q.

Employing the general properties of projective duality we can now immediately reconstruct the geometrically significant part of a duality mapping and obtain a series of beautiful and not obvious geometric theorems, examples of which we shall present. In what follows Q is a (non-degenerate) quadric in P^2 or P^3.

a) Let $Q \subset P^2$ and let p_1 and p_2 be two points on the quadric and p_3 the point of intersection of the tangents to Q at p_1 and p_2. According to the general duality principle, the point p_3 then corresponds to the straight line $\overline{p_1p_2}$, passing through p_1 and p_2, that is, to the projective span of p_1 and p_2.

We constrain the point p_3 to move along the straight line l; we draw from every point of the straight line two tangents to Q and connect pairs of tangent points. Then, all "chords" of Q so obtained will intersect at one point r, which corresponds to l by virtue of duality. We note once again that the proof does not require any calculations: this follows simply from the fact that according to the general principle of duality the projective span of the points $p_3, p_3', p_3'', \ldots$ is polar to the intersection of their dual straight lines, which are precisely the corresponding chords.

One point, however, deserves special mention. The intersecting pairs of tangents to points of Q may not sweep out the entire plane. For example, for an ellipse in $\mathbf{R}P^2$, as in Fig. 4 (of course, we have drawn only a piece of the affine map in $\mathbf{R}P^2$), we only obtain the *exterior* of the ellipse. How do we determine the straight lines corresponding to the interior of the ellipse ? The sketch in the figure suggests the answer: by virtue of the duality symmetry we must draw through the interior point r a pencil of chords of Q, and then construct the points of intersection of the tangents to Q at the opposite ends of these chords; they sweep out the straight line l dual to the point r.

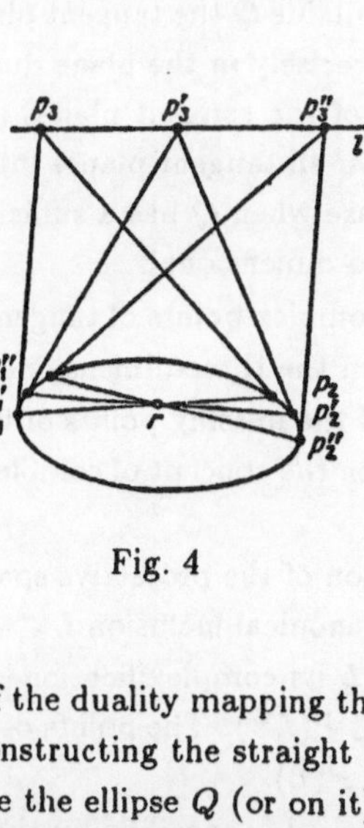

Fig. 4

However, the description of the duality mapping thus becomes inhomogeneous. We now have two recipes for constructing the straight line l polar to the point r.

1) If the point r lies outside the ellipse Q (or on it), draw two tangents from r to Q (or one) and connect the tangent points with the straight line l (or take the

tangent l).

2) If the point r lies inside the ellipse Q, draw all straight lines through r, and construct the points of intersection of the tangents at the two points of intersection of straight lines through r with Q. Their geometric locus will be the straight line dual to r.

The main point here is that the basic field $\mathbf{R}$ is not algebraically closed. If we had worked in $\mathbf{C}P^2$ instead, *both recipes could have been used*, and moreover, *for all points* $r \in \mathbf{C}P^2$. The real straight line l, lying wholly outside the real ellipse Q nevertheless *intersects* it, *but at two complex conjugate points*, and two complex conjugate tangents to Q at these points now intersect at a *real* point r inside Q. It is nevertheless possible to draw from the real point r inside Q *two complex conjugate tangents* to Q, through whose points of tangency passes a real straight line – this is l.

In this sense, the real projective geometry $\mathbf{R}P^2$ is only a piece of the geometry $\mathbf{C}P^2$, and the truly simple and symmetric duality theory exists in $\mathbf{C}P^2$, whereas $\mathbf{R}P^2$ reflects only its real part.

Classical projective geometry was largely concerned with clarifying the details of this beautiful world of configurations, consisting of quadrics, chords, and tangents and "invisible" complex points of tangency and intersection. Indeed, the entire quadric may not have real points, such as $x_0^2+x_1^2+x_2^2=0$. Nevertheless, the visible part of the duality appears in $\mathbf{R}P^2$.

b) We shall give one more illustration for the three-dimensional case. Consider the projective non-degenerate quadric Q in a three-dimensional projective space, and draw from the point r outside Q the tangent planes to Q. All points of tangency then lie in one plane, and precisely in the plane dual to r. The reason is once again the same: the intersection of the tangent planes is dual to the projective span of the points of tangency, and if all tangent planes intersect precisely at r (this should and can be proved in the case when Q has a sufficient number of points), then this projective span must be two-dimensional.

The comments about complex points of tangency and intersections made in the two-dimensional case hold in the three-dimensional case also.

A rigorous definition of the missing points of the projective space and quadrics in the case $\mathcal{K} = \mathbf{R}$ is based on the concept of complexification (see §12 of Chapter 1).

7.5. a) The complexification of the projective space $P(L)$ over $\mathbf{R}$ is the projective space $P(L^{\mathbf{C}})$ over $\mathbf{C}$. The canonical inclusion $L \subset L^{\mathbf{C}}$ makes it possible to associate to every $\mathbf{R}$-straight line in L its complexification — a $\mathbf{C}$-straight line in $L^{\mathbf{C}}$ which defines the inclusion $P(L) \subset P(L^{\mathbf{C}})$. The points of $P(L^{\mathbf{C}})$ are the "complex points" of the real projective space $P(L)$.

b) The isomorphism $L \to L^*$, specified by the scalar product g on L, induces a complexified isomorphism $L^{\mathbf{C}} \to (L^{\mathbf{C}})^*$. It is defined by the symmetric inner product $g^{\mathbf{C}}$ on $L^{\mathbf{C}}$, which defines the quadric $Q^{\mathbf{C}}$ and the projective duality in $P(L^{\mathbf{C}})$. The operation of complex conjugation, induced by the antilinear isomorphism $L^{\mathbf{C}} \to \bar{L}^{\mathbf{C}}$, an identity on $L \subset L^{\mathbf{C}}$, operates on $L^{\mathbf{C}}$ and $P(L^{\mathbf{C}})$. The points of $P(L)$ are the points of $P(L^{\mathbf{C}})$ that are invariant under complex conjugation; they are called real points. More generally, the projective subspaces of $P(L^{\mathbf{C}})$, transforming into themselves under complex conjugation, form a bijection with the projective subspaces of $P(L)$. We shall call such subspaces real. Then, two mappings establishing one-to-one bijections can be described as follows:

(real projective subspace in $P(L^{\mathbf{C}})$) $\to$(set of its real points in $P(L)$);

(projective subspace in $P(L)$) $\to$ (its complexification in $P(L^{\mathbf{C}})$).

c) The duality mapping in $P(L)$, defined with the help of g, is obtained from the duality mapping in $P(L^{\mathbf{C}})$ by restricting the latter to the system of real subspaces of $P(L^{\mathbf{C}})$ identified with the system of subspaces of $P(L)$, as in b).

Using the results of §12 of Chapter I, it is straightforward to verify all these assertions, and in the real coordinate system of $L^{\mathbf{C}}$, transferred from L, they are entirely tautological. The only serious aspect of the situation illustrated in the examples given above is the possibility that the real points will appear in invisible complex configurations, such as the points of intersection, lying inside an ellipse, of two imaginary tangents at two complex-conjugate points of this ellipse.

In the case of the basic field $\mathcal{K}$ differing from $\mathbf{R}$, the general functor that extends the main field (for example, up to algebraic closure of $\mathcal{K}$) must be used instead of complexification. The situation, however, is complicated somewhat by the fact that instead of the one mapping of complex conjugation the entire Galois group must be invoked in order to distinguish objects defined over the starting field (which are real in the case $\mathcal{K} = \mathbf{R}$).

§8. Projective Groups and Projections

8.1. Let L and M be two linear subspaces and $f : L \to M$ a linear mapping. If $\ker f = \{0\}$ then f maps any straight line from L into a uniquely determined straight line in M and hence induces the mapping $P(f) : P(L) \to P(M)$, called the projectivization of f. In particular, if f is an isomorphism, then $P(f)$ is called a projective isomorphism. When $\ker f \neq \{0\}$ the situation is more complicated: straight lines contained in $\ker f$, that is, consisting of the projective subspace $P(\ker f) \subset P(L)$, are mapped into zero, which does not determine any point in $P(M)$. Therefore, the projectivization $P(f)$ is determined only on the complement $U_f = P(L) \backslash P(\ker f)$. Both of these cases are important. But they lead in different directions, so that we shall study them separately. The most important geometric features of this situation are already revealed when $L = M$.

8.2. The projective group. Let $M = L$, and let f run through the group of linear automorphisms of the space L. The following assertions are obvious:

a) $P(\mathrm{id}_L) = \mathrm{id}_{P(L)}$;

b) $P(fg) = P(f)P(g)$.

In particular, all mappings $P(f)$ are bijective and $P(f^{-1}) = P(f)^{-1}$. Therefore, $P(f)$ runs through the group of mappings of $P(L)$ into itself, which is called the *projective group* of the space $P(L)$ and is denoted by $\mathrm{PGL}(L)$; the mapping P: $\mathrm{GL}(L) \to \mathrm{PGL}(L)$, $f \to P(f)$ is a surjective homomorphism of groups.

Every mapping $P(f)$ maps the projective subspaces of $P(L)$ into projective subspaces, preserving dimension and all incidence relations.

$\mathrm{PGL}(n)$ is written instead of $\mathrm{PGL}(\mathcal{K}^{n+1})$.

8.3. Proposition. *The kernel of a canonical mapping $P: GL(L) \to PGL(L)$ consists precisely of the homotheties. Therefore, $PGL(L)$ is isomorphic to the quotient group $GL(L)/\mathcal{K}^*$ where*

$$\mathcal{K}^* = \{a\,\mathrm{id}_L | a \in \mathcal{K} \backslash \{0\}\}.$$

Proof. By definition, $\ker P = \{f \in \mathrm{GL}(L) | P(f) = \mathrm{id}_{P(L)}\}$. Every homothety maps every straight line in L into itself. Therefore, $\mathcal{K}^* \subset \ker P$. Conversely, every element of $\ker P$ maps any straight line into itself and is therefore diagonalizable in any basis of L. But then all of its eigenvalues must be equal. Indeed, let $f(e_1) = \lambda_1 e_1$, $f(e_2) = \lambda_2 e_2$, where e_1, e_2 are linearly independent. Then the conditions $f(e_1 + e_2) = \mu(e_1 + e_2) = \lambda_1 e_1 + \lambda_2 e_2$ imply that $\lambda_1 = \mu = \lambda_2$. Hence f is a homothety, which proves the required result.

8.4. Coordinate representation of the mappings $P(f)$. If the linear mapping $f : L \to L$ is represented in terms of coordinates of the matrix A:

$$f(x_0, \ldots, x_n) = A \cdot [x_0, \ldots, x_n]$$

(product of a matrix A by the column $[x_0, \ldots, x_n]$), then $P(f)$ in appropriate homogeneous coordinates is represented by the same matrix A or any matrix proportional to it:

$$P(f)(x_0 : \ldots : x_n) \subset (\lambda A) \cdot [x_0, \ldots, x_n], \ \lambda \in \mathcal{K}^*.$$

If we study only points with $x_0 \neq 0$, whose projective coordinates can be chosen in the form $(1 : y_1 : \ldots : y_n)$, and also write the coordinates of the image of the point, then we arrive at linear-fractional formulas:

$$P(f)(1 : y_1 : \ldots : y_n) = (1 : y'_1 : \ldots : y'_n) =$$

$$= \left(a_{00} + \sum_{i=1}^{n} a_{i0} y_i : a_{01} + \sum_{i=1}^{n} a_{i1} y_i : \ldots : a_{0n} + \sum_{i=1}^{n} a_{in} y_i \right) =$$

$$\left(1 : \frac{a_{01} + \sum_{i=1}^{n} a_{i1} y_i}{a_{00} + \sum_{i=1}^{n} a_{i0} y_i} : \ldots : \frac{a_{0n} + \sum_{i=1}^{n} a_{in} y_i}{a_{00} + \sum_{i=1}^{n} a_{i0} y_i} \right).$$

($P(f)$, of course, has an analogous form on the set of points where $x_i \neq 0$, for arbitrary i.) These expressions become meaningless when the denominator vanishes, that is, at those points of the complement to the hyperplane $x_0 = 0$ that $P(f)$ maps into this hyperplane. If there are no such points, then in terms of affine coordinates $(y_1, \ldots y_n)$ in $P(L) \setminus \{x_0 = 0\}$ we obtain an affine mapping. The following result gives an invariant explanation of this.

8.5. Proposition. *Let $M \subset L$ be a subspace with codimension one, $P(M) \subset P(L)$ the corresponding hyperplane, and A_M its complement with the affine structure described in §6. We associate with any projective automorphism $P(f) : P(L) \to P(L)$ with the condition $f(M) \subset M$, its restriction to A_M. We obtain an isomorphism of a subgroup of $PGL(L)$, mapping $P(M)$ into itself, to* Aff *A. The linear part of the restriction of $P(f)$ to A_M is proportional to the restriction of f to M.*

Proof. We introduce an affine structure on A_M identifying A_M with the linear variety $m' + M \subset L$: every point in A_M is associated with the intersection of the corresponding straight line in L with $m' + M$. If $f(M) \subset M$, then the set of f with the same $P(f)$ contains a unique mapping f_0 for which $f_0(m' + M) = m' + M$. The restrictions of all of these mappings f_0 form a group of affine mappings of $m' + M$, since $m' + M$ is an affine subspace of L with its affine structure, while $f_0 : L \to L$ is linear and therefore affine. The linear part of such an f_0 evidently coincides with the restriction of f_0 on M. For any linear part there exists a corresponding f_0, and for a fixed linear part there exists an f_0 that maps any point in $m' + M$ into any other point. To see this, it is sufficient to choose a basis of L consisting of a basis of M and a vector in m', and then to apply the formulas of §3. Finally, if f acts like an identity on $m' + M$ and M, then $P(f) = \mathrm{id}_{P(L)}$, because f maps every straight line in L into itself. This completes the proof.

8.6. Action of the projective group on projective configurations. We shall call a finite ordered system of projective subspaces in $P(L)$ a projective configuration. We shall say that two configurations are projectively congruent if and only if one can be mapped into the other by a projective mapping of $P(L)$ into itself. Evidently, for this it is necessary and sufficient that the corresponding configurations of the linear subspaces of L be identically arranged in the sense of §5 of Chapter 1. Therefore we can immediately translate the results proved there into the projective language and obtain the following facts.

a) The group $\mathrm{PGL}(L)$ acts transitively on the set of projective subspaces with fixed dimension in $P(L)$, that is, all such subspaces are congruent (see §5.1 of Chapter 1).

b) The group $\mathrm{PGL}(L)$ acts transitively on the set of ordered pairs of projective subspaces of $P(L)$, each member of a pair and their intersections having fixed dimensions, that is, all such pairs are congruent (see §5.5 of Chapter 1).

c) The group $\mathrm{PGL}(L)$ acts transitively on the set of ordered n-tuples of projective subspaces $(P_1, \ldots, P_n)$ with fixed dimensions $\dim P_i$, which have the following property: for each i the subspace P_i does not intersect the projective span of $(P_1, \ldots, P_{i-1}, P_{i+1}, \ldots, P_n)$, that is, the smallest projective subspace containing this system.

Thus let $P_i = P(L_i)$, $L_i \subset L$. The projective span of $(P_1, \ldots, P_{i-1}, P_{i+1}, \ldots$ $\ldots, P_n)$, as it is easy to verify, coincides with $P(L_1 + \ldots + L_{i-1} + L_{i+1} + \ldots + L_n)$, and the condition that its intersection with $P(L_i)$ is empty means that $L_i \cap \sum_{j \neq i}^{n} L_j =$ $= \{0\}$. Theorem 5.8a of Chapter 1 implies that the sum $L_1 \oplus \ldots \oplus L_n$ is a direct sum and $\mathrm{GL}(L)$ acts transitively on such n-tuples of subspaces (choose a basis of L, extending the union of the bases of all L_i, and use the fact that $\mathrm{GL}(L)$ is transitive on the bases of L).

As a particular case ($\dim P_i = 0$ for all i), we have the following result: all collections of n points in $P(L)$ with the property that no point is contained in the projective span of the remaining points is projectively congruent.

d) The group $\mathrm{PGL}(L)$ acts transitively on the set of projective flags $P_1 \subset P_2 \subset$ $\subset \ldots \subset P_n$ in $P(L)$ of fixed length n and with fixed dimensions $\dim P_i$.

Indeed, any such flag is an image of the flag $L_1 \subset L_2 \subset \ldots$ in L; choose a basis of L, the first $\dim P_i + 1$ elements of which generate the subspace L_i for each i, and use once again the transitivity of the action of $\mathrm{GL}(L)$ on the bases.

In addition to these results, which are a direct consequence of the corresponding theorems for linear spaces, we shall analyse an interesting new case in which a nontrivial invariant relative to projective congruency appears for the first time: the classical "cross ratio of a quadruple of points on a projective line". Many of the arguments can be given in the case of arbitrary dimension, and we shall begin with a general definition.

8.7. Definition. *A system of points $p_1, \ldots, p_N$ in an n-dimensional projective space P is in general position if for all $m \leq \min\{N, n+1\}$ and all subsets $S \subset$ $\subset \{1, \ldots, N\}$ with cardinality m, the dimension of the projective span of the points $\{p_i | i \in S\}$ equals $m - 1$.*

We shall be especially interested in the cases $N = n+1$, $n+2$, and $n+3$.

a) $n+1$ *points in the general position.* Since no point of the system is contained in the projective span of the remaining points (otherwise the projective span of the entire system would have the dimension $n-1$ and not n), such configurations have already been studied in §8.6c; in particular, the projective group on them is transitive. We now want to call attention to the fact that a projective mapping that maps one system of $n+1$ points in general position into another is not defined uniquely.

Indeed, if $e_1, \ldots, e_{n+1}$ are non-zero vectors in $p_1, \ldots, p_{n+1}$ respectively, then $\{e_1, \ldots, e_{n+1}\}$ is a basis of L (where $P = P(L)$) and the group of projective mappings which leave all points p_i in place consists precisely of mappings of the form $P(f)$, where the f are diagonal in the basis $\{e_1, \ldots, e_{n+1}\}$. This remaining degree of freedom makes it possible to prove the transitivity of the action of $\mathrm{PGL}(L)$ on systems of $(n+2)$ points in general position.

b) $n+2$ *points in general position.* If the points $\{p_1, \ldots, p_{n+2}\}$ are in general position, then the points $\{p_1, \ldots, p_{n+1}\}$ are also in general position. As done in the preceding item, we choose a basis $\{e_1, \ldots, e_{n+1}\}$, $e_i \in p_i$. It defines a system of homogeneous coordinates in P. Let $(x_1 : \ldots : x_{n+1})$ be the coordinates of the point p_{n+2} in this basis. No coordinate x_i vanishes, otherwise the vector $(x_1, \ldots, x_{n+1})$ on the straight line p_{n+2} can be expressed linearly in terms of the vectors e_j, $1 \leq j \leq n+1$, $j \neq i$, whence it follows that the projective span of the $n+1$ points $\{p_i | i \neq j\}$ would have the dimension $n-1$ and not n. But the mapping $P(f)$ with $f = \mathrm{diag}(\lambda_1, \ldots, \lambda_{n+1})$ (in the basis $\{e_1, \ldots, e_{n+1}\}$) maps $(x_1 : \ldots : x_{n+1})$ into the point $(\lambda_1 x_1 : \ldots : \lambda_{n+1} x_{n+1})$, and leaves $p_1 \ldots, p_{n+1}$ in place. From here it follows that any point $(x_1 : \ldots : x_{n+1})$ (all $x_i \neq 0$) can be mapped into any other point $(y_1 : \ldots : y_{n+1})$ (all $y_i \neq 0$) by a unique projective mapping which leaves $p_1, \ldots, p_n$ in place.

We have thus established that all ordered systems of $n+2$ points in general position in P, where $\dim P = n$, are congruent and moreover, they form the main homogeneous space over the group $\mathrm{PGL}(L)$.

Adopting the passive rather than the active viewpoint, we can say that for any ordered system of points $\{p_1, \ldots, p_{n+2}\}$ there exists a unique system of homogeneous coordinates in P in which the coordinates of $p_1, \ldots, p_{n+2}$ have the form:

$$p_i = (1 : 0 : \ldots : 0),\ p_2 = (0 : 1 : 0 : \ldots : 0), \ldots,\ p_{n+1} = (0 : \ldots 0 : 1),$$

$$p_{n+2} = (1 : \ldots : 1).$$

We can say that this system is adapted to $\{p_1, \ldots, p_{n+2}\}$.

c) $n+3$ *points in general position.* Such configurations are no longer all congruent: if $\{p_1, \ldots, p_{n+3}\}$ and $\{p'_1, \ldots, p'_{n+3}\}$ are given, then we can find a unique projective mapping that maps p_i into p'_i for all $1 \leq i \leq n+2$, but p_{n+3} does or does not fall into p'_{n+3}, depending on the situation.

It is not difficult to describe the projective invariants of a system of $n+3$ points. Choose a system of homogeneous coordinates in P, in which the first $n+2$ points have the coordinates described in the preceding item. In it, the point p_{n+3} has the coordinates $(x_1 : \dots : x_{n+1})$, determined uniquely up to proportionality. Any projective automorphism P, applied simultaneously to the configuration $\{p_1, \dots, p_{n+3}\}$ and to the system of coordinates adapted to it, maps this configuration into another configuration and the system of coordinates into the adapted system of coordinates of the new configuration. Therefore, the coordinates $(x_1 : \dots : x_{n+1})$ of the last point will remain unchanged.

All preceding arguments can also be transferred with obvious alterations to the case when we have two n-dimensional projective spaces P and P', and configurations $\{p_1, \dots, p_N\} \subset P$ and $\{p_1, \dots, p'_N\} \subset P'$, and we are interested in the projective isomorphisms $P \to P'$ mapping the first configuration into the second. We summarize the results of the discussion in the following theorem.

8.8. Theorem. a) *Let P and P' be n-dimensional projective spaces and let $\{p_1, \dots, p_{n+2}\} \subset P$ and $\{p'_1, \dots, p'_{n+2}\} \subset P'$ be two systems of points in the general position. Then there exists a unique projective isomorphism $P \to P'$ which maps the first configuration into the second.*

b) *An analogous result holds for systems of $n+3$ points in general position if and only if the coordinates of the $(n+3)$rd point in the system, adapted to the first $(n+2)$ points, coincide for both configurations (of course, up to a scalar factor).*

8.9. The cross-ratio. Let us apply Theorem 8.8 to the case $n = 1$. We find, first of all, that if ordered triples of pairwise different points are given on two projective straight lines (this is the condition for the generality of position here), then there exists a unique projective isomorphism of straight lines mapping one triple into another.

Further let a quadruple of pairwise different points $\{p_1, p_2, p_3, p_4\} \subset P^1$ with coordinates in the adapted system $(1 : 0), (0 : 1), (1 : 1)$ and $(x_1 : x_2)$ be given. Then $x_2 \neq 0$. Let

$$[p_2, p_3, p_1, p_4] = x_1 x_2^{-1}.$$

This number is called the *cross-ratio* of the quadruple of points $\{p_i\}$. The unusual order is explained by the desire to maintain consistency with the classical definition: in an affine map, where $p_2 = \infty$, $p_3 = 0$, $p_1 = 1$, the coordinates of points in the brackets are positioned as follows: $[0, 1, \infty, x]$, where x is the cross-ratio of this quadruple.

The term "cross-ratio" itself originates from the following explicit formula for calculating the invariant $[x_1, x_2, x_3, x_4]$, where the $x_i \in \mathcal{K}$ are interpreted here as the coordinates of the points p_i in an arbitrary affine map P^1. According to the results of §8.4, the group PGL(1) in this map is represented by linear-fractional mappings of the form $x \mapsto \frac{ax+b}{cx+d}$, $ad - bc \neq 0$. Such a mapping, which maps (x_1, x_2, x_3) into

$(0, 1, \infty)$ has the form

$$x \mapsto \frac{x_1 - x}{x_3 - x} : \frac{x_1 - x_2}{x_3 - x_2}.$$

Substituting here $x = x_4$, we find

$$[x_1, x_2, x_3, x_4] = \frac{x_1 - x_4}{x_3 - x_4} : \frac{x_1 - x_2}{x_3 - x_2}.$$

Another classical construction related to Theorem 8.8a for $n = 1$ describes the representation of the symmetric group S_3 of linear-fractional mappings. According to this theorem, any permutation $\{p_1, p_2, p_3\} \mapsto \{p_{\sigma(1)}, p_{\sigma(2)}, p_{\sigma(3)}\}$ of three points on a projective straight line is induced by a unique projective mapping of this straight line.

In an affine map, where $\{p_1, p_2, p_3\} = \{0, 1, \infty\}$, these projective mappings are represented by linear-fractional mappings,

$$x \mapsto \{x, \frac{1}{x}, 1 - x, \frac{1}{1 - x}, \frac{x - 1}{x}, \frac{x}{x - 1}\}.$$

We shall now study projections.

8.10. Let a linear space L be represented in the form of the direct sum of two of its subspaces with dimension ≥ 1: $L = L_1 \oplus L_2$. We set $P = P(L), P_i = P(L_i)$.

As shown in §8.1, the linear projection $f : L \to L_2$, $f(l_1 + l_2) = l_2$, $l_1 \in L_i$ induces the mapping

$$P(f) : P \backslash P_1 \to P_2,$$

which we shall call a *projection from the centre of* P_1 *to* P_2. In order to describe the entire situation in purely projective terms, we note the following.

a) $\dim P_1 + \dim P_2 = \dim P - 1$ and $P_1 \cap P_2 = \emptyset$. Conversely, any configuration (P_1, P_2) with these properties originates from a unique direct decomposition $L = L_1 \oplus L_2$.

b) If $a \in P_2$, then $P(f)a = a$; if $a \in P \backslash (P_1 \cup P_2)$, then $P(f)a$ is determined as the point of intersection with P_2 of a unique projective straight line in P, intersecting P_1 and P_2 and passing through a.

Indeed, the case $a \in P_2$ is obvious. If $a \neq P_1 \cup P_2$, then in terms of the space L the required result is formulated thus: through any straight line $L_0 \subset L$, not lying in L_1 and L_2 there passes a unique plane, intersecting L_1 and L_2 along straight lines, and its intersection with L_2 coincides with its projection on L_2. Indeed, one plane with this property exists: it is spanned by the projection of L_0 onto L_1 and L_2 respectively. The existence of two such planes would imply the existence of two different decompositions of the non-zero vector $l_0 \in L_0$ into a sum of two vectors from L_1 and L_2 respectively, which is impossible, because $L = L_1 \oplus L_2$.

Since the described projective construction of the mapping $P \to P \backslash P_1$ is lifted to a linear mapping, we immediately find that for any subspace $P' \subset P \backslash P_1$ the restriction of the projection $P' \to P_1$ is a projective mapping, that is, it has the form $P(g)$, where g is some linear mapping of the corresponding vector spaces.

In the important particular case, when P_1 is a point and P_2 is a hyperplane, the mapping of the projection from the centre of P_1 onto P_2 maps the point a into its image in P_2, visible by an observer from P_1. Therefore the relationship between a figure and its projection, in this case, is also called a perspectivity. For example, the projection from a straight line to the straight line in P^3 is intuitively less obvious (see Figs. 5, 6):

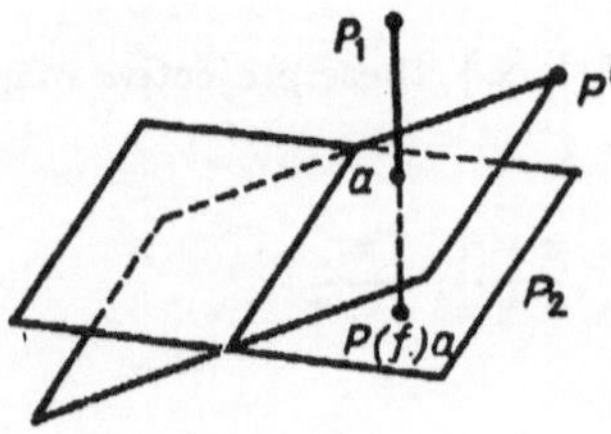

Fig. 5

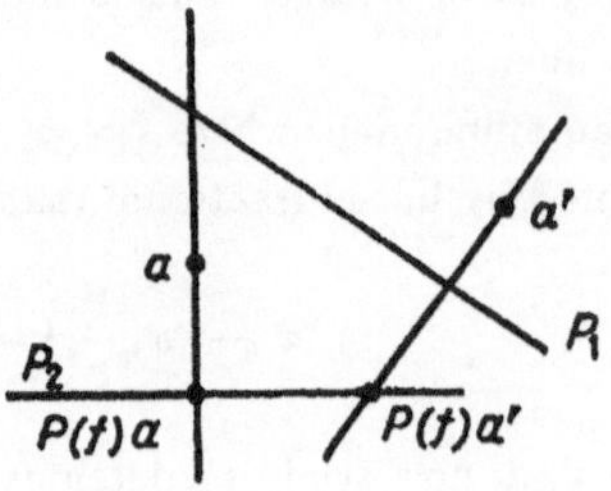

Fig. 6

An important property of projections, which should be kept in mind, is the following: *if $P' \subset P \backslash P_1$, then the projection from the centre of P_1 defines a projective isomorphism of P' and its image in P_2.* Indeed, in the language of linear spaces, this means that the projection $f : L_1 \oplus L_2 \to L_2$ induces an isomorphism of M with $f(M)$, where $M \subset L$ is any subspace with $L_1 \cap M = \{0\}$. This is so because $L_1 = \ker f$.

8.11. Behaviour of a projection near the centre. We shall restrict ourselves below to the analysis of projections from the point $p_1 = a \in P$ and we shall try to understand what happens to points located near the centre. In the case $\mathcal{K} = \mathbf{R}$ or $\mathbf{C}$, when we can indeed talk about the proximity of points, the picture is as follows: continuity of the projection breaks down at the point a, because points b, which are arbitrarily close to a, but approach a "from different sides", are projected into points p_2 far away from one another. It is precisely this property of a projection that is the basis for its applications to different questions concerning the "resolution of singularities". If some "figure" (algebraic variety, vector field) with an unusual structure near the point a is contained in P, then by projecting it from the point a we can stretch the neighbourhood of this point and see what happens in it in an enlarged scale; in addition, the magnification factor increases without bound as a is approached.

Although these applications refer to substantially non-linear situations (becoming trivial in linear models), it is worth while analysing the structure of the projection near its centre in somewhat greater detail, since this can be done within the framework of linear geometry.

8.12. We introduce in $P(L)$ a projective system of coordinates, in which the centre of the projection is the point $(0,\dots,0,1)$, while $P_2 = P^{n-1}$ consists of points $(x_0 : x_1 : \dots : x_{n-1} : 0)$; to achieve this we must select a basis of L which is a union of bases of L_1 and L_2 (since the centre is a point, $\dim L_1 = 1$).

It is not difficult to see that the point $(x_0 : \dots : x_n)$ is then projected into $(x_0 : \dots : x_{n-1} : 0)$. The complement A of P_2 is equipped with an affine system of coordinates

$$(y_0,\dots,y_{n-1}) = (x_0/x_n,\dots,x_{n-1}/x_n)$$

with the origin O at the centre of the projection. Consider the direct product $A \times P_2 = A \times P^{n-1}$ and in it the graph Γ_0 of the mapping of the projection, which we recall, is defined only on $A\backslash\{0\}$. This graph consists of pairs of points with the coordinates

$$((x_0/x_n,\dots,x_{n-1}/x_n),\ (x_0 : \dots : x_{n-1})),$$

where not all $x_0,\dots,x_{n-1}$ equal zero at the same time. We enlarge the graph Γ_0, adding to it above the point $O \in A$ the set $\{0\} \times P^{n-1} \subset A \times P^{n-1}$,

$$\Gamma = \Gamma_0 \cup \{0\} \times P^{n-1},$$

following geometric intuition, according to which for the projection from zero the centre "is mapped into the entire space P^{n-1}".

The set Γ has a number of useful properties.

a) Γ consists precisely of the pairs and points

$$((y_0,\dots,y_{n-1}),\ (x_0 : \dots : x_{n-1})),$$

that satisfy the system of algebraic equations

$$y_i x_j - y_j x_i = 0;\ i,j = 0,\dots,n-1.$$

Indeed, these equations mean that all minors of the matrix $\begin{pmatrix} x_0 & \dots & x_{n-1} \\ y_0 & \dots & y_{n-1} \end{pmatrix}$ equal zero, so that the rank of the matrix equals one (because the first row is non-zero), and hence the second row is proportional to the first one. If the coefficient of proportionality is not zero, then we obtain a point in Γ_0 and if it is zero, then we obtain a point in $\{0\} \times P^{n-1}$.

b) The mapping $\Gamma \to A$; $((y_0, \ldots, y_{n-1}), (x_0 : \ldots : x_{n-1})) \to (y_0, \ldots, y_{n-1})$ is a bijection everywhere except in the fibre over the point $\{0\}$. In other words, Γ is obtained from A by "sewing on" instead of one point, an entire projective space P^{n-1}. We say that Γ is obtained from A by blowing up the points, or by a σ-process centred at the point O. The inverse image in Γ of each straight line in A passing through the point O intersects the sewed-on projective space P^{n-1} also at one point, but a unique point for each straight line. In the case $\mathcal{K} = \mathbf{R}, n = 2$ one can imagine an affine map of the sewed-on projective space P_1 as the axis of the screw of a meat grinder Γ, which makes a half revolution over its infinite length (Fig. 7).

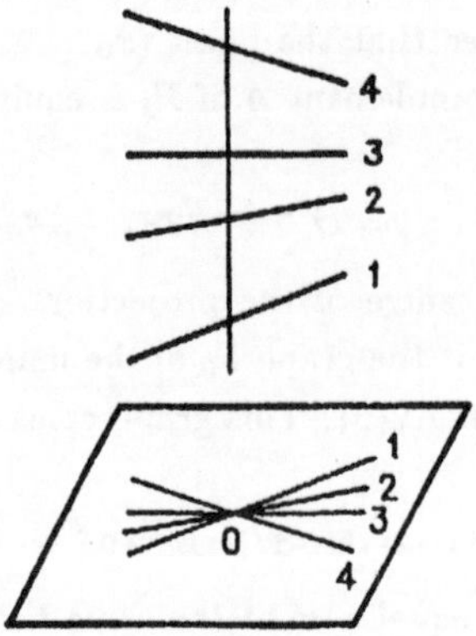

Fig. 7

The real application of a projection to the study of a singularity at a point $O \in A$ involves the transfer of the figure of interest from A to Γ and the study of the geometry of its inverse image near the sewed-on space P^{n-1}. In so doing, for example, the inverse image of an algebraic variety will be algebraic because of the fact that Γ is represented by algebraic equations.

§9. Desargues' and Pappus' Configurations and Classical Projective Geometry

9.1. Classical synthetic projective geometry was largely concerned with the study of families of subspaces in a projective space with an incidence relation; the properties of this relation can be placed at the foundation of the axiomatics, and we then arrive at the modern definition of the spaces $P(L)$ and the field of scalars $\mathcal{K}$

so that L and $\mathcal{K}$ will appear as derivative structures. In this construction, two configurations – Desargues' and Pappus' – play an important role. We shall introduce and study them within the framework of our definitions, and we shall then briefly describe their role in the synthetic theory.

9.2. Desargues' configuration. Let S be a set of points in a projective space. We denote by the symbol $\overline{S}$ its projective span. We shall study in three-dimensional projective space the ordered sextuple of points $(p_1, p_2, p_3; q_1, q_2, q_3)$. It is assumed that the points are pairwise different and that $\overline{p_1p_2p_3}$ and $\overline{q_1q_2q_3}$ are planes. Furthermore, let the straight lines $\overline{p_1q_1}$, $\overline{p_2q_2}$ and $\overline{p_3q_3}$ intersect at a single point r, different from p_i and q_j. In other words, the "triangles" $p_1p_2p_3$ and $q_1q_2q_3$ are "perspectives" and each of them is a projection of the other from the centre r, if they lie in different planes. Then for any pair of indices $\{i, j\} \subset \{1, 2, 3\}$ the straight lines $\overline{p_ip_j}$ and $\overline{q_iq_j}$ do not coincide, otherwise we would have $p_i = q_i$, because p_i and q_i are the points of intersection of these straight lines with the straight line $\overline{p_iq_i}$. In addition, the straight lines $\overline{p_ip_j}$ and $\overline{q_iq_j}$ lie in the common plane $\overline{rp_ip_j}$. Therefore, they intersect at a point which we shall denote by s_k, where $\{i, j, k\} = \{1, 2, 3\}$: this is the point of intersection of the continuations of pairs of corresponding sides of the triangles $p_1p_2p_3$ and $q_1q_2q_3$.

Desargues' theorem, which we shall prove in the next section, asserts that the three points s_1, s_2, s_3 lie on the same straight line. The configuration, consisting of the ten points p_i, q_i, s_k, r and the ten straight lines connecting them, shown in Fig. 8, is called Desargues' configuration. Each of its straight lines contains exactly three of its points, and exactly three of its straight lines pass through each of its points. The reader should verify that it is intrinsically symmetric (in the sense that the group of permutations of its points and straight lines preserving the incidence relation is transitive both on the points and on the straight lines).

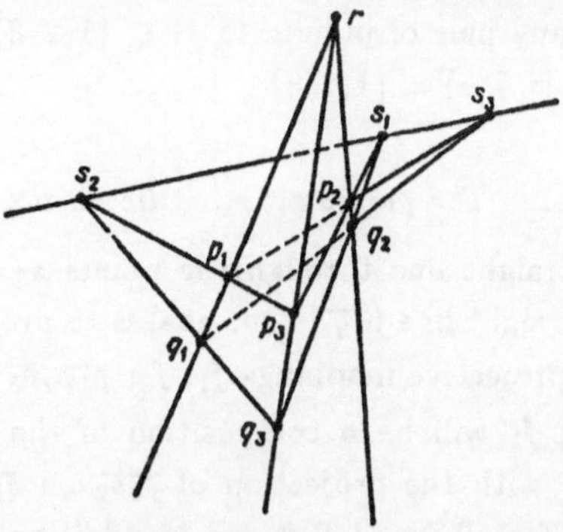

Fig. 8

9.3. Desargues' theorem. *Under the conditions stated above, the points s_1, s_2, s_3 lie on the same straight line.*

Proof. We shall analyse two cases, depending on whether or not the planes $\overline{p_1p_2p_3}$ and $\overline{q_1q_2q_3}$ coincide.

a) $\overline{p_1p_2p_3} \neq \overline{q_1q_2q_3}$ ("three-dimensional Desargues' theorem"). In this case, the planes $\overline{p_1p_2p_3}$ and $\overline{q_1q_2q_3}$ intersect along a straight line, and it is not difficult to verify that s_1, s_2, s_3 lie on it. Indeed, the point s_1, for example, lies on the straight lines $\overline{p_2p_3}$ and $\overline{q_2q_3}$, which in their turn lie in the planes $\overline{p_1p_2p_3}$ and $\overline{q_1q_2q_3}$ and, therefore, in their intersection.

b) $\overline{p_1p_2p_3} = \overline{q_1q_2q_3}$ ("two-dimensional Desargues' theorem"). In this case, we select a point r' in space not lying in the plane $\overline{p_1p_2p_3}$, and we connect it by straight lines with the points r, p_i, p_j. The straight line $\overline{p_1q_1}$ and hence the point r lie in the plane $\overline{r'p_1q_1}$. We draw in it through r a straight line not passing through the points r' and p_1, and denote its intersection with the straight lines $\overline{r'p_1}$, $\overline{r'q_1}$, by p_1', q_1' respectively. The triples (p_1', p_2, p_3) and (q_1', q_2, q_3) now lie in different planes; otherwise the common plane containing them would contain the straight lines $\overline{p_2p_3}$ and $\overline{q_2q_3}$ and would therefore coincide with $\overline{p_1p_2p_3}$ but this is impossible, because p_1', q_1' do not lie in this initial plane. In addition, the straight lines $\overline{p_1'q_1'}$, $\overline{p_2q_2}$ and $\overline{p_3q_3}$ pass through the point r. The three-dimensional Desargues' theorem implies that the points $\overline{p_1'p_2} \cap \overline{q_1'q_2}$, $\overline{p_1'p_3} \cap \overline{q_1'q_3}$ and $\overline{p_2p_3} \cap \overline{q_2q_3}$ lie on the same straight line. But if these points are projected from r' into the plane $\overline{p_1p_2p_3}$, then precisely the points s_3, s_2, s_1 respectively are obtained, because r' projects (p_1', p_2, p_3) into (p_1, p_2, p_3) and (q_1', q_2, q_3) into (q_1, q_2, q_3) and hence the sides of each of these triangles into the corresponding sides of the starting triangles.

This completes the proof.

9.4. Pappus' configuration. We shall examine in the projective plane two different straight lines and two triples of pairwise different points p_1, p_2, p_3 and q_1, q_2, q_3 in them. For any pair of indices $\{i, j\} \subset \{1, 2, 3\}$ we construct the point $s_k = \overline{p_iq_j} \cap \overline{q_ip_j}$, where $\{i, j, k\} = \{1, 2, 3\}$.

9.5. Pappus' theorem. *The points s_1, s_2, s_3 lie on the same straight line.*

Proof. We draw a straight line through the points s_3, s_2 and denote by s_4 its intersection with the straight line $\overline{p_1q_1}$. Our goal is to prove that s_1 lies on it.

We construct two projective mappings f_1, f_2: $\overline{p_1p_2p_3} \to \overline{q_1q_2q_3}$.

The first mapping f_1 will be a composition of the projection of $\overline{p_1p_2p_3}$ on $\overline{s_2s_3}$ from the point q_1 with the projection of $\overline{s_3s_2}$ on $\overline{q_1q_2q_3}$ from the point p_1. Obviously, $f_1(p_i) = q_i$ for all $i = 1, 2, 3$ and in addition, $f_1(t_1) = t_2$ where $t_1 = \overline{p_1p_2p_3} \cap \overline{q_1q_2q_3}$, $t_2 = \overline{s_4s_3s_2} \cap \overline{q_1q_2q_3}$ (see Fig. 9.)

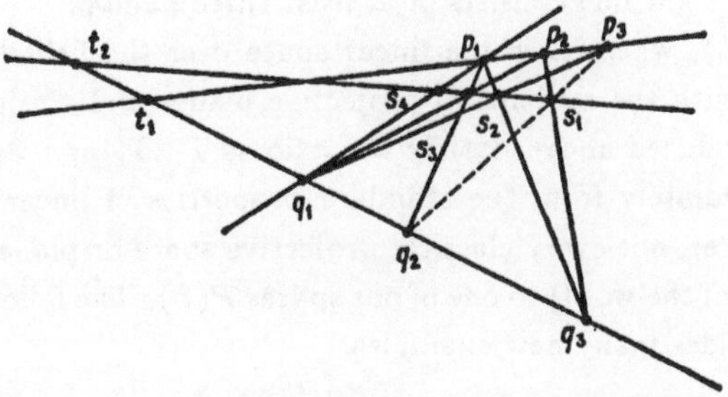

Fig. 9

The second mapping f_2 will be a composition of the projection of $\overline{p_1p_2p_3}$ on $\overline{s_3s_2}$ from the point q_2 with the projection of $\overline{s_3s_2}$ on $\overline{q_1q_2q_3}$ from the point p_2. This composition transfers p_1 into q_1, p_2 into q_2, and t_1 into t_2.

Since f_1 and f_2 operate identically on triples (t_1, p_1, p_2) they must be the same mapping. In particular, $f_1(p_3) = f_2(p_3)$. But $f_1(p_3) = q_3$. Hence $f_2(p_3) = q_3$. This assertion has the following geometric interpretation: if s_1' denotes the intersection $\overline{q_2p_3} \cap \overline{s_3s_2}$, then the straight line $\overline{p_2q_3}$ passes through s_1'. But then $s_1' = \overline{q_2p_3} \cap \overline{p_2q_3} = s_1$. Hence s_1 lies on $\overline{s_2s_3}$, which is what we were required to prove.

9.6. Classical axioms of three-dimensional projective space and the projective plane. The classical three-dimensional projective space is defined as a set whose elements are called points and which is equipped with two systems of subsets, whose elements are called straight lines and planes, respectively. In addition, the following axioms must hold.

T_1. Two different points belong to the same straight line.

T_2. Three different points not lying on the same straight line belong to the same plane.

T_3. A straight line and a plane have a common point.

T_4. The intersection of two planes contains a straight line.

T_5. There exist four points, not lying on the same plane, such that any three of the points do not lie on the same straight line.

T_6. Every straight line consists of not less than three points.

The classical projective plane is defined as a set whose elements are called points, equipped with a system of subsets whose elements are called straight lines. In addition, the following axioms must hold.

P_1. Two different points belong to the same straight line.

P_2. The intersection of two straight lines is not empty.

P_3. There exist three points which do not lie on the same straight line.

P_4. Every straight line consists of at least three points.

The sets $P(L)$, where L is the linear space over the field $\mathcal{K}$ with dimension 4 or 3, together with the systems of projective planes and straight lines in them, as they were introduced above, satisfy the axioms T_1–T_6 and P_1–P_4 respectively; this follows immediately from the standard properties of linear spaces proved in Chapter 1. However, not every classical projective space or plane is isomorphic (in the obvious sense of the word) to one of our spaces $P(L)$. The following fundamental construction provides many new examples.

9.7. Linear and projective spaces over division rings. A *division ring* (or a not necessarily commutative field) is an associative ring K, the set of whose non-zero elements forms a group under multiplication (not necessarily commutative). All fields are division rings, but the converse is not true. For example, the ring of classical quaternions is a division ring but not a field.

The additive group L together with the binary multiplication law $K \times L \to L$: $(a, l) \mapsto al$ is called a (left) *linear space* over the division ring K, if the conditions of Definition 1.2 of Chapter 1 hold. A significant part of the theory of linear spaces over fields can be transferred almost without any changes to linear spaces over division rings. This refers, in particular, to the theory of dimension and basis and the theory of subspaces, including the theorem about the dimension of intersections. This enables the construction, based on every division ring K and linear space L over it, of a projective space $P(L)$ consisting of straight lines in L and the system of its projective subspaces $P(M)$, where $M \subset L$ runs through the linear subspaces of different dimensions. When $\dim_K L = 4$ or 3 these objects satisfy all axioms T_1–T_6 and P_1–P_4 respectively.

9.8. Role of Desargues' theorem. It turns out, however, that there exist classical projective spaces which are not isomorphic even to any plane of the form $P(L)$, where L is a three-dimensional projective space over some division ring. The reason for this lies in the fact that in projective spaces of the form $P(L)$ Desargues' theorem is true, as before, while there exist non-Desarguian planes in which the theorem is not true. We shall formulate without proof the following result.

9.9. Theorem. *The following three properties of a classical projective plane are equivalent:*

a) *Desargues' two-dimensional theorem holds in it.*

b) *It can be embedded in a classical projective space.*

c) *There exists a linear three-dimensional space L over some division ring K, determined uniquely up to isomorphism, such that our plane is isomorphic to $P(L)$.*

The implication b) $\Rightarrow$ a) is established by direct verification of the fact that the proof of the three-dimensional Desargues' theorem employs only the axioms

T_1–T_6. The implication c) $\Rightarrow$ b) follows from the fact that L can be embedded in a four-dimensional linear space over the same division ring.

Finally, the implication a) $\Rightarrow$ c), which is the most subtle point of the proof, is established by direct construction of the division ring in a Desarguian projective plane. Namely, at first, with the help of the geometric construction of projections from the centre, the concept of a projective mapping of projective straight lines in the plane is introduced. Then it is proved that for two ordered triples of points lying on two straight lines there exists a unique projective mapping of one straight line into another. Finally, a straight line D with the triple of points p_0, p_1 and p_2 is fixed, the set K is defined as $D\backslash\{p_2\}$ with zero p_0 and unity p_1 and the laws of addition and multiplication in K are introduced geometrically with the help of projective transformations. In the verifications of the axioms of the division ring, substantial use is made of Desargues' theorem, which in this context arises as *Desargues' axiom* P_5.

9.10. Role of Pappus' theorem. Pappus' theorem may not hold even in Desarguian planes. Calling the corresponding assertion *Pappus' axiom* P_6, we can formulate the following theorem, which we shall also state without proof.

9.11. Theorem. a) *If Pappus' axiom holds in a classical projective plane, then the plane is Desarguian.*

b) *A Desarguian classical plane satisfies Pappus' axiom if and only if the associated division ring is commutative, that is, this plane is isomorphic to* $P(L)$, *where* L *is a three-dimensional linear space over a field.*

Further details and proofs which we have omitted can be found in the book "Foundations of Projective Geometry" by R. Hartshorne (W.A. Benjamin, Inc., N.Y. (1967)).

§10. The Kähler Metric

10.1. If L is a unitary linear space over **C**, then a special metric, called the Kähler metric, in honour of the mathematician E. Kähler, who discovered its important generalizations, can be introduced on the projective space $P(L)$. It plays an especially important role in complex algebraic geometry and implicitly in quantum mechanics also, because spaces such as $P(L)$, as explained in Chapter 2, are spaces of the states of quantum-mechanical systems.

This metric is invariant under projective transformations of $P(L)$ that have the form $P(f)$, where f is a unitary mapping of L into itself. It is introduced in the following manner. Let $p_1, p_2 \in P(L)$. The points p_1, p_2 correspond to two great circles on the unit sphere $S \subset L$, as shown in §6.3c. Then the Kähler distance

$d(p_1, p_2)$ equals the distance between these circles in the Euclidean spherical metric of S, that is, the length of the shortest arc of the great circle on S connecting two points in the inverse images of p_1 and p_2.

The main purpose of this section is to prove the following two formulas for $d(p_1, p_2)$.

10.2. Theorem. a) *Let $l_1, l_2 \in L$, $|l_1| = |l_2| = 1$; and let $p_1, p_2 \in P(L)$ be straight lines $\mathbf{C}l_1$ and $\mathbf{C}l_2$. Then*

$$d(p_1, p_2) = \cos^{-1} |(l_1, l_2)|,$$

where (l_1, l_2) is the inner product in L.

b) *Let an orthonormal basis, relative to which a homogeneous coordinate system is defined in $P(L)$, be chosen in L. Let two close-lying points $p_1, p_2 \in P(L)$ be given by their coordinates $(y_1, \ldots, y_n)$ and $(y_1 + dy_1, \ldots, y_n + dy_n)$ in the affine map U_0 (see §6.3a). Then the square of the distance between them, up to a third order infinitesimal in dy_i, equals*

$$\frac{\sum_{i=1}^n |dy_i|^2}{1 + \sum_{i=1}^n |y_i|^2} - \frac{\left|\sum_{i=1}^n y_i \overline{dy_i}\right|^2}{(1 + \sum_{i=1}^n |y_i|^2)^2}.$$

Proof. a) In a Euclidean space the distance between two points on the unit sphere equals the length of the arc connecting their great circle which lies between 0 and π, that is, the Euclidean angle, or the $\cos^{-1}$ of the Euclidean inner product of the radii. The Euclidean structure on L corresponding to the starting unitary structure is given by the inner product $\mathrm{Re}(l_1, l_2)$. Since we must find the minimum distance between the points on the great circles $(e^{i\phi} l_1)$, $(e^{i\psi} l_2)$, while the $\cos^{-1}$ is a decreasing function, we must find ϕ and ψ such that for given l_1, l_2 the quantity $\mathrm{Re}(e^{i\phi} l_1, e^{i\psi} l_2)$ assumes the maximum possible value. But it does not exceed $|(l_1, l_2)|$ and for suitable ϕ and ψ reaches this value: if $\phi = -\arg(l_1, l_2)$ then $(e^{i\phi} l_1, l_2) = |(l_1, l_2)|$. Therefore, finally

$$d(p_1, p_2) = \cos^{-1} |(l_1, l_2)|.$$

b) Let

$$R = \left(1 + \sum_{i=1}^n |y_i|^2\right)^{1/2}, \quad R + dR = \left(1 + \sum_{i=1}^n |y_i + dy_i|^2\right)^{1/2}.$$

Then the inverse images of the points $(y_1, \ldots y_n)$ and $(y_1 + dy_1, \ldots, y_n + dy_n)$ on S will be the points

$$l_1 = \left(\frac{1}{R}, \frac{y_1}{R}, \ldots, \frac{y_n}{R}\right), \; l_2 = \left(\frac{1}{R + dR}, \frac{y_1 + dy_1}{R + dR}, \ldots, \frac{y_n + dy_n}{R + dR}\right).$$

Therefore

$$(l_1, l_2) = \frac{1}{R(R+dR)}\left(1 + \sum_{i=1}^{n} y_i(\bar{y}_i + \overline{dy}_i)\right) = \frac{R^2 + \sum_{i=1}^{n} y_i\overline{dy}_i}{R(R+dR)}$$

and

$$|(l_1, l_2)|^2 = \frac{R^4 + R^2\sum_{i=1}^{n}(y_i\overline{dy}_i + \bar{y}_i dy_i) + \left|\sum_{i=1}^{n} y_i\overline{dy}_i\right|^2}{R^2(R+dR)^2}.$$

Furthermore,

$$(R+dR)^2 = 1 + \sum_{i=1}^{n}|y_i + dy_i|^2 = R^2 + \sum_{i=1}^{n}(y_i\overline{dy}_i + \bar{y}_i dy_i + |dy_i|^2).$$

Therefore, up to a third-order infinitesimal in dy_i,

$$|(l_1, l_2)|^2 = 1 - \frac{\sum_{i=1}^{n}|dy_i|^2}{R^2} + \frac{\left|\sum_{i=1}^{n} y_i\overline{dy}_i\right|^2}{R^4} + \ldots$$

On the other hand, if $\phi = \cos^{-1}|(l_1, l_2)|$, then up to ϕ^4 for small ϕ we have

$$|(l_1, l_2)|^2 = (\cos\phi)^2 = \left(1 - \frac{\phi^2}{2} + \ldots\right)^2 = 1 - \phi^2 + \ldots$$

Comparison of these formulas completes the proof.

§11. Algebraic Varieties and Hilbert Polynomials

11.1. Let $P(L)$ be an n-dimensional projective space over a field $\mathcal{K}$ with a fixed system of homogeneous coordinates. We have already repeatedly encountered the projective subspaces of $P(L)$ and quadrics, which are defined by the following systems of equations, respectively

$$\sum_{i=0}^{n} a_{ik}x_i = 0,\ k = 1, \ldots, m,$$

or

$$\sum_{i,j=0}^{n} a_{ij}x_i x_j = 0,\ a_{ij} = a_{ji}.$$

More generally, we shall study an arbitrary *homogeneous polynomial*, or form, of degree $m \geq 1$:

$$F(x_0, \ldots, x_n) = \sum_{i_0+\ldots+i_n=m} a_{i_0\ldots i_n} x_0^{i_0} \ldots x_n^{i_n}.$$

Although it does not determine functions on $P(L)$, the set of points with homogeneous coordinates $(x_0 : \ldots : x_n)$ for which $F = 0$ is determined uniquely. It is called an *algebraic hypersurface* (of degree m), defined by the equation $F = 0$.

More generally, the set of points in $P(L)$ satisfying the system of equations

$$F_1 = F_2 = \ldots = F_k = 0,$$

where the F_i are forms (possibly of different degree), is called an *algebraic variety*, determined by this system of equations.

The study of algebraic varieties in a projective space is one of the basic goals of algebraic geometry. Of course, the general algebraic variety is a substantially non-linear object, so that, like in other geometric disciplines, methods of linearization of non-linear problems play an important role in algebraic geometry.

In this section we shall introduce one such method, which dates back to Hilbert and which gives important information about the algebraic variety $V \subset P(L)$ with minimum preliminary preparation. The idea of the method is to form a correspondence between the algebraic variety V and a countable series of linear spaces $\{I_m(V)\}$ and to study its dimension as a function of m. Namely, let $I_m(V)$ be the space of forms of degree m that vanish on V.

We shall show that there exists a polynomial with rational coefficients $Q_V(m)$, such that $\dim I_m(V) = Q_V(m)$ for all sufficiently large m. The coefficients of the polynomial Q_V are the most important invariants of V. We shall actually establish a much more general result, but in order to formulate and prove it we must introduce several new concepts.

11.2. Graded linear spaces. We fix once and for all the main field of scalars $\mathcal{K}$. We shall call a linear space L together with its fixed decomposition into a direct sum of subspaces $L = \bigoplus_{i=0}^{\infty} L_i$ *a graded linear space over* $\mathcal{K}$. This sum is infinite, but every element $l \in L$ can be represented uniquely as a finite sum $l = \sum_{i=0}^{\infty} l_i$, $l_i \in L_i$, in the sense that all but a finite number of l_i vanish. The vector l_i is called the *homogeneous component of* l of degree i; if $l \in L_i$, then l is called a *homogeneous element of degree* i.

Example: The ring of polynomials $A^{(n)}$ of independent variables $x_0, \ldots, x_n$ can be decomposed as a linear space into the direct sum $\bigoplus_{i=0}^{\infty} A_i^{(n)}$, where $A_i^{(n)}$ consists of homogeneous polynomials of degree i. We note that if the x_i are interpreted as coordinate functions on the linear space L and the elements $A^{(n)}$ are interpreted as polynomial functions on this space, then the linear invertible substitutions of coordinates preserve homogeneity and degree.

Another example: $I = \bigoplus_{m=0}^{\infty} I_m(V)$ where V is some algebraic variety. Obviously $I \subset A^{(n)}$ and $I_m(V) = A_m^{(n)} \cap I$.

More generally, a *graded subspace* of M of the graded space $L = \bigoplus_{i=0}^{\infty} L_i$ is a linear subspace with the following property: $M = \bigoplus_{i=0}^{\infty}(M \cap L_i)$. An obvious equivalent condition is that all homogeneous components of any element M are themselves elements of M.

If $M \subset L$ is a pair consisting of a graded space and a graded subspace of it, then the quotient space L/M also exhibits a natural grading. Namely, consider the natural linear mapping

$$\bigoplus_{i=0}^{\infty} L_i/M_i \to L/M : \sum_{i=0}^{\infty}(l_i + M_i) \mapsto \left(\sum_{i=0}^{\infty} l_i\right) + M$$

(the sums on the right are finite). It is surjective, so that any element $\sum_{i=0}^{\infty} l_i$, $l_i \in L$ is the image of the element $\sum_{i=0}^{\infty}(l_i + M_i)$. It is injective, because if $\sum_{i=0}^{\infty} l_i + M = M$, then $\sum_{i=0}^{\infty} l_i \in M$ and $l \in M$ by virtue of the homogeneity of M. Therefore this mapping is an isomorphism, and we can define the grading of L/M by setting $(L/M)_i = L_i/M_i$.

The family of graded subspaces of L is closed relative to intersections and sums, and all standard isomorphisms of linear algebra have obvious graded versions.

11.3. Graded rings. Let A be a graded linear space over $\mathcal{K}$, which is at the same time a commutative $\mathcal{K}$-algebra with unity, multiplication in which obeys the condition

$$A_i A_j \subset A_{i+j}.$$

Then A is said to be a *graded ring* (more precisely, a graded $\mathcal{K}$-algebra). Since $\mathcal{K}A_i \subset A_i$, we have $\mathcal{K} \subset A_0$. A very important example is the ring of polynomials $A^{(n)}$; in them, of course, $A_0^{(n)} = \mathcal{K}$.

11.4. Graded ideals. An *ideal* I in an arbitrary commutative ring A is a subset, forming an additive subgroup A and closed under multiplication on the elements of A: *if* $f \in I$ *and* $a \in A$, *then* $af \in I$. *A graded ideal in a graded ring* A is an ideal which, like the $\mathcal{K}$-subspace of A, is graded, that is, $I = \bigoplus_{m=0}^{\infty} I_m$, $I_m = I \cap A_m$. The basic example is the ideals $I_m(V)$ of algebraic varieties in polynomial rings $A^{(n)}$. The standard construction of ideals is as follows. Let $S \subset A$ be any subset of elements. Then the set of all finite linear combinations $\left\{\sum_{s_i \in S} a_i s_i | a_i \in A\right\}$ is an ideal in A *generated by the set* S. The set S is called a *system of generators* of this ideal. If the ideal has a finite number of generators, then it is said to be *finitely generated*. In the graded case it is sufficient to study sets S consisting only of homogeneous elements; the ideals generated by them are then automatically graded. Indeed, the homogeneous component of degree j of any linear combination $\sum a_i s_i$ will also be a linear combination $\sum a_i^{(k_i)} s_i$, where $a_i^{(k_i)}$ is the homogeneous component of a_i of degree $k_i = j - \deg s_i$ ($\deg s_i$ is the degree of s_i). Therefore it is contained in the ideal generated by S. If the graded ideal is finitely generated, then it contains a finite system of homogeneous generators: it consists of the homogeneous components of the elements of the starting system.

To simplify the proofs the concept of a graded ideal must be generalized and graded modules must also be studied. This is the last of the list of concepts which we shall require.

11.5. Graded modules. A *module* M over a commutative ring A, or an A-module, is an additive group equipped with the operation $A \times M \to M : (a, m) \mapsto$ $\mapsto am$, which is associative ($(ab)m = a(bm)$ for all $a, b \in A$, $m \in M$) and distributive with respect to both arguments:

$$(a+b)m = am + bm, \; a(m+n) = am + an.$$

In addition we require that $lm = m$ for all $m \in M$, where 1 is unity in A.

If A is a field, then M is simply the linear space over A. We can say that the concept of a module is the extension of the concept of a linear space to the case when the scalars form only a ring (see "Introduction to Algebra", §3 of Chapter 9).

If A is a graded $\mathcal{K}$-algebra, then a *graded A-module* M is an A-module which is a graded linear space over $\mathcal{K}$, $M = \bigoplus_{i=0}^{\infty} M_i$ and such that

$$A_i M_j \subset M_{i+j}$$

for all $i, j \geq 0$.

Examples.

a) A is a graded A-module.

b) Any graded ideal in A is a graded A-module.

If M is a graded A-module, then any graded subspace $N \subset M$ closed under multiplication on the elements of A, is itself a graded A-module and a submodule of M. One can construct based on any system of homogeneous elements $S \subset M$ a graded submodule generated by it, consisting of all finite linear combinations $\sum a_i s_i$, $a_i \in A$, $s_i \in S$. If it coincides with M, then S is called a homogeneous system of generators of M. A module which has a finite system of generators is said to be finitely generated. If a graded module has any finite system of generators, then it also has a finite system of homogeneous generators: the homogeneous components of the elements of the starting system.

The study of all possible modules, and not only ideals, in our problem gives great freedom of action . Multiplication by elements $a \in A$ in the quotient module is introduced by the formula

$$a(m+N) = am + N.$$

In the graded case the grading on M/N is determined by the previous formula $(M/N)_i = M_i/N_i$. The verification of the correctness of this definition is trivial.

The elements of the theory of direct sums, submodules, and quotient modules are formally identical to the corresponding results for linear spaces.

We can now take up the proof of the basic results of this section.

11.6. Theorem. *Let M be an arbitrary finitely generated module over a ring of polynomials $A^{(n)} = \mathcal{K}[x_0, \ldots, x_n]$ of a finite number of variables. Then any submodule $N \subset M$ is finitely generated.*

Proof. We shall divide the proof into several steps. We shall use the standard terminology: *a module, each submodule of which is finitely generated, is called a noetherian module* (in honour of Emmy Noether).

a) *A module M is noetherian if and only if any infinite chain of increasing submodules $M_1 \subset M_2 \subset \ldots$ in M stabilizes: there exists an a_0 such that $M_a = M_{a+1}$ for all $a \geq a_0$.*

Indeed, let M be noetherian. Set $N = \bigcup_{i=1}^{\infty} M_i$. Let $n_1, \ldots, n_k$ be a finite system of generators of N. For all $1 \leq j \leq k$ there exists $i(j)$ such that $n_j \in M_{i(j)}$. We set $a_0 = \max\{i(j) | j = 1, \ldots, k\}$. Then M_a contains $n_1, \ldots n_k$ for all $a \geq a_0$ and therefore $M_a = N$.

Conversely, assume that any ascending chain of submodules in M terminates. We shall construct a system of generators of the submodules $n \subset M$ inductively: let $n_1 \in N$ be any element; if $n_1, \ldots, n_i \in N$ have already been constructed, then we denote by $M_i \subset N$ the submodule generated by them and for $N \neq M_i$ we choose n_{i+1} from $N \backslash M_i$. This process terminates after a finite number of steps, otherwise the chain $M_1 \subset \ldots \subset M_i \subset \ldots$ would not stabilize.

b) *If the submodule $N \subset M$ is noetherian and the quotient module M/N is noetherian, then M is noetherian; the converse holds as well.*

In fact, let $M_1 \subset M_2 \subset \ldots$ be a chain of submodules of M. Let a_0 be such that the chains $M_1 \cap N \subset M_2 \cap N \subset \ldots$ and $(M_1 + N)/N \subset (M_2 + N)/N \subset \ldots$ stabilize when $a \geq a_0$. Then the chain $M_1 \subset M_2 \subset \ldots$ also stabilizes when $a \geq a_0$. The converse is obvious.

c) *The direct sum of a finite number of noetherian modules is noetherian.*

For let $M = \bigoplus_{i=0}^{n} M_i$, where the M_i are noetherian. We proceed by induction on n. The case $n = 1$ is trivial. For $n \geq 2$, the module M contains a submodule isomorphic to M_n with quotient isomorphic to $\bigoplus_{i=0}^{n-1} M_i$. Both modules are noetherian so that M is noetherian by virtue of b).

d) *The ring $A^{(n)}$ is noetherian as a module over itself. In other words, every ideal in $A^{(n)}$ is finitely generated.*

This is the main special case of a theorem first proved by Hilbert. It is proved by induction on n. The case $n = -1$, that is, $A^{(-1)} = \mathcal{K}$ is trivial. For any ideal I in the field $\mathcal{K}$ is either $\{0\}$ or the whole of $\mathcal{K}$. If $a \in I$, $a \neq 0$, then $b = (ba^{-1})a \in I$ for all $b \in \mathcal{K}$. The induction step is based on regarding $A^{(n)}$ as $A^{(n-1)}[x_n]$. Let

$I^{(n)} \subset A^{(n)}$ be an ideal. We represent each element of $I^{(n)}$ as a polynomial in powers of x_n with coefficients in $A^{(n-1)}$. The set of all the leading coefficients of such polynomials is an ideal $I^{(n-1)}$ in $A^{(n-1)}$. By the induction hypothesis it has a finite number of generators $\phi_1, \ldots, \phi_m$. We assign to each generator ϕ_i the element $f_i = \phi_i x_n^{d_i} + \ldots$ in $I^{(n)}$, where the dots denote terms of lower powers in x_n. We set $d = \max_{1 \le i \le m}\{d_i\}$. The polynomials $f_1, \ldots, f_m$ generate an ideal $I \subset I^{(n)}$ in $A^{(n)}$.

Now let $f = \phi x^s +$ (lower degree terms) be any element of $I^{(n)}$. By definition, $\phi \in I^{(n-1)}$, so that $\phi = \alpha_1\phi_1 + \ldots + \alpha_m\phi_m$. If $s \ge d$, then the polynomial $f - \sum \alpha_i f_i x^{s-d_i}$ belongs to $I^{(n)}$ and its degree is less than s. By acting in similar fashion we obtain the expression $f = g + h$, where $h \in I$ and g is a polynomial in $I^{(n)}$ of degree less than d.

All the polynomials in $I^{(n)}$ of degree $< d$ form a submodule J of the $A^{(n-1)}$-module generated by the finite system $\{1, x_n, \ldots, x_n^{d-1}\}$. By the induction hypothesis that $A^{(n-1)}$ is noetherian and from c), we see that J is finitely generated.

We have proved that $I^{(n)} = I + J$ is a sum of two finitely generated modules. Therefore the ideal $I^{(n)}$ is finitely generated.

We can now complete the proof of the theorem without difficulty.

Let the module M over $A^{(n)}$ have a finite number of generators $m_1, \ldots, m_k$. Then there exists a surjective homomorphism of $A^{(n)}$ modules

$$\underbrace{A^{(n)} \oplus \ldots \oplus A^{(n)}}_{k\text{times}} \to M : (f_1, \ldots, f_k) \mapsto \sum_{i=1}^{k} f_i m_i.$$

The module $A^{(n)} \oplus \ldots \oplus A^{(n)}$ is neotherian by virtue of items d) and c). Therefore, its quotient module M is noetherian.

11.7. Corollary. *Any infinite system of equations $F_i = 0$, $i \in S$, where F_i are polynomials in $A^{(n)}$, is equivalent to some finite subsystem of itself.*

Proof. Let I be the ideal, generated by all the F_i. It has a finite system of generators $\{G_j\}$. We consider a finite subset $S_0 \subset S$ such that all G_j are expressed linearly in terms of F_i, $i \in S_0$. Then the system of equations $F_i = 0$, $i \in S_0$, is equivalent to the starting system, that is, it has the same set of solutions.

11.8. Hilbert polynomials and Poincaré series of a graded module. Now let M be a graded finitely generated module over the ring $A^{(n)}$. Then all the linear spaces $M_k \subset M$ are finite-dimensional over $\mathcal{K}$. Indeed if $\{a_i\}$ is a homogeneous $\mathcal{K}$-basis of $A_0^{(n)} + \ldots + A_k^{(n)}$ and $\{m_j\}$ is a finite system of generators of M, then M_k as a linear space is generated by a finite number of elements $a_i m_j$ with $\deg a_i + \deg m_j = k$.

Let $d_k(M) = \dim M_k$. The formal power series in the variable t

$$H_M(t) = \sum_{k=0}^{\infty} d_k(M) t^k$$

is called the *Poincaré series* of the module M.

11.9. Theorem. a) *Under the conditions of the preceding section there exists a polynomial $f(t)$ with integer coefficients such that*

$$H_M(t) = \frac{f(t)}{(1-t)^{n+1}}.$$

b) *Under the same conditions there exists a polynomial $P(k)$ with rational coefficients and a number N such that*

$$d_k(M) = P(k) \text{ for all } k \geq N.$$

Proof. We first derive the second assertion from the first one. We set $f(t) = \sum_{i=0}^{N} a_i t^i$ and equate the coefficients of t^k on the right and left sides of the identity

$$H_M(t) = f(t)(1-t)^{-(n+1)}.$$

We obtain, taking into account the fact that $(1-t)^{-(n+1)} = \sum_{k=0}^{\infty} \binom{n+k}{n} t^k$:

$$d_k(M) = \sum_{i=0}^{\min(k,N)} a_i \binom{n+k-i}{n}.$$

For $k \geq N$ the expression on the right is a polynomial of k with rational coefficients.

We now prove the assertion a) by induction on n. It is convenient to set $A^{(-1)} = \mathcal{K} = A_0^{(-1)}$; $A_i^{(-1)} = \{0\}$ for all $i \geq 1$. The finitely generated graded module over $A^{(-1)}$ is simply a finite-dimensional vector space over $\mathcal{K}$ represented in the form of the direct sum $\sum_{k=1}^{N} M_k$. Its Poincaré series is the polynomial $\sum_{k=0}^{N} \dim M_k t^k$, so that the result is trivially true.

Now assume that it is proved for $A^{(n-1)}, n \geq 0$. We shall establish it for $A^{(n)}$. Let M be a finitely generated graded module over $A^{(n)}$. Let

$$K = \{m \in M | x_n m = 0\},\ C = M/x_n M.$$

Obviously, K and $x_n M$ are graded submodules of M; therefore C also has the structure of a graded $A^{(n)}$-module. But multiplication by x_n annihilates both K and C. Therefore, if we regard K and C as modules over the subring $A^{(n-1)} = \mathcal{K}[x_0, \ldots, x_{n-1}] \subset A^{(n)} = \mathcal{K}[x_0, \ldots, x_n]$, then any system of generators for them over $A^{(n)}$ will at the same time be a system of generators over $A^{(n-1)}$. According to Theorem 11.6, K is finitely generated over $A^{(n)}$ as a submodule of a finitely generated module. On the other hand, C is finitely generated over $A^{(n)}$, because if

$m_1, \ldots, m_p$ generate M, then $m_1 + x_n M, \ldots, m_p + x_n M$ generate C. Therefore, K and C are finitely generated over $A^{(n-1)}$, and the induction hypothesis is applicable to them. From the exact sequences of linear spaces over $\mathcal{K}$:

$$0 \longrightarrow K_m \longrightarrow M_m \xrightarrow{x_n} M_{m+1} \longrightarrow C_{m+1} \longrightarrow 0, \ m \geq 0,$$

it follows that

$$\dim M_{m+1} - \dim M_m = \dim C_{m+1} - \dim K_m.$$

Multiplying this equality by t^{m+1} and summing over m from 0 to ∞, we obtain

$$H_M(t) - \dim M_0 - tH_M(t) = H_C(t) - \dim C_0 - tH_K(t),$$

or, according to the induction hypothesis for K and C,

$$(1-t)H_M(t) = \dim M_0 - \dim C_0 + \frac{f_C(t)}{(1-t)^n} - \frac{tf_K(t)}{(1-t)^n},$$

where $f_C(t)$ and $f_K(t)$ are polynomials with integer coefficients. Obviously the required result follows from here.

11.10. The dimension and degree of an algebraic variety. Now let $V \subset P^n$ be an algebraic variety, corresponding to the ideal $I(V)$. We consider the Hilbert polynomial $P_V(k)$ of the quotient module $A^{(n)}/I(V)$:

$$P_V(k) = \dim A_k^{(n)}/I_k(V) \text{ for all } k \geq k_0.$$

It is not difficult to see that $P_{P^n}(k) = \binom{n+k}{n}$, so that $\deg P_{P^n}(k) = n$. Therefore, $\deg P_V \leq n$. The number $d = \deg P_V$ is called the *dimension* of the variety V. We represent the highest order coefficient in $P_V(k)$ in the form $e\frac{k^d}{d!}$. It can be shown that e is an integer, which is called the degree of the variety V. The dimension and the degree are the most important characteristics of the "size" of the algebraic variety. They can be defined purely geometrically: if the field $\mathcal{K}$ is algebraically closed, then the d-dimensional variety of degree e intersects a "sufficiently general" projective space $P^{n-d} \subset P^n$ of complementary dimension at precisely e different points. We shall not prove this theorem.

In conclusion, we note that after Hilbert's discovery, the question of how the values of the Hilbert polynomial $P_V(k)$ for those integer values of k for which $P_V(k) \neq \neq \dim I_k(V)$ (in particular, negative k) should be interpreted remained open for almost half a century. It was solved only in the 1950's when the cohomology theory of coherent sheaves was developed, and it became clear that for any k, $P_V(k)$ is an alternating sum of the dimensions of certain spaces of cohomologies of the variety

V. Hilbert's polynomials of any finitely generated graded modules were interpreted in an analogous manner.

EXERCISES

1. Prove that the Hilbert polynomial of the projective space P^n does not depend on the dimension m of the projective space P^m in which P^n is embedded: $P^n \subset P^m$.

2. Calculate the Hilbert polynomial of the module $A^{(n)}/FA^{(n)}$, where F is a form of degree e.

CHAPTER 4

Multilinear Algebra

§1. Tensor Products of Linear Spaces

1.1. The last chapter of our book is devoted to the systematic study of the multilinear constructions of linear algebra. The main algebraic tool is the concept of a tensor product, which is introduced in this section and is later studied in detail. Unfortunately, the most important applications of this formalism fall outside the purview of linear algebra itself: differential geometry, the theory of representations of groups, and quantum mechanics. We shall consider them only briefly in the final sections of this book.

1.2. Construction. Consider a finite family of vector spaces $L_1, \dots, L_p$ over the same field of scalars $\mathcal{K}$. We recall that a mapping $L_1 \times \dots \times L_p \to L$, where L is another space over $\mathcal{K}$, is said to be multilinear if it is linear with respect to each argument $l_i \in L_i$, $i = 1, \dots, p$, with the other arguments held fixed.

Our first goal is to construct a *universal* multilinear mapping of the spaces $L_1, \dots, L_p$. Its image is called the tensor product of these spaces. The precise meaning of the assertion of universality is explained below in Theorem 1.3. The construction consists of three steps.

a) *The space $\mathcal{M}$.* This is the set of all functions with finite support on $L_1 \times \dots \times L_p$ with values in $\mathcal{K}$, that is, set-theoretic mappings $L_1 \times \dots \times L_p \to \mathcal{K}$, which vanish at all except for a finite number of points of the set $L_1 \times \dots \times L_p$. It forms a linear space over $\mathcal{K}$ under the usual operations of pointwise addition and multiplication by a scalar.

A basis of it consists of the delta functions $\delta(l_1, \dots, l_p)$, equal to 1 at the point $(l_1, \dots, l_p) \in L_1 \times \dots \times L_p$ and 0 elsewhere. Omitting the symbol δ, we may assume that $\mathcal{M}$ consists of formal finite linear combinations of the families $(l_1, \dots, l_p) \in$ $\in L_1 \times \dots \times L_p$:

$$\mathcal{M} = \left\{ \sum a_{l_1 \dots l_p}(l_1, \dots, l_p) \middle| a_{l_1 \dots l_p} \in \mathcal{K} \right\}.$$

We note that if the field $\mathcal{K}$ is infinite and at least one of the spaces L_i is not zero-dimensional, then $\mathcal{M}$ is an infinite-dimensional space.

b) *The subspace* $\mathcal{M}_0$. By definition, it is generated by all vectors from $\mathcal{M}$ of the form

$$(l_1,\dots,l'_j+l''_j,\dots,l_p)-(l_1,\dots,l'_j,\dots,l_p)-(l_1,\dots,l''_j,\dots,l_p),$$

$$(l_1,\dots,al_j,\dots,l_p)-a(l_1,\dots,l_j,\dots,l_p),\ a\in\mathcal{K}.$$

c) *The tensor product* $L_1\otimes\dots\otimes L_p$. By definition

$$L_1\otimes\dots\otimes L_p=\mathcal{M}/\mathcal{M}_0,$$

$$l_1\otimes\dots\otimes l_p=(l_1,\dots,l_p)+\mathcal{M}_0\in L_1\otimes\dots\otimes L_p,$$

$$t:L_1\times\dots\times L_p\to L_1\otimes\dots\otimes L_p,\ t(l_1,\dots,l_p)=l_1\otimes\dots\otimes l_p.$$

Here $\mathcal{M}/\mathcal{M}_0$ is a quotient space in the usual sense of the word. The elements of $L_1\otimes\dots\otimes L_p$ are called tensors; $l_1\otimes\dots\otimes l_p$ are factorizable tensors. Since the families $(l_1,\dots,l_p)$ form a basis of $\mathcal{M}$, the factorizable tensors $l_1\otimes\dots\otimes l_p$ generate the entire tensor product $L_1\otimes\dots\otimes L_p$, but they are by no means a basis: there are many linear relations between them.

The basic property of tensor products is described in the following theorem:

1.3. Theorem. a) *The canonical mapping*

$$t:L_1\times\dots\times L_p\to L_1\otimes\dots\otimes L_p,\quad (l_1,\dots,l_p)\mapsto l_1\otimes\dots\otimes l_p,$$

is multilinear.

b) *The multilinear mapping t is universal in the following sense of the word: for any linear space M over the field* $\mathcal{K}$ *and any multilinear mapping* $s:L_1\times\dots\times L_p\to M$ *there exists a unique linear mapping* $f:L_1\otimes\dots\otimes L_n\to M$ *such that* $s=f\circ t$. *We shall say briefly that s operates through f.*

Proof. a) We must verify the following formulas:

$$l_1\otimes\dots\otimes(l'_j+l''_j)\otimes\dots\otimes l_p=$$

$$=l_1\otimes\dots\otimes l'_j\otimes\dots\otimes l_p+l_1\otimes\dots\otimes l''_j\otimes\dots\otimes l_p,$$

$$l_1\otimes\dots\otimes(al_j)\otimes\dots\otimes l_p=a(l_1\otimes\dots\otimes l_j\otimes\dots\otimes l_p),$$

that is, for example, for the first formula,

$$(l_1,\dots,l'_j+l''_j,\dots,l_p)+\mathcal{M}_0=[(l_1,\dots,l'_j,\dots,l_p)+\mathcal{M}_0]+$$

$$+[(l_1,\dots,l''_j,\dots,l_p)+\mathcal{M}_0].$$

Recalling the definition of a quotient space (§§6.2 and 6.3 of Chapter 1) and the system of generating subspaces of $\mathcal{M}_0$, described in §1.2b above, we immediately obtain these equalities from the definitions.

b) If f always exists, then the condition $s = f \circ t$ uniquely determines the value of f on the factorizable tensors:

$$f(l_1 \otimes \ldots \otimes l_p) = f \circ t(l_1, \ldots, l_p) = s(l_1, \ldots, l_p).$$

Since the latter generate $L_1 \otimes \ldots \otimes L_p$, the mapping f is unique.

To prove the existence of f we study the linear mapping $g : \mathcal{M} \to M$, which on the basis elements of $\mathcal{M}$ is determined by the formula

$$g(l_1, \ldots, l_p) = s(l_1, \ldots, l_p),$$

that is,

$$g(\sum a_{l_1 \ldots l_p}(l_1, \ldots, l_p)) = \sum a_{l_1 \ldots l_p} s(l_1, \ldots, l_p).$$

It is not difficult to verify that $\mathcal{M}_0 \subset \ker g$. Indeed,

$$g[(l_1, \ldots, l'_j + l''_j, \ldots, l_p) - (l_1, \ldots, l'_j, \ldots, l_p) - (l_1, \ldots, l''_j, \ldots, l_p)] =$$

$$= s(l_1, \ldots, l'_j + l''_j, \ldots, l_p) - s(l_1, \ldots, l'_j, \ldots, l_p) -$$

$$-s(l_1, \ldots, l''_j, \ldots, l_p) = 0$$

by virtue of the multilinearity of s. The fact that g annihilates the generators of $\mathcal{M}_0$ of the second type, associated with the multiplication of one of the components by a scalar, is verified analogously.

From here it follows that g induces the linear mapping

$$f : \mathcal{M}/\mathcal{M}_0 = L_1 \otimes \ldots \otimes L_p \to M$$

(see Proposition 6.8 of Chapter 1), for which

$$f(l_1 \otimes \ldots \otimes l_p) = s(l_1, \ldots, l_p).$$

This completes the proof.

We shall now present several immediate consequences of this theorem and the first applications of our construction.

1.4. Let $\mathcal{L}(L_1, \ldots, L_p;\ M)$ be the set of multilinear mappings $L_1 \times \ldots \times L_p$ in M. In Theorem 1.3 we constructed the mapping of sets

$$\mathcal{L}(L_1, \ldots, L_p;\ M) \to \mathcal{L}(L_1 \otimes \ldots \otimes L_p;\ M),$$

that associates with the multilinear mapping s, the linear mapping f with the property $s = f\circ t$. But the left and right sides are linear spaces over $\mathcal{K}$ (as spaces of functions with values in the vector space M: addition and multiplication by a scalar are carried out pointwise). From the construction it is obvious that our mapping is linear. Moreover, it is an isomorphism of linear spaces. Indeed, surjectivity follows from the fact that for any linear mapping $f : L_1\otimes\ldots\otimes L_p \to M$ the mapping $s = f\circ t$ is multilinear by virtue of the assertion of item a) of Theorem 1.3. Injectivity follows from the fact that if $s \neq 0$, then $f\circ t \neq 0$ and therefore $f \neq 0$. Finally, we obtain the canonical identification of the linear spaces

$$\mathcal{L}(L_1,\ldots,L_p;\ M) = \mathcal{L}(L_1\otimes\ldots\otimes L_p;\ M).$$

Thus the construction of the tensor product of spaces reduces the study of multilinear mappings to the study of linear mappings by means of the introduction of a new operation on the category of linear spaces.

1.5. Dimension and bases. a) *If at least one of the spaces $L_1,\ldots,L_p$ is a null space, then their tensor product is a null space.*

Indeed, let, for example, $L_j = 0$. Any multilinear mapping $f : L_1\times\ldots\times L_p \to M$ with fixed $l_i \in L_i$, $i \neq j$, is linear on L_j; but a unique linear mapping of a null space is itself a null mapping. Hence, $f = 0$ for all values of the arguments. In particular, the universal multilinear mapping $t : L_1\times\ldots\times L_p \to L_1\otimes\ldots\otimes L_p$ is a null mapping. But its image generates the entire tensor product. Hence the latter is zero-dimensional.

b) $\dim(L_1\otimes\ldots\otimes L_p) = \dim L_1 \ldots \dim L_p$.

If at least one of the spaces is a null space, this follows from the preceding result. Otherwise we argue as follows: the dimension of $L_1\otimes\ldots\otimes L_p$ equals the dimension of the dual space $\mathcal{L}(L_1\otimes\ldots\otimes L_p,\mathcal{K})$. In the preceding section we identified it with the space of multilinear mappings $\mathcal{L}(L_1\times\ldots\times L_p,\mathcal{K})$. We select in the spaces L_i the bases $\{e_1^{(i)},\ldots,e_{n_i}^{(i)}\}$. We form a correspondence between every multilinear mapping

$$f : L_1\times\ldots\times L_p \to \mathcal{K}$$

and a set of $n_1\ldots n_p$ scalars

$$F\left(e_{i_1}^{(1)},\ldots,e_{i_p}^{(p)}\right),\ 1\leq i_j\leq n_j,\ 1\leq j\leq p.$$

By virtue of the property of multilinearity this set uniquely defines f:

$$f\left(\sum_{i_1=1}^{n_1} x_{i_1}^{(1)}e_{i_1}^{(1)},\ldots,\sum_{i_p=1}^{p} x_{i_p}^{(p)}e_{i_p}^{(p)}\right) = \sum_{(i_1\ldots i_p)} x_{i_1}^{(1)}\ldots x_{i_p}^{(p)} f\left(e_{i_1}^{(1)},\ldots,e_{i_p}^{(p)}\right).$$

In addition, it can be arbitrary: the right side of the last formula defines a multilinear mapping of the vectors $(\vec{x}^{(1)}, \ldots, \vec{x}^{(p)})$ for any values of the coefficients. This shows that the dimension of the space of multilinear mappings $L_1 \times \ldots \times L_p \to \mathcal{K}$ equals $n_1 \ldots n_p = \dim L_1 \ldots \dim L_p$. This completes the proof.

c) *The tensor basis of* $L_1 \otimes \ldots \otimes L_p$. The preceding arguments also enable us to establish the fact that the tensor products $\{e_{i_1}^{(1)} \otimes \ldots \otimes e_{i_p}^{(p)}\}$ form a basis of the space $L_1 \otimes \ldots \otimes L_p$ (we assume that the dimensions of all spaces $L_j \geq 1$ and, for simplicity, are finite). Indeed, these tensor products generate $L_1 \otimes \ldots \otimes L_p$, because all factorizable tensors are expressed linearly in terms of them. Indeed, their number equals exactly the dimension of $L_1 \otimes \ldots \otimes L_p$.

1.6. Tensor products of spaces of functions. Let $S_1, \ldots, S_p$ be finite sets, and let $F(S_i)$ be the space of functions on S_i with values in $\mathcal{K}$. Then there exists a canonical identity

$$F(S_1 \times \ldots \times S_p) = F(S_1) \otimes \ldots \otimes F(S_p),$$

which associates with the function $\delta_{(s_1, \ldots, s_p)}$ the element $\delta_{s_1} \otimes \ldots \otimes \delta_{s_p}$ (see §1.7 of Chapter I). Since both of these families form a basis for their spaces, this is indeed an isomorphism. If $f_i \in F(S_i)$, then

$$f_1 \otimes \ldots \otimes f_p = \Big(\sum_{s_1 \in S_1} f_1(s_1)\delta_{s_1} \Big) \otimes \ldots \otimes \Big(\sum_{s_p \in S_p} f_p(s_p)\delta_{s_p} \Big)$$

transforms under this isomorphism into the function

$$(s_1, \ldots, s_p) \mapsto f_1(s_1) \ldots f_p(s_p),$$

that is, factorizable tensors correspond to "separate variables".

If $S_1 = \ldots = S_p = S$, then the tensor product of functions on S corresponds to the usual product of their values "at independent points of S".

It is precisely in this context that tensor products appear most often in functional analysis and physics. However, the algebraic definition of a tensor product is substantially different in functional analysis, because of the fact that the topology of the spaces is taken into account; in particular, it usually must be extended over different topologies.

1.7. Extension of the field of scalars. Let L be a linear space over the field $\mathbf{R}$, and let $L^{\mathbf{C}}$ be its complexification (see §12 of Chapter I). Since the field $\mathbf{C}$ can be regarded as a linear space over $\mathbf{R}$ (with the basis $1, i$), we can construct a linear space $\mathbf{C} \otimes L$ generated by the basis over $\mathbf{R}$, $1 \otimes e_1, \ldots, 1 \otimes e_n$, $i \otimes e_1, \ldots, i \otimes e_n$, where $\{e_1, \ldots, e_n\}$ is a basis of L. It is obvious that the $\mathbf{R}$-linear mapping

$$\mathbf{C} \otimes L \to L^{\mathbf{C}} : 1 \otimes e_j \mapsto e_j, \ i \otimes e_j \mapsto ie_j$$

determines an isomorphism of $\mathbf{C} \otimes L$ to $L^{\mathbf{C}}$.

More generally, let $\mathcal{K} \subset K$ be a field and a subfield of it, and L a linear space over $\mathcal{K}$. Regarding K at first as a linear space over $\mathcal{K}$, we construct the tensor product $K \otimes L$. Next, we introduce on it the structure of a linear space over K, defining multiplication by a scalar $a \in K$ by the formula

$$a(b \otimes l) = ab \otimes l; \; a, b \in K, \; l \in L.$$

To check the correctness of this definition we construct the space $\mathcal{M}$, freely generated by the elements of $K \times L$, and its subspace $\mathcal{M}_0$, as in §2 so that $K \otimes L = \mathcal{M}/\mathcal{M}_0$. We define multiplication by scalars from K in $\mathcal{M}$, setting on the basis elements

$$a(b, l) = (ab, l); \; a, b \in K, \; l \in L,$$

and extending this rule by $\mathcal{K}$-linearity to the remaining elements of $K \times L$. Direct verification shows that $\mathcal{M}$ transforms into the K-linear space, while $\mathcal{M}_0$ transforms into a subspace of it, so that $\mathcal{M}/\mathcal{M}_0 = K \otimes L$ also becomes a linear space over K. This is the general construction of the extension of the field of scalars mentioned in §12.15 of Chapter 1.

An important particular case is the case when for $K = \mathcal{K}$ the linear space $\mathcal{K} \otimes L$ over $\mathcal{K}$ is canonically isomorphic to L. This isomorphism transforms $a \otimes l$ into al.

§2. Canonical Isomorphisms and Linear Mappings of Tensor Products

2.1. Tensor multiplication has some of the algebraic properties of operations which are called multiplications in other contexts, for example, associativity. These properties, however, are formulated in their own peculiar manner because of the fact that tensor multiplication is an operation over objects belonging to a category. For example, the spaces $(L_1 \otimes L_2) \otimes L_3$ and $L_1 \otimes (L_2 \otimes L_3)$ *do not coincide*, as is obvious from a comparison of their construction: they are only related by a *canonically defined isomorphism*.

In this section we shall describe a number of such "elementary" isomorphisms, which are very useful in working with tensor products. We warn the reader, however, that we shall have to confine ourselves only to an introduction to the theory of canonical isomorphisms. The main question, the systematic study of which we shall omit, is their compatibility. Let us assume, for example, that we have two natural isomorphisms between some tensor products, constructed differently from several "elementary" natural isomorphisms. Are these isomorphisms necessarily the same? One can try to make a direct check in each specific case or one can attempt to construct a general theory, which turns out to be quite cumbersome. Analogous problems arise in connection with natural mappings which are not isomorphisms, for example, mappings such as symmetrization or contraction.

2.2. Associativity. Let $L_1, \ldots, L_p$ be linear subspaces over $\mathcal{K}$. We want to construct canonical isomorphisms between spaces of the type $(L_1 \otimes L_2)(\ldots \otimes L_p)$, obtained as the result of the tensor multiplication of $L_1, \ldots, L_p$ in groups of different order, established by parentheses. The most convenient method, which automatically guarantees compatibility, consists in constructing for each arrangement of parentheses the linear mapping $L_1 \otimes \ldots \otimes L_p \to (L_1 \otimes L_2)\,(\ldots \otimes L_p)$ with the help of the universal property from Theorem 1.3b, and checking that it is an isomorphism.

We shall study in detail the construction $L_1 \otimes L_2 \otimes L_3 \to (L_1 \otimes L_2) \otimes L_3$; the general case is completely analogous.

The mapping $L_1 \times L_2 \to L_1 \otimes L_2 : (l_1, l_2) \mapsto l_1 \otimes l_2$ is bilinear. Therefore the mapping $L_1 \times L_2 \times L_3 \to (L_1 \otimes L_2) \otimes L_3 : (l_1, l_2, l_3) \mapsto (l_1 \otimes l_2) \otimes l_3$ is trilinear. Hence it can be constructed through the unique linear mapping $L_1 \otimes L_2 \otimes L_3 \to (L_1 \otimes L_2) \otimes L_3$. According to the construction itself, the latter mapping transforms $l_1 \otimes l_2 \otimes l_3$ into $(l_1 \otimes l_2) \otimes l_3$. Choosing bases of the spaces L_1, L_2, and L_3 and using the results of §1.5, we find that this mapping transforms a basis into another basis, and is therefore an isomorphism.

Finally, *the product* $(L_1 \otimes L_2)\,(\ldots \otimes L_p)$ *with any arrangement of the parentheses can be identified with* $L_1 \otimes L_2 \otimes \ldots \otimes L_p$, *simply by omitting all parentheses; on the elements* $(l_1 \otimes l_2)\,(\ldots \otimes l_p)$ *this identification operates according to the same rule.* We can therefore write $(l_1 \otimes l_2) \otimes l_3 = l_1 \otimes l_2 \otimes l_3 = l_1 \otimes (l_2 \otimes l_3)$ etc.

2.3. Commutativity. Let σ be any permutation of the numbers $1, \ldots, p$. We shall determine the system of isomorphisms

$$f_\sigma : L_1 \otimes \ldots \otimes L_p \to L_{\sigma(1)} \otimes \ldots \otimes L_{\sigma(p)}$$

with the property $f_{\sigma\tau} = f_\sigma \circ f_\tau$ for any σ and τ. To this end, we note the mapping

$$L_1 \times \ldots \times L_p \to L_{\sigma(1)} \otimes \ldots \otimes L_{\sigma(p)} : (l_1, \ldots l_p) \mapsto l_{\sigma(1)} \otimes \ldots \otimes l_{\sigma(p)}$$

is multilinear. Therefore, according to Theorem 1.3b, it is constructed through the mapping $f_\sigma : L_1 \otimes \ldots \otimes L_p \to L_{\sigma(1)} \otimes \ldots \otimes L_{\sigma(p)}$. It operates on the products of vectors in an obvious manner, permuting the cofactors, and an examination of its action on the tensor product of the bases of $L_1, \ldots, L_p$ shows that this is an isomorphism. The property $f_{\sigma\tau} = f_\sigma \circ f_\tau$ is obvious.

We have thus determined the action of the symmetric groups S_p *on* $L_1 \otimes \ldots \otimes L_p$. In the case when all spaces L_i are different, the isomorphisms f_σ can be used to identify uniquely $L_1 \otimes \ldots \otimes L_p$ with $L_{\sigma(1)} \otimes \ldots \otimes L_{\sigma(p)}$. In this sense, tensor multiplication is commutative. However, it is dangerous to write this identity as an equality without indicating f_σ explicitly (as we did for associativity), if the set of spaces $L_1, \dots, L_p$ contains identical spaces.

2.4. Duality. There is a canonical isomorphism

$$L_1^* \otimes \dots \otimes L_p^* \to (L_1 \otimes \dots \otimes L_p)^*.$$

To construct it, we associate with every element $(f_1, \dots, f_p) \in L_1^* \times \dots \times L_p^*$, the multilinear function $f_1(l_1) \dots f_p(l_p)$ from $(l_1, \dots, l_p) \in L_1 \times \dots \times L_p$. According to Theorem 1.3b, this mapping is constructed through the mapping $L_1^* \otimes \dots \otimes L_p^* \to$ (the space of multilinear functions on $L_1 \times \dots \times L_p$). The latter space, by virtue of the construction in §1.4, is identified with the space $\mathcal{L}(L_1 \otimes \dots \otimes L_p, \mathcal{K}) = (L_1 \otimes \dots \otimes L_p)^*$. We have thus constructed the mapping sought. To show that it is an isomorphism we note that the dimensions of the spaces $(L_1 \otimes \dots \otimes L_p)^*$ and $L_1^* \otimes \dots \otimes L_p^*$ are the same (we confine ourselves to finite-dimensional L_i). It is therefore sufficient to verify that our mapping is surjective. But its image contains the functions $f_1(l_1) \dots f_p(l_p)$, where the f_i run through some basis of L_i^* and as in §1.5, it is not difficult to verify that they form a basis of $\mathcal{L}(L_1, \dots, L_p; \mathcal{K}) = (L_1 \otimes \dots \otimes L_p)^*$.

The identification of $(L_1 \otimes \dots \otimes L_p)^*$ with $L_1^* \otimes \dots \otimes L_p^*$ with the help of the isomorphism described is usually harmless.

2.5. Isomorphism of $\mathcal{L}(L, M)$ with $L^* \otimes M$. We consider the bilinear mapping

$$L^* \times M \to \mathcal{L}(L, M) : (f, m) \mapsto [l \mapsto f(l)m],$$

where $f \in L^*$, $l \in L$, $m \in M$. The bilinearity of the expression $f(l)m$ with respect to f and m and its linearity with respect to l are both obvious. Repeated application of the universality property shows that it corresponds to the linear mapping

$$L^* \otimes M \to \mathcal{L}(L, M).$$

We choose in L, M the bases $\{l_1, \dots, l_a\}$, $\{m_1, \dots, m_b\}$ and in L^* the dual basis $\{l^1, \dots, l^a\}$. The element $l^i \otimes m_j$ of the tensor product of the bases of L^* and M transforms into a linear mapping that transforms the vector $l_k \in L$ into $l^i(l_k)m_j = \delta_{ik}m_j$. The $b \times a$ matrix of this linear mapping has a one at the location (ji) and zeros elsewhere. Since these matrices form a basis of $\mathcal{L}(L, M)$, the mapping constructed transforms a basis into a basis and is an isomorphism.

Let us analyse the important case $L = M$. Here

$$\mathcal{L}(L, L) = L^* \otimes L.$$

The space of endomorphisms $\mathcal{L}(L, L)$ contains a distinguished element: the identity mapping id_L. Its image in $L^* \otimes L$, as is evident from the preceding arguments, equals

$$\sum_{k=1}^{a} l^k \otimes l_k,$$

where $\{l_k\}, \{l^k\}$ is a pair of dual bases in L and L^*. Thus the formula for this element has the same form in all pairs of dual bases.

In addition, the space of the endomorphisms $\mathcal{L}(L, L)$ is equipped with a canonical linear functional — the trace: $\mathrm{Tr} : \mathcal{L}(L, L) \to \mathcal{K}$. From the preceding arguments it follows that the trace of a mapping, with which is associated the element $l^i \otimes l_j$ equals δ_{ij} (look at the matrix), so that associated with the general element of the tensor product $L^* \otimes L$ is the number

$$\sum_{i,j=1}^{a} a_i^j l^i \otimes l_j \mapsto \sum_{i=1}^{a} a_i^i.$$

This linear functional $L^* \otimes L \to \mathcal{K}$ is called a *contraction*. Later we shall give the definition of a contraction in a more general context.

2.6. The isomorphism of $\mathcal{L}(L \otimes M, N)$ with $\mathcal{L}(L, \mathcal{L}(M, N))$. The space $\mathcal{L}(L \otimes M, N)$ is isomorphic to the space of bilinear mappings $L \times M \to N$. Each such bilinear mapping $f : (l, m) \to f(l, m)$ with the first argument l fixed is a linear mapping $M \to N$; this mapping is a linear function of l. Thus we obtain the canonical linear mapping

$$\mathcal{L}(L \otimes M, N) = \mathcal{L}(L, M; N) \to \mathcal{L}(L, \mathcal{L}(M, N)).$$

An argument with bases of L, M, N, analogous to that given in the preceding section, shows that it is an isomorphism (as always, the space is assumed to be finite-dimensional).

(This identification is an important example of the general-categorical concept of "adjoint functors, in the sense of Kan".)

2.7. The tensor product of linear mappings. Let $L_1, \ldots, L_p$ and $M_1, \ldots, M_p$ be two families of linear spaces, and $f_i : L_i \to M_i$ a linear mapping. Then one can construct the linear mapping

$$f_1 \otimes \ldots \otimes f_p : L_1 \otimes \ldots \otimes L_p \to M_1 \otimes \ldots \otimes M_p,$$

called the tensor product of the f_i and uniquely characterized by the simple property

$$(f_1 \otimes \ldots \otimes f_p)(l_1 \otimes \ldots \otimes l_p) = f_1(l_1) \otimes \ldots \otimes f_p(l_p)$$

for all $l_i \in L_i$. Noting that the mapping

$$L_1 \times \ldots \times L_p \to M_1 \otimes \ldots \otimes M_p : (l_1, \ldots, l_p) \mapsto f_1(l_1) \otimes \ldots \otimes f_p(l_p)$$

is multilinear. The existence of the above tensor product can be proved by the same standard application of Theorem 1.3b.

If all f_i are isomorphisms, then $f_1 \otimes \ldots \otimes f_p$ is also an isomorphism.

2.8. Contraction and raising of indices. With the help of this construction we can give a general definition of the contraction "with respect to pairs or several pairs of indices". Assume that we have a tensor product $L_1 \otimes \ldots \otimes L_p$, and in addition for some two indices $i, j \in \{1, \ldots, p\}$ we have $L_i = L^*, L_j = L$. *The contraction with respect to the indices* i, j is the linear mapping

$$L_1 \otimes \ldots \otimes L_p \to \bigotimes_{\substack{k=1\\k\neq i,j}}^{p} L_k,$$

which is obtained as a composition of the following linear mappings:

a) f_σ, where σ is a permutation of the indices $\{1, \ldots, p\}$ carrying i into 1 and j into 2 and preserving the order of the remaining indices:

$$f_\sigma : L_1 \otimes \ldots \otimes L_p \to L_i \otimes L_j \otimes \Big(\bigotimes_{\substack{k=1\\k\neq i,j}}^{p} L_k\Big) = L^* \otimes L \otimes \Big(\bigotimes_{\substack{k=1\\k\neq i,j}}^{p} L_k\Big).$$

b) The contraction of the first two factors, tensor-multiplied by the identity mapping of the remaining factors:

$$L^* \otimes L \otimes \Big(\bigotimes_{\substack{k=1\\k\neq i,j}}^{p} L_k\Big) \to \mathcal{K} \otimes \Big(\bigotimes_{\substack{k=1\\k\neq i,j}}^{p} L_k\Big).$$

c) The identification

$$\mathcal{K} \otimes \Big(\bigotimes_{\substack{k=1\\k\neq i,j}}^{p} L_k\Big) \simeq \bigotimes_{\substack{k=1\\k\neq i,j}}^{p} L_k.$$

If we have several pairs of indices $(i_1, j_1), \ldots, (i_r, j_r)$ such that $L_{i_k} = M_k^*$, $L_{j_k} = M_k$, then this construction can be repeated several times for all pairs successively. The resulting linear mapping is a contraction with respect to these pairs of indices. It depends on the pairs themselves, but not on the order in which the contractions with respect to them are carried out. It can happen that $\{1, \ldots, p\} = \{i_1, j_1, \ldots, i_r, j_r\}$. Then a *complete contraction* is obtained.

We again consider the tensor product $L_1 \otimes \ldots \otimes L_p$ and we shall assume that the isomorphism $g : L_i \to L_i^*$ is given for the ith space (in applications it is most often constructed with the help of a non-degenerate symmetric bilinear form on L_i). Then the linear mapping

$$\mathrm{id} \otimes \ldots \otimes g \otimes \ldots \otimes \mathrm{id} : L_1 \otimes \ldots \otimes L_i \otimes \ldots \otimes L_p \to$$

$$\to L_1 \otimes \ldots \otimes L_i^* \otimes \ldots \otimes L_p$$

is called the "lowering of the ith index", and the reverse mapping is called the "raising of the ith index". This terminology will be explained in the next section.

Both constructions, contractions and raising/lowering of an index, are most often used in the case $L_i = L$ or L^*, when an orthogonal structure is given on L. There are a large number of linear mappings, coupling the spaces $L^{*\otimes p} \otimes L^{\otimes q}$, which are constructed as compositions of the raising and lowering of indices and contractions. These mappings play a large role in Riemannian geometry, where with their help (and analytical operations of differentiation type) important differential-geometric invariants are constructed.

2.9. Tensor multiplication as an exact functor. We fix a linear space M and consider the mapping of the category of finite linear spaces into itself: $L \mapsto L \otimes M$ on objects and $f \mapsto f \otimes \mathrm{id}_M$ on morphisms. From the definitions it is easy to see that $\mathrm{id}_L \mapsto \mathrm{id}_{L\otimes M}$ and

$$f \circ g \mapsto f \circ g \otimes \mathrm{id}_M = (f \otimes \mathrm{id}_M) \circ (g \otimes \mathrm{id}_M).$$

This mapping is therefore a *functor*, which is called the functor of tensor multiplication on M.

We shall show that if the sequence $0 \longrightarrow L_1 \xrightarrow{f} L \xrightarrow{g} L_2 \longrightarrow 0$ is exact, then the sequence

$$0 \longrightarrow L_1 \otimes M \xrightarrow{f\otimes \mathrm{id}_M} L \otimes M \xrightarrow{g\otimes \mathrm{id}_M} L_2 \otimes M \longrightarrow 0$$

is also exact. This property is called the *exactness of the functor of tensor multiplication*. Like the exactness of the functor $\mathcal{L}$, it *breaks down* in the category of modules, and this breakdown is an important object of study in homological algebra (cf. the discussion in §14 of Chapter 1).

Exactness is most easily checked by choosing bases of L_1, L, and L_2 that are adapted to f and g in such a way that $\{e_1, \ldots, e_a\}$ is a basis of L_1, $\{f(e_1), \ldots, f(e_a);$ $e'_{a+1}, \ldots, e'_{a+b}\}$ is a basis of L, and $\{g(e'_{a+1}), \ldots, g(e'_{a+b})\}$ is a basis of L_2. Choosing, in addition, the basis $\{e''_1, \ldots, e''_a\}$ of the space M we find that the tensor products of the bases

$$\{e_i \otimes e''_j\},\ \{f(e_i) \otimes e''_j, e'_k \otimes e''_j\},\ \{g(e'_k) \otimes e''_j\}$$

are adapted to $f \otimes \mathrm{id}_M$, $g \otimes \mathrm{id}_M$ in the same sense of the word.

§3. The Tensor Algebra of a Linear Space

3.1. Let L be some finite-dimensional linear space over the field $\mathcal{K}$. Any element of the tensor product

$$T_p^q(L) = \underbrace{L^* \otimes \ldots \otimes L^*}_{p} \otimes \underbrace{L \otimes \ldots \otimes L}_{q}$$

is called a *tensor of type* (p, q) *and rank* (or *valency*) $p + q$ *on* L. It is also called a *p-covariant and q-contravariant mixed tensor.* The first two chapters of this book were actually devoted to the study of the following tensors of low rank.

a) It is convenient to set $T_0^0(L) = \mathcal{K}$, that is, to call scalars tensors of rank 0.

b) $T_1^0(L) = L^*$, that is, tensors of the type (1,0) are linear functionals on L. Tensors of the type (0,1) are simply vectors from L.

c) $T_1^1(L) = L^* \otimes L$. In §2.5 we identified $L^* \otimes L$ with the space $\mathcal{L}(L, L)$. Therefore tensors of the type (1,1) "are" linear operators on L.

d) $T_2^0(L) = L^* \otimes L^*$. In §2.4 we identified $L^* \otimes L^*$ with $(L \otimes L)^*$, or with the bilinear mappings $L \times L \to \mathcal{K}$. Thus tensors of the type (2,0) "are" inner products on L. In §2.5 we identified $L^* \otimes L^*$ with $\mathcal{L}(L^{**}, L^*) \simeq \mathcal{L}(L, L^*)$. In this identification the inner product on L is associated with the linear mapping $L \to L^*$, which corresponds to an interpretation of this inner product as a function of one of its arguments with the second argument fixed. Thus the tensor constructions given in Section 2 generalize the constructions of Chapter 2.

e) We present one more example: the structure tensor of a $\mathcal{K}$-algebra. Here by a $\mathcal{K}$-algebra we mean the linear space L together with the bilinear operation of multiplication $L \times L \to L : (l, m) \to lm$, not necessarily commutative or even associative, so that, for example, the Lie algebras fall within this definition.

According to Theorem 1.3b, multiplication can be defined just like the linear mapping $L \otimes L \to L$. In §2.5 the space $\mathcal{L}(L \otimes L, L)$ was identified with $(L \otimes L)^* \otimes L$, or, using in addition §2.4 and associativity, with $L^* \otimes L^* \otimes L$. Therefore the specification of the structure of a $\mathcal{K}$-algebra on the space L is equivalent to the specification of a tensor of the type (2,1), called the *structure tensor* of this algebra.

3.2. Tensor multiplication. In accordance with §2.4, we can identify the space $T_p^q(L)$ with $(L^{\otimes p} \otimes (L^*)^{\otimes q})^*$ and then with the space of multilinear mappings

$$f : \underbrace{L \times \ldots, \times L}_{p} \times \underbrace{L^* \times \ldots \times L^*}_{q} \to \mathcal{K}.$$

Two such multilinear mappings of types (p, q) and (p', q') can be tensorially multiplied, yielding as a result the multilinear mapping of type $(p + p', q + q')$:

$$(f \otimes g)(l_1, \ldots, l_p; l_1', \ldots, l_{p'}'; l_1^*, \ldots, l_q^*; l_1'^*, \ldots, l_{q'}'^* =$$

$$= f(l_1, \dots, l_p); l_1^{*}, \dots, l_q^{*}) g(l'_1, \dots, l'_{p'}; l_1'^{*}, \dots, l_{q'}'^{*})$$

where $l_i, l'_j \in L$, $l_i^*, l_i'^* \in L^*$. This definition immediately reveals the bilinearity of tensor multiplication with respect to its arguments:

$$(af_1 + bf_2) \otimes g = a(f_1 \otimes g) + b(f_2 \otimes g);$$

$$f \otimes (ag_1 + bg_2) = a(f \otimes g_1) + b(f \otimes g_2),$$

and also its associativity:

$$(f \otimes g) \otimes h = f \otimes (g \otimes h).$$

However, it is not commutative: $f \otimes g$, generally speaking, is not the same thing as $g \otimes f$.

If tensors are not interpreted as multilinear mappings, then tensor multiplication can be defined with the help of the permutation operations from §2.3, taking into account associativity, as the mapping

$$f_\sigma : \underbrace{L^* \otimes \dots \otimes L^*}_{p} \otimes \underbrace{L \otimes \dots \otimes L}_{q} \otimes \underbrace{L^* \otimes \dots \otimes L^*}_{p'} \otimes \underbrace{L \otimes \dots \otimes L}_{q'} \to$$

$$\to \underbrace{L^* \otimes \dots \otimes L^*}_{p+p'} \otimes \underbrace{L \otimes \dots \otimes L}_{q+q'}.$$

where σ permutes the third group of p' indices into the location after the first group of p indices, preserving their relative order as well as the relative order of the remaining indices. In this variant, the bilinearity of tensor multiplication is equally obvious, and its associativity becomes an identity between permutations, which the reader will find it easier to find out for himself than to follow long but banal explanations.

3.3. Tensor algebra of the space L. We set

$$T(L) = \bigoplus_{p,q=1}^{\infty} T_p^q(L)$$

(the direct sum of linear spaces). This infinite-dimensional space, together with the operation of tensor multiplication in it, defined in the preceding section, is called a tensor algebra of the space L.

We note that it is sometimes important to study over the field of complex numbers an extended tensor algebra, which is a direct sum of the spaces $L^{\otimes p} \otimes$ $\otimes L^{*\otimes q} \otimes \bar{L}^{\otimes p'} \otimes \bar{L}^{*\otimes q'}$. For example, the sesquilinear form on L as a tensor is contained in $L^* \otimes \bar{L}^*$. Because of lack of space we shall not make a systematic study of this construction.

§4. Classical Notation.

4.1. In classical tensor analysis the tensor formalism is described in terms of coordinates. This description is widely used in the physics and geometry literature, and this language must be given its due: it is compact and flexible. In this section we shall introduce it and show how the different constructions described above are expressed in terms of it.

4.2. Bases and coordinates. Let L be a finite-dimensional linear space. We choose a basis $\{e_1, \ldots, e_n\}$ of L and we specify the vectors in L by their coordinates $(a^1, \ldots, a^n)$ in this basis: $\sum_{i=1}^n a^i e_i$.

We choose in L^* the dual basis $\{e^1, \ldots, e^n\}$, $(e^i, e_j) = \delta_j^i = 0$ if $i \neq j$ and 1 if $i = j$, and we specify vectors in L^* by the coordinates $(b_1, \ldots, b_n) : \sum_{j=1}^n b_j e^j$. The arrangement of the indices in both cases is chosen so that pairs of identical indices, one of which is a superscript and the other is a subscript, appear in the summation.

We construct in $L^{*\otimes p} \otimes L^{\otimes q}$ the tensor product of the bases under study

$$\{e^{i_1} \otimes \ldots \otimes e^{i_p} \otimes e_{j_1} \otimes \ldots \otimes e_{j_q} | 1 \leq i_k \leq n,\ 1 \leq j_l \leq n\}.$$

Any tensor $T \in T_p^q(L)$ is given in it by its coordinates $T_{i_1 \ldots i_p}^{j_1 \ldots j_q}$:

$$T = \sum T_{i_1 \ldots i_p}^{j_1 \ldots j_q} e^{i_1} \otimes \ldots \otimes e^{i_p} \otimes e_{j_1} \otimes \ldots \otimes e_{j_q}.$$

We note that here the summation once again extends over pairs of identical indices, one of which is a superscript and the other a subscript. This is such a characteristic feature of the classical formalism that the summation sign is conventionally dropped in all cases when such summation is presupposed.

In particular, under this convention the vectors in L are written in the form $a^j e_j$, while the functionals are written in the form $b_i e^i$. The inner product of L^* and L, that is, the value of the functional $b_i e^i$ on the vector $a^j e_j$, is written as $a^i b_i$ or $b_i a^i$.

Moreover, we can simplify the notation even more by omitting the vectors e_i and e^i themselves. Then the elements of L are written in the form a^j, the elements of L^* are written in the form b_i, and the general tensor $T \in T_p^q(L)$ is written in the form $T_{i_1 \ldots i_p}^{j_1 \ldots j_q}$. In other words, in the classical notation for the tensor T:

the coordinates or the *components* of T in the tensor basis $L^{*\otimes p} \otimes L^{\otimes q}$, enumerated as elements of the tensor basis, *are indicated explicitly*; the numbers are compound indices; the covariant part of the index $(i_1, \ldots, i_p)$ is written as a subscript, while the contravariant part $(j_1, \ldots, j_q)$ is written as a supercript;

the choice of the starting basis $\{e_1, \ldots, e_n\}$ of L, according to which the dual basis $\{e^1, \ldots, e^n\}$ in L^* and then the tensor bases in all the spaces $T_p^q(L)$ are constructed is presupposed.

Sometimes it is convenient to study tensors in spaces where the factors of L and L^* appear in a different order than the one which we adopted, for example, $L \otimes L^*$ instead of $L^* \otimes L$ or $L \otimes L^* \otimes L \otimes L$. This is indicated with the help of a "block arrangement" of compound indices in the tensor coordinates. For example, the tensor $T \in L \otimes L^*$ can be specified by its components which are denoted by $T_i^{\ j}$, while $T \in L \otimes L^* \otimes L \otimes L$ can be specified by the components $T_{i_1}{}^{j_1}{}_{i_2 i_3}$.

4.3. Some important tensors. These include the following:

a) *The metric tensor* g_{ij}. According to our notation, it lies in $T_2^0(L)$, and by virtue of §3.1d it can represent the inner product on L. Its value on the pair of vectors a^i, b^j equals $\sum g_{ij} a^j b^i$ or simply $g_{ij} a^i b^j$. Thus the components of the metric tensor are the elements of the Gram matrix of the starting basis of L relative to the corresponding inner product.

b) *The matrix* A_j^i. This is an element of $T_1^1(L)$, that is, by virtue of §3.1c, a linear mapping of L into itself. It transforms the vector a^j into the vector with the ith coordinate $\sum A_j^i a^j$ or simply $A_j^i a^j$. The tensor of rank $p+q$ can be thought of as a "$p+q$-dimensional matrix", and the standard matrices can be thought of as two-dimensional matrices.

c) *The Kronecker tensor* δ_j^i. This is an element of $T_1^1(L)$, representing the identity mapping of L into itself.

d) *The structure tensor of an algebra.* According to §3.1e, it lies in $T_2^1(L)$ and is therefore written in terms of components in the form γ_{ij}^k. It defines bilinear multiplication in L according to the formula

$$a^i \cdot b^j = c^k = \gamma_{ij}^k a^i b^j.$$

The complete notation is as follows:

$$\Big(\sum_i a^i e_i\Big)\Big(\sum_j b^j e_j\Big) = \sum_k \Big(\sum_{i,j} \gamma_{ij}^k a^i b^j\Big) e_k.$$

4.4. Transformation of the components of a tensor under a change of basis in L. Let A_j^i be the matrix describing the change of basis in L: $e'_k = A_k^i e_i$; let B_j^i be the matrix of the transformation from the basis $\{e^k\}$, dual to $\{e_k\}$, to the basis $\{e'^k\}$ dual to $\{e'_k\}$. It is easy to verify that $B = (A^t)^{-1}$. This matrix is said to be *contragradient* to A.

The coordinates a'^j in the basis $\{e'_j\}$ of a vector initially defined in terms of the coordinates a^i in the basis $\{e_i\}$ will be $B_k^j a^k$.

Analogously, the coordinates b_i in the basis $\{e'^i\}$ of the functional (or "covector"), initially defined by the coordinates b_i in the basis $\{e^i\}$, will be $A_j^k b_k$.

To find the coordinates $T'^{j_1\dots j_q}_{i_1\dots i_p}$ in the primed tensor basis of the tensor initially defined by the coordinates $T^{j_1\dots j_q}_{i_1\dots i_p}$, it is now sufficient to note that they transform just like the coordinates of the tensor product of q-vectors and p-covectors, that is

$$T'^{j_1\dots j_q}_{i_1\dots i_p} = A^{l_1}_{i_1}\dots A^{l_p}_{i_p}B^{j_1}_{k_1}\dots B^{j_q}_{k_q}T^{k_1\dots k_q}_{l_1\dots l_p}.$$

One should not forget that on the right side summation over repeated indices is presumed.

In the classical exposition this formula is used as a basis for the *definition* of tensors.

Namely, *a tensor on an n-dimensional space of type (p,q) is a mapping T which associates with each basis of L a family of n^{p+q} components — the scalars $T^{j_1\dots j_q}_{i_1\dots i_p}$, and in addition the correspondence is such that under a transformation of the basis by means of the matrix A the components of the tensor transform according to the formulas written out above.*

4.5. Tensor constructions in terms of coordinates.

a) *Linear combinations of tensors of the same type.* Here the formulas are obvious:

$$(aT+bT')^{j_1\dots j_q}_{i_1\dots i_p} = aT^{j_1\dots j_q}_{i_1\dots i_p} + bT'^{j_1\dots j_q}_{i_1\dots i_p}.$$

b) *Tensor multiplication.* According to Definition 3.2

$$(T\otimes T')^{j_1\dots j_q j'_1\dots j'_{q'}}_{i_1\dots i_p i'_1\dots i'_{p'}} = T^{j_1\dots j_q}_{i_1\dots i_p}T'^{j'_1\dots j'_{q'}}_{i'_1\dots i'_{p'}}.$$

In particular, a factorizable tensor has the components $T_{i_1}\dots T_{j_p}T^{j_1}\dots T^{j_q}$.

c) *Permutations.* Let σ be a permutation of $1,\dots,p$, τ a permutation of $1,\dots,q$, and $f_{\sigma,\tau}: T^q_p(L)\to T^q_p(L)$ the linear mapping corresponding to these permutations, as in §2.3. Then for any $T\in T^q_p(L)$ we have

$$[f_{\sigma,\tau}(T)]^{j_1\dots j_q}_{i_1\dots i_p} = T^{j_{\tau^{-1}(1)}\dots j_{\tau^{-1}(q)}}_{i_{\sigma^{-1}(1)}\dots i_{\sigma^{-1}(p)}}.$$

d) *Contraction.* Let $a\in\{1,\dots,p\}$, $b\in\{1,\dots,q\}$. As in §2.8, there exists a mapping $T^q_p(L)\to T^{q-1}_{p-1}(L)$ which "annihilates" the ath L^* factor and the bth L factor with the help of the contraction mapping $L^*\otimes L\to\mathcal{K}$, which is the standard inner product of vectors and functionals: $(b_i)\otimes(a^j)\mapsto b_ia^i$. Therefore, denoting by T' the tensor T contracted over a pair of indices (ath subscript and bth superscript), we obtain:

$$T'^{j_1\dots j_{b-1}j_{b+1}\dots j_q}_{i_1\dots i_{a-1}i_{a+1}\dots i_p} = T^{j_1\dots j_{b-1}kj_{b+1}\dots j_q}_{i_1\dots i_{a-1}ki_{a+1}\dots i_p}$$

(summation over k on the right side). Iterating this construction, we obtain a definition of contraction over several pairs of indices.

We have already verified that many formulas in tensor algebra are written in terms of tensor multiplication and subsequent contraction with respect to one or several pairs of indices. We repeat them here for emphasis:

$$g_{ij}a^i b^j \text{ is the contraction of } ((g_{ij}) \otimes (a^k) \otimes (b^l)).$$

The inner product

$$b_i a^i \text{ is the contraction of } ((b_i) \otimes (a^j)).$$

The coordinates of a tensor in a new basis or, from the "active viewpoint", the image of the tensor under a linear transformation of the base space:

$$T'^{j_1 \dots j_q}_{i_1 \dots i_p} \text{ is the contraction of } ((A \otimes \dots \otimes A) \otimes (B \otimes \dots \otimes B) \otimes T).$$

Multiplication in algebra:

$$a^i \cdot b^j \text{ is the contraction of } ((\text{structural tensor }) \otimes (a^i) \otimes (b^j)).$$

One more example — matrix multiplication:

$$(A^i_j)(B^j_k) = (A^i_j B^j_k) \text{ is the contraction of } (A^i_j \otimes B^j_k).$$

We remind the reader once again that in order to define the contraction completely the indices with respect to which it is carried out must be specified; in the examples presented above this is either obvious or it is obvious from the complete formulas presented previously.

In general, we can say that the operation of contraction in the classical language of tensor algebra plays the same unifying role as does the operation of matrix multiplication in the language of linear algebra. In §4 of Chapter 1 we underscored the fact that different types of set-theoretic operations are described in a unified manner with the help of matrix multiplication. This remark is even more pertinent to tensor algebra and contraction, combined with tensor multiplication.

e) *Raising and lowering of indices.* According to Definition 2.8c, the raising of the ath index and the lowering of the bth index are the linear transformations

$$T^q_p(L) \to T^{q+1}_{p-1}(L),\ T^q_p(L) \to T^{q-1}_{p+1}(L),$$

induced by some isomorphisms $g : L^* \to L$ or $g^{-1} : L \to L^*$: the ath L^* factor in the product $L^{*\otimes p} \otimes L^{\otimes q}$ must be replaced by an L factor or, correspondingly, the bth L factor must be replaced by L^*.

In accordance with the conventions stated at the end of §4.2 the components of the tensors obtained must be written in the form

$$T_{i_1 \dots i_{a-1}}{}^{i_a}{}_{i_{a+1} \dots i_p}{}^{j_1 \dots j_q},\ T_{i_1 \dots i_p}{}^{j_1 \dots j_{b-1}}{}_{j_b}{}^{j_{b+1} \dots j_q}.$$

If it is agreed that the mapping raising (lowering) an index is followed by a permutation mapping, which shifts the new L factor to the right and the L^* factor to the left until it adjoins the old factors, then the previous notation for the components can be retained.

As we already pointed out, the isomorphisms $g : L^* \to L$ and $g^{-1} : L \to L^*$ originate in applications most often from the symmetric, non-degenerate, bilinear form g_{ij} on L. Since it is itself a tensor, the operation of raising and lowering indices can be applied to it also. We shall describe this formalism in greater detail.

The form g_{ij} forms a correspondence between the vector a^i and the linear functional

$$b^j \mapsto \sum g_{ij} a^i b^j.$$

The coordinates of this functional in a dual basis in L^* are $g_{ij}a^i$ (summation over i) or, in view of the symmetry, $g_{ij}a^j$. In other words, *the lowering of the (single) upper index of the tensor a^i with the help of the metric tensor g_{ij}* yields the tensor

$$a_i = g_{ij}a^j.$$

From here we obtain immediately the general formula for lowering any number of indices in a factorizable tensor and then, by linearity, in any tensor:

$$T_{i_1 \dots i_p j_1 \dots j_r}{}^{j_{r+1} \dots j_q} = g_{j_1 j_1'} \dots g_{j_r j_r'} T_{i_1 \dots i_p}{}^{j_1' \dots j_r' j_{r+1} \dots j_q}.$$

In particular, we can employ it to calculate the tensor g^{ij}, obtained from g_{ij} by raising the indices. Indeed

$$g_{ij} = g_{ik} g_{jl} g^{kl}.$$

We interpret the right side here as a formula for the (i,j)th element of the matrix obtained by multiplying the matrix (g_{ik}) by the matrix $(\sum_l g_{jl} g^{kl})$. Since the matrix (g_{ik}) also appears on the left side, obviously

$$g_{jl} g^{kl} = \delta_j^k,$$

that is, *the matrix (g^{kl}) is the inverse of the matrix (g_{ij}) (symmetry taken into account)*. This calculation shows that g_i^j is the Kronecker tensor.

Hence the general formula for raising indices has the form

$$T_{i_1 \dots i_b}{}^{i_{b+1} \dots i_p j_1 \dots j_q} = g^{i_{b+1} i_{b+1}'} \dots g^{i_p i_p'} T_{i_1 \dots i_p}{}^{j_1 \dots j_q}.$$

If we want to lower (or raise) other sets of indices, the formulas can be modified in an obvious manner.

§5. Symmetric Tensors

5.1. Let L be a fixed linear space and $T_0^q(L) = L^{\otimes q}$, $q \geq 1$. In §2.3 we showed that with every permutation σ from the group S_q of permutations of the numbers $1, \ldots, q$ we can associate a linear transformation $f_\sigma : T_0^q(L) \to T_0^q(L)$, that operates on factorizable tensors according to the formula

$$f_\sigma(l_1 \otimes \ldots \otimes l_q) = l_{\sigma(1)} \otimes \ldots \otimes l_{\sigma(q)}.$$

We call the tensor $T \in T_0^q(L)$ *symmetric*, if $f_\sigma(T) = T$ for all $\sigma \in S_q$. Obviously, symmetric tensors form a linear subspace in $T_0^q(L)$. It is convenient to regard all scalars as symmetric tensors. With the identification of §3.1d, the symmetric tensors from $T_0^2(L^*)$ correspond to symmetric bilinear forms on L.

We denote by $S^q(L)$ the subspace of symmetric tensors in $T_0^q(L)$. We now construct the projection operator $S : T_0^q(L) \to T_0^q(L)$, whose image is $S^q(L)$, assuming that the characterisitic of the base field vanishes or at least is not a factor of $q!$. It is called the symmetrization mapping. In classical notation, $T^{(i_1 \ldots i_q)}$ is written instead of $S(T)$.

5.2. Proposition. *Let*

$$S = \frac{1}{q!} \sum_{\sigma \in S_q} f_\sigma : T_0^q(L) \to T_0^q(L).$$

Then $S^2 = S$ *and* $\operatorname{im} S = S^q(L)$.

Proof. Obviously, the result of symmetrization of any tensor is symmetric, so that $\operatorname{im} S \subset S^q(L)$. Conversely, on symmetric tensors symmetrization is an identity operation, so that if $T \in S^q(L)$, then $T = S(T)$. This shows at the same time that $\operatorname{im} S = S^q(L)$ and $S^2 = S$.

5.3. Let $\{e_1, \ldots, e_n\}$ be a basis of the space L. Then the factorizable tensors $e_{i_1} \otimes \ldots \otimes e_{i_q}$ form a basis of $T_0^q(L)$, and their symmetrizations $S(e_{i_1} \otimes \ldots \otimes e_{i_q})$ generate $S^q(L)$. We introduce the notation

$$S(e_{i_1} \otimes \ldots \otimes e_{i_q}) = e_{i_1} \ldots e_{i_q}.$$

The formal product $e_{i_1} \ldots e_{i_q}$ does not change under a permutation of the indices, and we can choose the notation $e_1^{a_1} \ldots e_n^{a_n}$ as the canonical notation for such symmetric tensors, where $a_i \geq 0$, $a_1 + \ldots + a_n = q$; here the number a_i shows how many times the vector e_i appears in $e_{i_1} \otimes \ldots \otimes e_{i_q}$.

5.4. Proposition. *The tensors* $e_1^{a_1} \ldots e_n^{a_n} \in S^q(L)$, $a_1 + \ldots + a_n = q$, *form a basis of the space* $S^q(L)$, *which can thus be identified with the space of homogeneous polynomials of degree* q *of the elements of the basis of* L.

Proof. We need only verify that the tensors $e_1^{a_1} \dots e_n^{a_n}$ are linearly independent in $T_0^q(L)$. If

$$\sum c_{a_1 \dots a_n} e_1^{a_1} \dots e_n^{a_n} = 0,$$

then

$$S\left(\sum c_{a_1 \dots a_n} \underbrace{e_1 \otimes \dots \otimes e_1}_{a_1} \otimes \dots \otimes \underbrace{e_n \otimes \dots \otimes e_n}_{a_n}\right) = 0.$$

Collecting all similar terms on the left side, it is easy to verify that the coefficients in front of the elements of the tensor basis of the space $T_0^q(L)$ are scalars $c_{a_1 \dots a_n}$, multiplied by integers consisting of products of prime numbers $\leq q$. Since the characteristic of $\mathcal{K}$ is, by definition, greater than $q!$, it follows from the fact that these coefficients vanish that all the $c_{a_1 \dots a_n}$ vanish.

5.5. Corollary. $\dim S^q(L) = \binom{n+q-1}{q}$.

5.6. Let $S(L) = \bigoplus_{q=1}^{\infty} S^q(L)$. The definition of §5.4 implies that $S(L)$ can be identified with the space of all polynomials of the elements of the basis of L. On this space there exists a structure of an algebra in which multiplication is the standard multiplication of polynomials. It is not clear immediately, however, whether or not this multiplication depends on the choice of the starting basis. For this reason we introduce it invariantly. Since all the $S^q(L)$ will have to be considered simultaneously in what follows, we assume that the characteristic of $\mathcal{K}$ equals zero.

5.7. Proposition. *We introduce on the space $S(L)$ bilinear multiplication according to the formula*

$$T_1 T_2 = S(T_1 \otimes T_2),\ f \in S^p(L),\ g \in S^q(L).$$

It transforms $S(L)$ into a commutative, associative algebra over the field $\mathcal{K}$. In the representation of symmetric tensors in the form of polynomials of the elements of the basis of L this multiplication is the same as the multiplication of polynomials.

Proof. We first verify that for any tensors $T_1 \in T_0^p(L)$, $T_2 \in T_0^q(L)$, the formula

$$S(S(T_1) \otimes T_2) = S(T_1 \otimes S(T_2)) = S(T_1 \otimes T_2)$$

holds. Indeed

$$S(T_1) \otimes T_2 = \frac{1}{p!} \sum_{\sigma \in S_p} f_\sigma(T_1) \otimes T_2,$$

whence

$$S(S(T_1) \otimes T_2) = \frac{1}{p!} \sum_{\sigma \in S_p} S(f_\sigma(T_1) \otimes T_2).$$

But $S(f_\sigma(T_1) \otimes T_2) = S(T_1 \otimes T_2)$ for any $\sigma \in S_p$. This is obvious for factorizable tensors T_1 and T_2, and follows for other tensors by virtue of linearity. Therefore the sum on the right side consists of $p!$ terms $S(T_1 \otimes T_2)$, so that

$$S(S(T_1) \otimes T_2) = S(T_1 \otimes T_2).$$

The second equality is established analogously.

From here it is easy to derive the fact that on symmetric tensors the operation $(T_1, T_2) \mapsto S(T_1 \otimes T_2) = T_1 T_2$ is associative. Indeed

$$(T_1 T_2) T_3 = S(S(T_1 \otimes T_2) \otimes T_3) = S(T_1 \otimes T_2 \otimes T_3)$$

and analogously

$$T_1(T_2 T_3) = S(T_1 \otimes S(T_2 \otimes T_3)) = S(T_1 \otimes T_2 \otimes T_3).$$

In addition, it is commutative: the formula $S(T_1 \otimes T_2) = S(T_2 \otimes T_1)$ is obvious for factorizable tensors and follows for other tensors by virtue of linearity.

It follows from these assertions that

$$(e_1^{a_1} \dots e_n^{a_n})(e_1^{b_1} \dots e_n^{b_n}) = e_1^{a_1+b_1} \dots e_n^{a_n+b_n},$$

which completes the proof.

5.8. The algebra $S(L)$ constructed above is called the *symmetric algebra* of L.

The elements of the algebra $S(L^*)$ can be regarded as polynomial functions on the space L with values in the field $\mathcal{K}$: we associate with an element $f \in L^*$, the element itself as a functional on L, and with the product of elements in $S(L^*)$ and their linear combination we associate the product and linear combination of the corresponding functions. It is not entirely obvious that the different elements of $S(L^*)$ are distinguished also as functions on L. We leave the question to the reader as an exercise. For symmetric algebras over finite fields, which we shall introduce below, this is no longer the case: for example, the function $x^p - x$ vanishes identically in the field $\mathcal{K}$ of p elements.

5.9. Second definition of a symmetric algebra. In the definition which we adopted for a symmetric algebra with the help of the operator S, it is necessary to divide by factorials. This is impossible to do over fields with a finite characteristic and in the theory of modules over rings, where the formalism of tensor algebra also exists and is very useful. We shall therefore briefly describe a different definition of a symmetric algebra of the space L, in which it is realized not as a subspace, but rather as a quotient space of $T_0(L) = \bigoplus_{p=0}^{\infty} T_0^p(L)$.

To this end, we study the *two-sided ideal* I in the tensor algebra $T_0(L)$, generated by all elements of the form

$$T - f_\sigma(T),\ T \in T_0^p(L),\ \sigma \in S_p,\ p = 1, 2, 3, \dots$$

It consists of all possible sums of such tensors, multiplied tensorially on the left and right by any elements from $T_0(L)$. It is easy to see that $I = \bigoplus_{i=1}^{\infty} I^p$, where $I^p = I \cap T_0^p(L)$, that is, this ideal is graded.

We set

$$\tilde{S}(L) = T_0(L)/I$$

as a quotient space. The same argument as that used in §11 of Chapter 3 shows that

$$\tilde{S}(L) = \bigoplus_{p=0}^{\infty} \tilde{S}^p(L),\ \tilde{S}^p(L) = T_0^p(L)/I^p.$$

Since I is an ideal, multiplication in $S(L)$ can be introduced according to the formula

$$(T_1 + I)(T_2 + I) = T_1 \otimes T_2 + I.$$

It is bilinear and associative, since this is true for tensor multiplication. In addition, it is commutative, because if T_1 and T_2 are factorizable, then $T_2 \otimes T_1 = f_\sigma(T_1 \otimes T_2)$ for an appropriate permutation σ and hence $T_1 \otimes T_2 - T_2 \otimes T_1 \in I$. Thus $S(L)$ is a commutative, associative $\mathcal{K}$-algebra. It can be shown that the natural mapping $L \to \tilde{S}(L) : l \mapsto l + I$ is an embedding and that the elements of $\tilde{S}(L)$ can be uniquely represented in terms of any basis of the space L as polynomials of this basis. The elements of $\tilde{S}^p(L)$ correspond to homogeneous polynomials of degree p.

If the characteristic of $\mathcal{K}$ equals zero, the composition mapping

$$S(L) \to T_0(L) \to \tilde{S}(L)$$

is a grade-preserving isomorphism of algebras. Since $\tilde{S}(L)$ exists in more general situations, for algebraic purposes it is convenient to introduce the symmetric algebra precisely in this manner.

§6. Skew-Symmetric Tensors and the Exterior Algebra of a Linear Space

6.1. In the same situation as in §5.1, we shall call a tensor $T \in T_0^q(L)$ *skew-symmetric* (or *antisymmetric*) if $f_\sigma(T) = \epsilon(\sigma)T$, where $\epsilon(\sigma)$ is the sign of the permutation σ, for all $\sigma \in S_q$. Obviously, skew-symmetric tensors form a linear subspace in $T_0^q(L)$. It is convenient to regard all scalars as being skew-symmetric and symmetric tensors simultaneously. With the identification of §3.1d, skew-symmetric tensors from $T_0^2(L^*)$ correspond to symplectic bilinear forms on L.

We denote by $\Lambda^q(L)$ the subspace of skew-symmetric tensors in $T_0^q(L)$. By analogy with §5, we construct a linear projection operator $A\colon T_0^q(L) \to T_0^q(L)$, whose image is $\Lambda^q(L)$. As in §5, we assume for the time being that the characteristic of the field of scalars is not a factor of $q!$. The projector A will be called *antisymmetrization* or *alternation*. In classical notation $T^{[i_1 \ldots i_q]}$ is written instead of $A(T)$.

6.2. Proposition. *Let*

$$A = \frac{1}{q!} \sum_{\sigma \in S_q} \epsilon(\sigma) f_\sigma : T_0^q(L) \to T_0^q(L).$$

Then $A^2 = A$ *and* $\operatorname{im} A = \Lambda^q(L)$.

Proof. We first verify that the result of alternation of any tensor is skew-symmetric. Indeed, since f_σ and $\epsilon(\sigma)$ are multiplicative with respect to σ and $\epsilon(\sigma)^2 = 1$, we have

$$f_\sigma(AT) = f_\sigma \left(\frac{1}{q!} \sum_{\tau \in S_q} \epsilon(\tau) f_\tau(T) \right) = \frac{1}{q!} \sum_{\tau \in S_q} \epsilon(\tau) f_{\sigma\tau}(T) =$$

$$= \epsilon(\sigma) \frac{1}{q!} \sum_{\tau \in S_q} \epsilon(\sigma\tau) f_{\sigma\tau}(T) = \epsilon(\sigma) AT.$$

Further, A is a projection operator because

$$A^2 = \frac{1}{(q!)^2} \sum_{\sigma, \tau \in S_q} \epsilon(\sigma\tau) f_{\sigma\tau} = \frac{1}{q!} \sum_{\rho \in S_q} \epsilon(\rho) f_\rho = A.$$

Indeed, any element $\rho \in S_q$ can be represented in precisely $q!$ ways in the form of the product $\sigma\tau$: we choose an arbitrary σ, and we find τ from the equality $\tau = \sigma^{-1}\rho$.

From here, as in Proposition 5.1, it follows that $\operatorname{im} A = \Lambda^q(L)$.

6.3. Let $\{e_1, \ldots, e_n\}$ be a basis of the space L. Then the factorizable tensors $e_{i_1} \otimes \ldots \otimes e_{i_q}$ form a basis of $T_0^q(L)$ and their antisymmetrizations $A(e_{i_1} \otimes \ldots \otimes e_{i_q})$ generate $\Lambda^q(L)$. We introduce the notation

$$A(e_{i_1} \otimes \ldots \otimes e_{i_q}) = e_{i_1} \wedge \ldots \wedge e_{i_q}$$

(the symbol $\wedge$ denotes "exterior multiplication").

We now note that unlike the symmetric case, the transposition of any two vectors in $e_{i_1} \wedge \ldots \wedge e_{i_q}$ changes the sign of this product, because this tensor is antisymmetric. This implies two results: a) $e_{i_1} \wedge \ldots \wedge e_{i_q} = 0$, *if* $i_a = i_b$ *for some* a *and* b, *provided that* $\operatorname{char} \mathcal{K} \neq 2$.

b) *The space* $\Lambda^q(L)$ *is generated by tensors of the form* $e_{i_1} \wedge \ldots \wedge e_{i_q}$, *where* $1 \le i_1 < i_2 < \ldots < i_q \le n$. From here, in particular, it follows immediately that $\Lambda^m(L) = 0$ for $m > n = \dim L$.

The next result parallels Proposition 5.4.

6.4. Proposition. *The tensors* $e_{i_1} \wedge \ldots \wedge e_{i_q} \in \Lambda^q(L)$ *for* $q \le n$, $1 \le i_1 < i_2 < \ldots < i_q \le n$ *form a basis of the space* $\Lambda^q(L)$.

Proof. We need only verify that these tensors are linearly independent in $T_0^q(L)$. If

$$\sum c_{i_1 \ldots i_q} e_{i_1} \wedge \ldots \wedge e_{i_q} = 0,$$

then

$$A(\sum e_{i_1 \ldots i_q} e_{i_1} \otimes \ldots \otimes e_{i_p}) = 0.$$

But since the indices $i_1, \ldots, i_q$ are all different and they are arranged in increasing order, as a result of the permutation of the indices we obtain in the sum on the left a linear combination of different elements of the tensor basis $T_0^q(L)$ with coefficients of the form $\pm\frac{1}{q!}c_{i_1 \ldots i_q}$. This sum can vanish only if all the $c_{i_1 \ldots i_{qn}}$ vanish.

6.5. Corollary. $\dim \Lambda^q(L) = \binom{n}{q}$, $\dim \bigoplus_{q=0}^n \Lambda^q(L) = 2^n$.

6.6. Let $\Lambda(L) = \bigoplus_{q=0}^n \Lambda^q(L)$. By analogy with the symmetric case we introduce on the space of antisymmetric tensors the operation of exterior multiplication, and we show that it transforms $\Lambda(L)$ into an associative algebra, called the *exterior algebra*, or the *Grassmann algebra*, of the space L.

6.7. Proposition. *The bilinear operation*

$$T_1 \wedge T_2 = A(T_1 \otimes T_2);\ T_1 \in \Lambda^p(L),\ T_2 \in \Lambda^q(L),$$

on $\Lambda(L)$ *is associative,* $T_1 \wedge T_2 \notin L^{p+q}(L)$ *and* $T_2 \wedge T_1 = (-1)^{pq} T_1 \wedge T_2$ *(this property is sometimes called skew-commutativity).*

In particular, the subspace $\Lambda^+(L) = \bigoplus_{q=0}^{[n/2]} \Lambda^{2q}(L)$ *is a central subalgebra of* $\Lambda(L)$.

Proof. By analogy with the symmetric case we first verify that for all $T_1 \in$ $\in T_0^p(L)$, $T_2 \in T_0^q(L)$ the formulas

$$A(A(T_1) \otimes T_2) = A(T_1 \otimes A(T_2)) = A(T_1 \otimes T_2)$$

hold. Indeed,

$$A(T_1) \otimes T_2 = \frac{1}{p!} \sum_{\sigma \in S_p} \epsilon(\sigma) f_\sigma(T_1) \otimes T_2,$$

whence

$$A(A(T_1)\otimes T_2) = \sum_{\sigma\in S_p} \epsilon(\sigma)A(f_\sigma(T_1)\otimes T_2).$$

Consider the embedding $S_p \to S_{p+q}$, $\sigma \mapsto \tilde{\sigma}$, where

$$\tilde{\sigma}(i) = \begin{cases} \sigma(i) & \text{for } 1 \le i \le p, \\ i & \text{for } i > p. \end{cases}$$

Obviously, $f_\sigma(T_1)\otimes T_2 = f_{\tilde{\sigma}}(T_1\otimes T_2)$, and in addition A and $f_{\tilde{\sigma}}$ commute, so that

$$Af_{\tilde{\sigma}}(T_1\otimes T_2) = f_{\tilde{\sigma}}A(T_1\otimes T_2) = \epsilon(\tilde{\sigma})A(T_1\otimes T_2) = \epsilon(\sigma)A(T_1\otimes T_2).$$

Therefore

$$A(A(T_1)\otimes T_2) = \frac{1}{p!}\sum_{\sigma\in S_p} \epsilon^2(\sigma)A(T_1\otimes T_2) = A(T_1\otimes T_2).$$

The second equality is proved analogously. Now the associativity of exterior multiplication can be checked just as in the symmetric case:

$$(T_1\wedge T_2)\wedge T_3 = A(A(T_1\otimes T_2)\otimes T_3) = A(T_1\otimes T_2\otimes T_3),$$

$$T_1\wedge(T_2\wedge T_3) = A(T_1\otimes A(T_2\otimes T_3)) = A(T_1\otimes T_2\otimes T_3).$$

The equality $A(T_1\otimes T_2) = (-1)^{pq}A(T_2\otimes T_1)$ with $T_1\in T_0^p(L)$, $T_2\in T_0^q(L)$ follows from the fact that $T_2\otimes T_1 = f_\sigma(T_1\otimes T_2)$, where σ is a permutation consisting of a product of pq transpositions: the cofactors in T_2 must be transposed one at a time to the left of T_1, exchanging them with the left neighbours from T_1.

6.8. Second definition of an exterior algebra. As in the symmetric case, our definition of an exterior algebra has the drawback that it requires division by factorials. A second definition, which does not have this drawback and which realizes $\Lambda(L)$ as a quotient space rather than a subspace of $T_0(L)$, is constructed by complete analogy with the symmetric case.

Consider the two-sided ideal J in the algebra $T_0(L)$, generated by all elements of the form

$$T - \epsilon(\sigma)f_\sigma(T),\ T\in T_0^p(L),\ \sigma\in S_p,\ p = 1,2,3\ldots$$

It is very easy to verify that $J = \bigoplus_{p=0}^{\infty} J^p$, where $J^p = J\cap T_0^p(L)$, that is, this is a graded ideal. Let $\tilde{\Lambda}(L) = T_0(L)/J$ as a quotient space. Then

$$\tilde{\Lambda}(L) = \bigoplus_{p=0}^{\infty}\tilde{\Lambda}^p(L),\ \tilde{\Lambda}^p(L) = T_0^p(L)/J^p.$$

Since J is an ideal, multiplication can be introduced in $\tilde{\Lambda}(L)$ according to the formula

$$(T_1 + J) \wedge (T_2 + J) = T_1 \otimes T_2 + J.$$

It is bilinear and associative, since this is true for tensor multiplication. Moreover, it is skew-commutative, because for $T_1 \in T_0^p(L)$, and $T_2 \in T_0^q(L)$ we have $T_1 \otimes T_2 - -(-1)^{pq} T_2 \otimes T_1 \in J$.

It is not difficult to verify that *the algebra constructed in this manner is isomorphic to the Clifford algebra of the space L with zero inner product, introduced in §15 of Chapter 2.* Indeed, the mapping $\sigma : L \to \tilde{\Lambda}(L)$, $\sigma(l) = l + J$ satisfies the condition $\sigma(l)^2 = \sigma(l) \wedge \sigma(l) = 0$ for all l, because $\sigma(l) \wedge \sigma(l) = -\sigma(l) \wedge \sigma(l)$. Hence, according to Theorem 15.2 of Chapter 2, there exists a unique homomorphism of $\mathcal{K}$-algebras $C(L) \to \tilde{\Lambda}(L)$, such that σ coincides with the composition $L \xrightarrow{\rho} C(L) \longrightarrow \tilde{\Lambda}(L)$, where ρ is the canonical mapping. Since L generates $T_0(L)$ as an algebra, $\sigma(L)$ generates $\tilde{\Lambda}(L)$, so that $C(L) \to \tilde{\Lambda}(L)$ is surjective. We know that $\dim C(L) = 2n$. Therefore to check the fact that this is an isomorphism it is sufficient to verify that $\dim \tilde{\Lambda} = 2n$. This can be done by establishing the fact that a basis of $\tilde{\Lambda}^q(L)$ is formed by elements of the form $e_{i_1} \wedge \ldots \wedge e_{i_q}$, $1 \leq i_1 < \ldots < i_q \leq n$, where $\{e_1, \ldots, e_n\}$ is a basis of L. We omit this verification.

As in the symmetric case, if the characteristic of $\mathcal{K}$ equals zero, the composite mapping

$$\Lambda(L) \to T_0(L) \to \tilde{\Lambda}(L)$$

is also an isomorphism of grade-preserving algebras.

Since $\tilde{\Lambda}(L)$ is defined in more general situations, for algebraic purposes the exterior algebra is introduced precisely by this method. In application to differential geometry or analysis, where $\mathcal{K} = \mathbf{R}$ or $\mathbf{C}$ our starting definition can be used.

6.9. Exterior multiplication and determinants. Let L be an n-dimensional space. According to Corollary 6.5, the space $\Lambda^n(L)$ is one-dimensional: it is the maximum non-zero exterior power of L.

According to §2.7, any endomorphism $f : L \to L$ induces endomorphisms of tensor powers

$$f^{\otimes p} = \underbrace{f \otimes \ldots \otimes f}_{p} : T_0^p(L) \to T_0^p(L).$$

It is easy to verify that $f^{\otimes p}$ commutes with the alternation operator A and therefore transforms $\Lambda^p(L)$ into $\Lambda^p(L)$. It is natural to denote the restriction of $f^{\otimes p}$ to $\Lambda^p(L)$ by $f^{\wedge p}$. In particular, for $p = n$ the mapping $f^{\wedge p} : \Lambda^n(L) \to \Lambda^n(L)$ must be a multiplication by a scalar $d(f)$, because $\Lambda^n(L)$ is one-dimensional.

6.10. Theorem. *In the notation used above, $d(f) = \det f$.*

Proof. We choose a basis $e_1, \ldots, e_n$ of the space L and write f as a matrix in this basis:

$$f(e_j) = \sum_{i=1}^{n} a_j^i e_i.$$

The exterior product $e_1 \wedge \ldots \wedge e_n$ is a basis of $\Lambda^n(L)$, and the number $d(f)$ is found from the equality

$$f^{\wedge n}(e_1 \wedge \ldots \wedge e_n) = d(f) e_1 \wedge \ldots \wedge e_n.$$

But

$$f^{\wedge n}(e_1 \wedge \ldots \wedge e_n) = A(f(e_1) \otimes \ldots \otimes f(e_n)) = f(e_1) \wedge \ldots \wedge f(e_n) =$$

$$= \left(\sum_{i_1=1}^{n} a_1^{i_1} e_{i_1} \right) \wedge \ldots \wedge \left(\sum_{i_n=1}^{n} a_n^{i_n} e_{i_n} \right).$$

According to the multiplication table in an exterior algebra,

$$a_1^{i_1} e_{i_1} \wedge a_2^{i_2} e_{i_2} \wedge \ldots \wedge a_n^{i_n} e_{i_n} =$$

$$= \begin{cases} \epsilon(\sigma) a_1^{i_1} \ldots a_n^{i_n} e_1 \wedge \ldots \wedge e_n, & \text{if } \{i_1, \ldots, i_n\} = \{1, \ldots, n\}, \\ 0 & \text{otherwise,} \end{cases}$$

where σ is the permutation transferring i_k into k, $1 \leq k \leq n$. Therefore the full sum of the coefficients $\epsilon(\sigma) a_1^{i_1} \ldots a_n^{i_n}$ coincides with the standard formula for the determinant $\det(a_j^i)$, which completes the proof.

6.11. Corollary. *The vectors $e'_1, \ldots, e'_n \in L$ are linearly dependent, if and only if $e'_1 \wedge \ldots \wedge e'_n = 0$.*

Indeed, let $f : L \to L$ be an endomorphism, transferring e_i into e'_i, where $\{e_1, \ldots, e_n\}$ is a basis of L. Then the linear dependence of $\{e'_i\}$ is equivalent to the fact that $\det f = 0$, that is, $e'_1 \wedge \ldots \wedge e'_n = 0$.

6.12. Factorizable vectors. The elements $T \in \Lambda^p(L)$ are called *p-vectors.* We shall call a p-vector of T factorizable, if there exist vectors $e_1, \ldots, e_p \in L$, such that $T = e_1 \wedge \ldots \wedge e_p$. For any p-vector of T we call the set

$$\operatorname{Ann} T = \{e \in L | e \wedge T = 0\}$$

its annihilator. Obviously, $\operatorname{Ann} T$ is a subspace of L.

6.13. Theorem. *Let T_1, T_2 be factorizable p- and q-vectors respectively, and let L_1 and L_2 be their annihilators. Then*

a) *$L_1 \supset L_2$ if and only if T_2 is a factor of T_1, that is $T_1 = T \wedge T_2$ for some $T \in \Lambda^{p-q}(L)$.*

b) $L_1 \cap L_2 = \{0\}$ *if and only if* $T_1 \wedge T_2 \neq 0$.

c) *If* $L_1 \cap L_2 = \{0\}$, *then* $L_1 + L_2 = \mathrm{Ann}(T_1 \wedge T_2)$.

Proof. a) If $x \wedge T_2 = 0$, then $x \wedge (T \wedge T_2) = \pm T \wedge (x \wedge T_2) = 0$, so that the fact that T_2 is a factor of T_1 implies that $L_2 \subset L_1$.

To prove the converse, we calculate the annihilator of the p-vector $e_1 \wedge \ldots \wedge e_p$. If $e_1, \ldots, e_p$ are linearly dependent, then one of the vectors e_i, for example, e_1, can be expressed as a linear combination of the remaining vectors, and then

$$e_1 \wedge \ldots \wedge e_p = \left(\sum_{i=2}^{p} a^j e_i \right) \wedge e_2 \wedge \ldots \wedge e_n = 0.$$

We shall assume that $e_1 \wedge \ldots \wedge e_p$ differs from zero and then show that $\mathrm{Ann}(e_1 \wedge \ldots \wedge e_p)$ coincides with the linear span of the vectors $e_1 \ldots, e_p$. It is clear that this linear span is contained in the annihilator, because

$$e_j \wedge (e_1 \wedge \ldots \wedge e_p) =$$

$$= \pm (e_j \wedge e_j) \wedge (e_1 \wedge \ldots \wedge e_{j-1} \wedge e_{j+1} \wedge \ldots \wedge e_p) = 0.$$

We extend the linearly independent system of vectors $\{e_1, \ldots, e_p\}$ to a basis $\{e_1, \ldots, e_p, e_{p+1}, \ldots, e_n\}$ of the space L and we shall show that if $\sum_{i=1}^{n} a^i e_i \in \mathrm{Ann}(e_1 \wedge \ldots \wedge e_p)$, then $a^i = 0$ for $i > p$. Indeed

$$\left(\sum_{i=1}^{n} a^i e_i \right) \wedge (e_1 \wedge \ldots \wedge e_p) = \sum_{i=p+1}^{n} a^i e_i \wedge e_2 \wedge \ldots \wedge e_p,$$

and the $(p+1)$-vectors $e_i \wedge e_1 \wedge \ldots \wedge e_p$, $p+1 \leq i \leq n$, are linearly independent.

Now let $L_1 \supset L_2$, $T_1 = e_1 \wedge \ldots \wedge e_p$, $T_2 \leq e'_1 \wedge \ldots \wedge e'_q$. Since the linear span of $\{e_1, \ldots, e_p\}$ contains the linear span of $\{e'_1, \ldots, e'_q\}$ we can select in the former a basis of the form $\{e'_1, \ldots, e'_q, e'_{q+1}, \ldots, e'_p\}$ and express e_j as a linear combination in terms of this basis. For T_1 we obtain the expression $ae'_1 \wedge \ldots \wedge e'_q \wedge e'_{q+1} \wedge \ldots \wedge e'_p$, where a is the determinant of the transformation from the primed basis to the unprimed basis. Therefore T_2 is a factor of T_1.

b), c). If $(e_1 \wedge \ldots \wedge e_p) \wedge (e'_1 \wedge \ldots \wedge e'_q) \neq 0$, then the vectors $\{e_1, \ldots, e_p, e'_1, \ldots, e'_q\}$ are linearly independent. Therefore, the linear spans of $\{e_1, \ldots, e_p\}$ and $\{e'_1, \ldots, e'_q\}$, that is, the annihilators of T_1 and T_2 intersect only at zero. This argument is obviously reversible. The characterization of the annihilator of a factorizable p-vector, given in the preceding section, proves the last assertion of the theorem.

6.14. Corollary. *Consider the mapping*

Ann: *(factorizable non-zero p-vectors to within a scalar factor)* $\to$

$\to$ *(p-dimensional subspaces of L).*

It is bijective.

Proof. It is obvious that if two non-zero factorizable vectors are proportional, then they have the same annihilators. Therefore the mapping described above is correctly defined. Any p-dimensional subspace $L_1 \subset L$ lies in the image of the mapping, because if $\{e_1, \ldots, e_p\}$ is a basis of L_1, then $L_1 = \mathrm{Ann}(e_1 \wedge \ldots \wedge e_p)$. Finally, Theorem 6.13a implies that this mapping is injective: if $\mathrm{Ann}\, T_1 = \mathrm{Ann}\, T_2$, then $T_1 = T \wedge T_2$ and T is an O-vector, that is a scalar.

6.15. Grassmann varieties. The *Grassmann* variety, or a *Grassmannian* $\mathrm{Gr}(p, L)$ is the set of all p-dimensional linear subspaces of the space L. In the case $p = 1$ the projective space $P(L)$, which we have studied in detail, is obtained. Corollary 6.14 enables us to realize $\mathrm{Gr}(p, L)$ for any p as a subset of the projective space $P(\Lambda^p(L))$.

Indeed the mapping inverse to Ann gives the embedding

$$\mathrm{Ann}^{-1} : \mathrm{Gr}(p, L) \to P(\Lambda^p(L)).$$

We shall write it out in a more explicit form. We select a basis $\{e_1, \ldots, e_n\}$ of L, and we consider the linear span of the p vectors

$$\sum_{i=1}^{n} a_j^i e_i; \; j = 1, \ldots, p.$$

The $\binom{n}{p}$ p-vectors $\{e_{i_1} \wedge \ldots \wedge e_{i_p} | 1 \le i_1 < \ldots < i_p \le n\}$ form a basis of $\Lambda^p(L)$. The mapping Ann^{-1} establishes a correspondence between our linear span and the straight line in $\Lambda^p(L)$ generated by the p-vector

$$\left(\sum_{i_1=1}^{n} a_1^{i_1} e_{i_1}\right) \wedge \ldots \wedge \left(\sum_{i_p=1}^{n} a_p^{i_p} e_{i_p}\right).$$

The homogeneous coordinates of the corresponding point in $P(\Lambda^p(L))$ are coefficients of the expansion of this p-vector in terms of $\{e_{i_1} \wedge \ldots \wedge e_{i_p}\}$:

$$\bigwedge_{j=1}^{p} \left(\sum_{j=1}^{n} a_j^i e_i\right) = \sum_{1 \le i_1 < \ldots < i_p \le n} \Delta^{i_1 \ldots i_p} e_{i_1} \wedge \ldots \wedge e_{i_p}.$$

Exactly the same calculation as that performed in the proof of Theorem 6.10 shows that $\Delta^{i_1 \ldots i_p}$ equals the minor of the matrix (a_j^i) , formed by the rows with the numbers $i_1, \ldots, i_p$. At least one of these minors differs from zero precisely when the rank of the matrix (a_j^i) has the highest possible value p, that is, when the linear span of our p vectors is indeed p-dimensional.

The vector $(\ldots : \Delta^{i_1\ldots i_p} : \ldots)$ is called the vector of the *Grassmann coordinates* of the p-dimensional subspace spanned by

$$\sum_{i=1}^{n} a_j^i e_i, \quad j = 1, \ldots, p.$$

It is clear from this construction that in order to characterize the image of $\mathrm{Gr}(p, L)$ in $P(\Lambda^p(L))$ we must have a criterion for factorizability of p-vectors. For this reason, we shall now study this problem.

6.16. Theorem. a) *A non-zero p-vector T is factorizable if and only if* $\dim \mathrm{Ann}\, T = p$; *for other non-zero p-vectors,* $\dim \mathrm{Ann}\, T < p$.

b) *Select a basis* $\{e_1, \ldots, e_n\}$ *of the space L and represent any p-vector T by the coefficients of its expansion in the basis* $\{e_{i_1} \wedge \ldots \wedge e_{i_p} | 1 \le i_1 < i_2 < \ldots < i_p \le n\}$ *of* $\Lambda^p(L)$:

$$T = \sum T^{i_1\ldots i_p} e_{i_1} \wedge \ldots \wedge e_{i_p}.$$

Then there exists a system of polynomial equations for $T^{i_1\ldots i_p}$ *with integer coefficients, depending only on n and p, such that the factorizability of T is equivalent to the fact that* $\{T^{i_1\ldots i_p}\}$ *is a solution of this system.*

Proof. We know that $\dim \mathrm{Ann}\, T = p$ for factorizable p-vectors from the proof of Theorem 6.10.

Let $\dim \mathrm{Ann}\, T = r$ and $\mathrm{Ann}\, T$ be generated by the vectors $e_1, \ldots, e_r$. We extend them to a basis $\{e_1, \ldots, e_n\}$ of L and set

$$T = \sum T^{i_1\ldots i_p} e_{i_1} \wedge \ldots \wedge e_{i_p}.$$

The condition $e_i \wedge T = 0$ for all $i = 1, \ldots, r$ means that $T^{i_1\ldots i_p} = 0$, if $\{1, \ldots, r\} \not\subset \{i_1, \ldots, i_p\}$. It follows immediately from here that if $T \neq 0$, then $r \le p$ and that $e_1 \wedge \ldots \wedge e_r$ is a factor of T. Therefore for $r = p$, the p-vector T is proportional to $e_1 \wedge \ldots \wedge e_r$ and is therefore factorizable.

b) Using this criterion we can now write the condition for factorizability of T as the requirement that the following linear system of equations for the unknowns $x^1, \ldots, x^n \in \mathcal{K}$ have a p-dimensional space of solutions

$$\left(\sum_{i=1}^{n} x^i e_i\right) \wedge \left(\sum T^{i_1\ldots i_p} e_{i_1} \wedge \ldots \wedge e_{i_p}\right) = 0.$$

It contains n unknowns and $\binom{n}{p+1}$ equations. Its matrix consists of integral linear combinations of $T^{i_1\ldots i_p}$. The rank of this matrix is always $\ge n - p$, because $\dim \mathrm{Ann}\, T \le p$. Therefore the condition of factorizability is equivalent to the fact that the rank must be $\le n-p$, that is, all its minors of order $(n-p+1)$ must vanish.

This is the system of equations for the Grassmann coordinates of the factorizable tensor sought above.

We shall present some examples and particular cases.

6.17. Proposition. *Any $(n-1)$-vector T is factorizable.*

Proof. Obviously $x \wedge T = f(x)e_1 \wedge \ldots \wedge e_n$, where $f(x)$ is a linear function on L; $\{e_1, \ldots, e_n\}$ is a fixed basis of L. Hence, $\dim \mathrm{Ann}\, T = \dim \ker f \geq n-1$. But if $T \neq 0$, then $f \neq 0$ so that $\dim \ker f = n-1$. Theorem 6.16a implies that T is factorizable.

In terms of the Grassmann varieties this means that there exists a bijection

$$\text{(hyperplane in L)} \to \mathrm{P}(\Lambda^{n-1}\mathrm{L}).$$

But the hyperplanes in L are the points $P(L^*)$. Therefore

$$P(L^* \simeq P(\Lambda^{n-1}(L))$$

(canonical isomorphism). We shall generalize this result below.

6.18. Proposition. *The non-zero bivector $T \in \Lambda^2(L)$ is factorizable if and only if $T \wedge T = 0$.*

Proof. Necessity is obvious. To prove sufficiency we perform induction on n, starting with the trivial case $n = 2$. Let $\{e_1, \ldots, e_{n+1}\}$ be a basis of L. Factoring T into $e_i \wedge e_j$, we can put T into the form $T = e_{n+1} \wedge T_1 + T_2$, where T_1 and T_2 can be decomposed into e_i and $e_i \wedge e_j$, $1 \leq i, j \leq n$. The condition $T \wedge T = 0$ implies that

$$T_2 \wedge T_2 + 2e_{n+1} \wedge T_1 \wedge T_2 = 0,$$

because $(e_{n+1} \wedge T_1) \wedge (e_{n+1} \wedge T_1) = 0$ and T_2 lies at the centre of $\Lambda(L)$. But $T_2 \wedge T_2$ cannot contain terms with e_{n+1}, so that

$$T_2 \wedge T_2 = e_{n+1} \wedge T_1 \wedge T_2 = 0.$$

Since $T_2 \wedge T_2 = 0$, by the induction hypothesis T_2 is factorizable. Since $T_1 \wedge T_2$ does not contain terms with e_{n+1}, we have $T_1 \wedge T_2 = 0$. Hence T_1 is contained in the two-dimensional annihilator of T_2, and $T_2 = T_1' \wedge T_1$. Therefore

$$T = e_{n+1} \wedge T_1 + T_1' \wedge T_1 = (e_{n+1} + T_1') \wedge T_1,$$

which completes the proof.

This result once again gives information about Grassmann varieties, but this time about $\mathrm{Gr}(2, L)$:

6.19. Corollary. *The canonical mapping $Gr(2,L) \to P(\Lambda^2(L))$ identifies for $n \geq 3$ the Grassmannian of planes in L with the intersection of quadrics in $P(\Lambda^2(L))$.*

Proof. Planes in L correspond to directly factorizable 2-vectors of $\Lambda^2(L)$. The condition of factorizability of the 2-vector $\sum_{i_1<i_2} T^{i_1 i_2} e_{i_1} \wedge e_i$, according to Proposition 6.18, has the form

$$\left(\sum_{i_1<i_2} T^{i_1 i_2} e_{i_1} \wedge e_{i_2}\right) \wedge \left(\sum_{j_1<j_2} t^{j_1 j_2} e_{j_1} \wedge e_{j_2}\right) = 0,$$

that is

$$\sum T^{i_1 i_2} T^{j_1 j_2} \epsilon(i_1, i_2, j_1, j_2) = 0,$$

where each sum on the left side corresponds to one quadruple of indices $1 \leq k_1 < < k_2 < k_3 < k_4 \leq n$ and $\epsilon(i_1, i_2, j_1, j_2)$ is the sign of the permutation of the set $\{i_1, i_2, j_1, j_2\} = \{k_1, k_2, k_3, k_4\}$, arranging this quadruple in increasing order.

In particular, for $n = 4$ we obtain one equation:

$$T^{12}T^{34} - T^{13}T^{24} + T^{14}T^{23} = 0.$$

In other words, $\mathrm{Gr}(2, \mathcal{K}^4)$ is a four-dimensional quadric in $P(\Lambda^2(\mathcal{K}^4)) = P(\mathcal{K}^6)$. It is called the *Plücker* quadric.

6.20. Exterior multiplication and duality. Let $\dim L = n$. According to Corollary 6.5 and the well-known symmetry of binomial coefficients

$$\dim \Lambda^p(L) = \binom{n}{p} = \binom{n}{n-p} = \dim \Lambda^{n-p}(L)$$

for all $1 \leq p \leq n$. This suggests that there should exist between $\Lambda^p(L)$ and $\Lambda^{n-p}(L)$ either a canonical isomorphism or a canonical duality. Except for a slight detail, the second assertion is correct.

Consider the operation of exterior multiplication

$$\Lambda^p(L) \times \Lambda^{n-p}(L) \to \Lambda^n(L) : (T_1, T_2) \to T_1 \wedge T_2.$$

Since it is bilinear, it defines the linear mapping

$$\Lambda^p(L) \to \mathcal{L}(\Lambda^{n-p}L, \Lambda^n L) \cong (\Lambda^{n-p}(L))^* \otimes \Lambda^n L$$

(the latter is an isomorphism – a particular case of the one described in §2.5). The kernel of this mapping is a null kernel. Indeed, let $\{e_1, \ldots e_n\}$ be a basis of L. We set $T_1 \wedge T_2 = (T_1, T_2)e_1 \wedge \ldots \wedge e_n$, where $T_1 \in \Lambda^p(L)$, $T_2 \in \Lambda^{n-p}(L)$. Obviously, (T_1, T_2) is a bilinear inner product of $\Lambda^p(L)$ and $\Lambda^{n-p}(L)$. We construct in $\Lambda^p(L)$ and $\Lambda^{n-p}(L)$ bases of factorizable p-vectors and $(n-p)$-vectors $\{e_{i_1} \wedge \ldots \wedge e_{i_p}\}$,

$\{e_{j_1} \wedge \ldots \wedge e_{j_{n-p}}\}$, $1 \le i_1 < \ldots < i_p \le n$, $1 \le j_1 < \ldots < j_{n-p} \le n$. We identify $\Lambda^p(L)$ and $\Lambda^{n-p}(L)$ with the help of the linear mapping that associates with the p-vector $e_{i_1} \wedge \ldots \wedge e_{i_p}$ the $(n-p)$-vector $e_{j_1} \wedge \ldots \wedge e_{j_{n-p}}$, for which $\{i_1, \ldots, i_p, j_1, \ldots, j_{n-p}\} = \{1, \ldots, n\}$. Then (T_1, T_2) will be an inner product on $\Lambda^p(L)$ with a diagonal Gram matrix of the form $\mathrm{diag}(\pm 1, \ldots, \pm 1)$. It is non-degenerate; in particular, its left kernel equals zero.

So we have constructed the canonical isomorphisms

$$\Lambda^p(L) \to (\Lambda^{n-p}(L))^* \otimes \Lambda^n(L).$$

For $p = n-1$ we obtain $\Lambda^{n-1}(L) \to L^* \otimes \Lambda^n(L)$, which explains the isomorphism $P(\Lambda^{n-1}(L)) \to P(L^*)$ in §18: the tensor product of L^* with the one-dimensional space $\Lambda^n(L)$ "does not change" the set of straight lines.

In the next section we shall continue the study of the relationship between exterior multiplication and duality, introducing into the analysis the exterior algebra $\Lambda(L^*)$.

§7. Exterior Forms

7.1. Let L be a finite-dimensional linear space over a field $\mathcal{K}$, and let L^* be its dual.

The elements of the pth exterior power $\Lambda^p(L^*)$ are called *exterior p-forms on the space L*. In particular, exterior 1-forms are simply linear functionals on L. For arbitrary p, two variants of this result can be established.

7.2. Theorem. *The space $\Lambda^p(L^*)$ is canonically isomorphic to*

a) $(\Lambda^p(L))^*$, *that is the space of linear functionals on p-vectors;*

b) *the space of skew-symmetric p-linear mappings $F : \underbrace{L \times \ldots \times L}_{p} \to \mathcal{K}$, that is, mappings with the property*

$$F(l_{\sigma(1)}, \ldots, l_{\sigma(p)}) = \epsilon(\sigma) F(l_1, \ldots, l_p)$$

for all $\sigma \in S_p$.

Proof. According to our definition

$$\Lambda^p(L^*) \subset L^* \otimes \ldots \otimes L^* = T_0^p(L^*).$$

In §2.4 we identified $T_0^p(L^*)$ with the space of all p-linear functions on L^*. In this identification the exterior forms become skew-symmetric p-linear mappings of L. Indeed, it is sufficient to verify this for factorizable forms. For them we have

$$(f_1 \wedge \ldots \wedge f_p)(l_1, \ldots, l_p) = A(f_1 \otimes \ldots \otimes f_p)(l_1, \ldots, l_p) =$$

$$=\frac{1}{p!}\sum_{\tau\in S_p}\epsilon(\tau)f_{\tau(1)}(l_1)\dots f_{\tau(p)}(l_p).$$

Therefore

$$(f_1\wedge\dots\wedge f_p)(l_{\sigma(1)},\dots,l_{\sigma(p)})=\frac{1}{p!}\sum_{\tau\in S_p}\epsilon(\tau)f_{\tau(1)}(l_{\sigma(1)})\dots f_{\tau(p)}(l_{\sigma(p)})=$$

$$=\frac{1}{p!}\sum_{\tau\in S_p}\epsilon(\tau\sigma)f_{\tau\sigma(1)}(l_{\sigma(1)})\dots f_{\tau\sigma(p)}(l_{\sigma(p)})=$$

$$=\epsilon(\sigma)(f_1\wedge\dots\wedge f_p)(l_1,\dots,l_p).$$

We have thus constructed the linear embedding $\Lambda^p(L^*)\to$(skew-symmetric p-linear forms on L). To verify that it is an isomorphism it is sufficient to establish that the dimension of the right side equals

$$\dim\Lambda^p(L^*)=\binom{n}{p}.$$

But if a basis $\{e_1,\dots e_n\}$ of L is chosen, any skew-symmetric p-linear form F on L is uniquely determined by its values $F(e_{i_1},\dots,e_{i_p})$, $1\le i_1<\dots<i_p\le n$, and they can be arbitrary. Therefore the dimension of the space of such forms equals $\binom{n}{p}$. This proves assertion b) of the theorem.

To prove assertion a) we identify $\underbrace{L^*\otimes\dots\otimes L^*}_{p}$ with $(\underbrace{L\otimes\dots\otimes L}_{p})^*$, once again with the help of the construction used in §2.4, and we restrict each element of $\Lambda^p(L^*)$ (as a linear function on $L\otimes\dots\otimes L$) to the subspace of p-vectors of $\Lambda^p(L)$. We obtain the linear mapping $\Lambda^p(L^*)\to(\Lambda^p(L))^*$. Since the dimensions of the spaces on the left and right sides are equal, it is sufficient to verify that it is surjective. The space of linear functionals on $\Lambda^p(L)$ is generated by functionals of the form $\frac{1}{p!}\delta_I$, where

$$I=\{i_1,\dots,i_p\}\subset\{1,\dots,n\},\ \delta_I(e_{i_1}\wedge\dots\wedge e_{i_p})=1,$$

$$\delta_I(e_{j_1}\wedge\dots\wedge e_{j_p})=0,$$

if $\{j_1,\dots,j_p\}\ne I$. We assert that this functional is the image of a p-form $e^{i_1}\wedge\dots\wedge e^{i_p}\in\Lambda^p(L^*)$, where, as usual $\{e^i\}$ indicates the basis which is the dual of the basis $\{e_i\}$. Indeed, the value of $e^{i_1}\wedge\dots\wedge e^{i_p}$ by $e_{j_1}\wedge\dots\wedge e_{j_p}$ equals

$$A(e^{i_1}\otimes\dots\otimes e^{i_p})(A(e_{j_1}\otimes\dots\otimes e_{j_p}))=$$

$$=\frac{1}{(p!)^2}\sum_{\sigma,\tau\in S_p}\epsilon(\sigma\tau)e^{i_{\sigma(1)}}(e_{j_{\tau(1)}})\dots e^{i_{\sigma(p)}}(e_{j_{\tau(p)}}).$$

Only the terms for which $i_{\sigma(1)} = j_{\tau(1)}, \ldots, i_{\sigma(p)} = j_{\tau(p)}$, on the right side differ from zero, so that the entire sum vanishes if $\{i_1, \ldots, i_p\} \neq \{j_1, \ldots, j_p\}$. If, however, these sets coincide, then the entire sum equals

$$\frac{1}{(p!)^2} \sum_{\sigma \in S_p} \epsilon(\sigma)^2 = \frac{1}{p!},$$

when $\{i_1, \ldots, i_p\} = \{j_1, \ldots, j_p\}$ are in increasing order. Therefore $e^{i_1} \wedge \ldots \wedge e^{i_p}$ as a functional on $\Lambda^p(L)$ equals $\frac{1}{p!}\delta_I$, which completes the proof.

7.3. Remarks. a) Our method for identifying $\Lambda^p(L^*)$ with $\Lambda^p(L))^*$ corresponds to the bilinear mapping

$$\Lambda^p(L^*) \times \Lambda^p(L) \to \mathcal{K},$$

which can be represented on arbitrary pairs of factorizable vectors in the form

$$(f^1 \wedge \ldots \wedge f^p, l_1 \wedge \ldots \wedge l_p) = \frac{1}{p!} \det(f^i(l_j)),$$

$f^i \in L^*$, $l_j \in L$. Indeed, both sides are multilinear and skew-symmetric separately in f^i and l_j; in addition, they coincide for

$$(f^1, \ldots, f^p) = (e^{i_1}, \ldots, e^{i_p}),\ (l_1, \ldots, l_p) = (e_{j_1}, \ldots, e_{j_p}),$$

as was verified in the preceding proof.

The factor $\frac{1}{p!}$ in this scalar product is sometimes omitted.

b) One of the identifications established in the theorem is sometimes adopted as the *definition* of an exterior power. For example, $\Lambda^p(L)$ is often introduced, especially in differential geometry, as the space of skew-symmetric p-linear functionals on L^*. The generality of this construction falls between that of the first and second definitions of the exterior power in §6; it is suitable for linear spaces over fields with finite characteristic as well as free modules over commutative rings. But for general modules, the extra dualization presents a difficulty, and the second definition is preferable.

A result analogous to Theorem 7.2 also holds for symmetric powers, and our preceding remark is pertinent to them also. In particular, $S^p(L)$ can be defined as the space of symmetric p-linear functionals on L^* for spaces over arbitrary fields and free modules over commutative rings. In the most general case, however, the correct definition of $S^p(L)$ is the definition given in §5.9.

7.4. The inner product. The bilinear mapping

$$L \times \Lambda^p(L^*) \to \Lambda^{p-1}(L^*) : (l, F) \mapsto i(l)F,$$

is called an *inner product*. It is defined as follows. Interpret $F \in \Lambda^p(L^*)$ as a skew-symmetric p-linear form on L, and $i(l)F$ analogously. Then by definition

$$i(l)F(l_1, \ldots, l_{p-1}) = F(l, l_1, \ldots, l_{p-1}).$$

Obviously, the right side is $(p-1)$-linear and skew-symmetric as a function of $l_1, \ldots, l_{p-1}$, and it is also bilinear as a function of F and l, so that the definition is correct. For $p = 0$ it is convenient to set $i(l)F = 0$. The notation $l \lrcorner F$ is also used instead of $i(l)F$.

§8. Tensor Fields

8.1. In this section we shall briefly describe the typical differential-geometric situations in which tensor algebra is used.

We consider some region $U \subset \mathbf{R}^n$ in a real coordinate space and the ring C of infinitely differentiable functions with real values on U. In particular, the coordinate functions x^i belong to C, $i = 1, \ldots, n$.

8.2. Definition. *Any linear mapping $X_a : C \to \mathbf{R}$ satisfying the condition*

$$X_a f = 0, \textit{if } f \textit{ is constant in some neighbourhood of } a,$$

and

$$X_a(fg) = X_a f \cdot g(a) + f(a) \cdot X_a g$$

is called a tangent vector X_a to U at the point $a \in U$.

If X_a, Y_a are tangent vectors at the point a, then any real linear combination of these vectors is also a tangent vector at this point:

$$(cX_a + dY_a)(fg) = cX_a(fg) + dY_a(fg) =$$

$$= cX_a f \cdot g(a) + cf(a)X_a g + dY_a f \cdot g(a) + df(a)Y_a g =$$

$$= (cX_a + dY_a)f \cdot g(a) + f(a)(cX_a + dY_a)g.$$

Therefore the tangent vectors form a linear space, which is denoted by T_a and is called the tangent space to U at the point a. The value of $X_a f$ is called the derivative of the function f along the direction of the vector X_a. It can be shown that the space T_a is n-dimensional.

8.3. Definition. *A set of tangent vectors $X = \{X_a \in T_a | a \in U\}$ such that for any function $f \in C$ the function on U*

$$a \mapsto X_a f$$

also belongs to C is called a vector field in the region U.

We denote this function by Xf. Obviously, a tangent field determines a linear mapping $X : C \to C$, which is a zero mapping on the constant functions and such that $X(fg) = Xf \cdot g + f \cdot Xg$ for all $f, g \in C$. Such mappings are called *derivations* of the ring C into itself.

Conversely, with every derivation $X : C \to C$ and point $a \in U$ there is associated a tangent vector X_a: $X_a f = (Xf)(a)$. This establishes a bijection between the vector fields on U and derivations of the ring C.

The sum of the vector fields $X+Y$, defined by the formula $(X+Y)f = Xf+Yf$ for all $f \in C$ is a vector field. The product fX, defined by the formula $(fX)g = = f(Xg)$, where X is a vector field and $f, g \in C$ is a vector field. In particular, any linear combination of vector fields $\sum_{i=1}^{m} f^i X_i$, $f^i \in C$ is a vector field.

8.4. Example. Let $\frac{\partial}{\partial x^i}$, $i = 1, \ldots, n$ be the classical partial differentiation operators. They are all vector fields on U. The following fundamental result, which we shall present without proof, is true.

8.5. Theorem. *Any vector field X in a connected region $U \subset \mathbf{R}^n$ can be uniquely represented in the form $\sum_{i=1}^{n} f^i \frac{\partial}{\partial x^i}$, where $x^1, \ldots, x^n$ are the coordinate functions on $\mathbf{R}^n$.*

8.6. In algebraic language this means that the set of all vector fields T in a connected region U is a free module of rank n over the commutative associative ring C of infinitely differentiable functions on U.

Free modules of finite rank over commutative rings form a category, whose properties are very close to those of the category of finite-dimensional spaces over a field. For them, in particular, the complete theory of duality and all constructions of tensor algebra from this part of the course are valid.

Another variant, which does not require the transfer of tensor algebra to rings and modules, but is instead predicated on the development of some geometric technique, consists of studying every vector field X as a collection of vectors $\{X_a | a \in U\}$, contained in a family of finite-dimensional spaces $\{T_a\}$. Then all required operations of tensor algebra can be constructed "pointwise", by defining, for example, $X \otimes Y$ as $\{X_a \otimes Y_a | a \in U\}$.

Both variants of the construction of the tensor algebra are completely equivalent; in the definitions presented below, we shall start from the first variant.

8.7. We denote by T^* the C-module of C-linear mappings

$$T^* = \mathcal{L}_C(T, C).$$

It consists of mappings $\omega : S \to C$ with the property

$$\omega\left(\sum_{i=1}^{m} f^i X_i\right) = \sum_{i=1}^{m} f^i \omega(X_i)$$

for all $X_i \in T$ and $f^i \in C$. Addition and multiplication by elements are performed by means of the standard formulas. The C-module T^* is often denoted by Ω^1 or $\Omega^1(U)$ and is called the module of (differential) 1-*forms* in the region U.

Every function $f \in C$ determines an element $df \in \Omega^1$ according to the formula

$$(df)(X) = Xf, \ X \in T.$$

It is called the *differential of the function* f. In particular, we can construct the differentials of the coordinate functions $dx^1, \ldots, dx^n \in \Omega^1$. The following proposition follows easily from Theorem 8.4.

8.8. Proposition. *Any 1-form $\omega \in \Omega^1$ can be uniquely represented as a linear combination $\sum_{i=1}^{n} f_i dx^i$.*

8.9. The elements of the tensor product of C-modules $T^* \underbrace{\otimes \ldots \otimes}_{p} T^* \otimes T \underbrace{\otimes \ldots \otimes}_{q} T$ are called *tensor fields* of type (p, q), or p-covariant and q-covariant tensor fields in the region U. In differential geometry, by the way, the word "field" is often omitted and tensor fields are simply called tensors.

It follows from Theorem 8.5 and Proposition 8.8 that any tensor of the type (p, q) is uniquely determined by its components $T_{i_1 \ldots i_p}^{j_1 \ldots j_q}$ according to the formula

$$T = \sum T_{i_1 \ldots i_p}^{j_1 \ldots j_q} dx^{i_1} \otimes \ldots \otimes dx^{i_p} \otimes \frac{\partial}{\partial x^{j_1}} \otimes \ldots \otimes \frac{\partial}{\partial x^{j_q}},$$

where i_k and j_l independently assume values from 1 to n. In the classical notation, all symbols on the right side are omitted except for the components $T_{i_1 \ldots i_p}^{j_1 \ldots j_q}$, and this symbol serves as the notation for the tensor. We emphasize once again that here $T_{i_1 \ldots i_p}^{j_1 \ldots j_q}$ are not numbers, but rather real infinitely differentiable functions on U.

8.10. Transformation of coordinates. The first contribution of analysis to the study of tensor fields is the possibility of performing non-linear transformations of coordinate functions in U: from $x^1, \ldots, x^n$ to $y^1, \ldots, y^n$, where $y^i = y^i(x^1, \ldots, x^n)$ are infinitely differentiable functions such that the inverse functions $x^j = x^j(y^1, \ldots, y^n)$ are defined and infinitely differentiable. The point is that the components of the vector fields and 1-forms in this case still transform linearly according to the classical formulas, only with coefficients which vary from point to point: according to the rule for differentiating a composite function we have

$$\frac{\partial}{\partial x^j} = \frac{\partial y^k}{\partial x^j} \frac{\partial}{\partial y^k}$$

(summation over k is implied on the right side), and also

$$dx^j = \frac{\partial x^j}{\partial y^k} dy^k$$

(same convention). Therefore the tensor $T^{j_1 \dots j_q}_{i_1 \dots i_p}$ in the new coordinates has the components

$$(T')^{j'_1 \dots j'_q}_{i'_1 \dots i'_q} = \frac{\partial x^{i_1}}{\partial y^{i'_1}} \dots \frac{\partial x^{i_p}}{\partial y^{i'_q}} \frac{\partial y^{j'_1}}{\partial x^{j_1}} \dots \frac{\partial y^{j'_q}}{\partial x^{j'_q}} T^{j_1 \dots j_p}_{i_1 \dots i_p}$$

(same convention with summation on the right side).

All algebraic constructions and language conventions of §4 can now be transferred to tensor fields.

We shall conclude this section with some examples of tensor fields which play an especially important role in geometry and physics.

8.11. The metric tensor. This tensor, denoted by g_{ij}, or in a more complete notation, by $\sum_{i,j=1}^n g_{ij} dx^i \otimes dx^j$, is assumed to be symmetric and non-degenerate at all points $a \in U$, that is, $\det(g_{ij}(a)) \neq 0$. It determines an orthogonal structure in every tangent space T_a, and the pairs (U, g_{ij}) (and also the generalizations to the case of manifolds consisting of several regions U "sewn together") form the basic object of study in Riemannian geometry, while in the case $n = 4$ and metrics with the signature (1,3) they are the basic object of study in the general theory of relativity.

The metric is used to measure the lengths of differentiable curves

$$\{x^1(t), \dots, x^n(t) | t_0 \leq t \leq t_1\},$$

the length being defined by the formula

$$\int_{t_0}^{t_1} \sqrt{g_{ij}(x^k(t)) \frac{\partial x^i}{\partial t} \frac{\partial x^j}{\partial t}} dt,$$

and also for raising and lowering the indices of tensor fields.

8.12. Exterior forms and volume forms. The elements of $\Lambda^p(\Omega^1)$, that is, skew-symmetric tensors of the type $(p, 0)$, are called exterior p-forms in U, while the exterior n-forms are called *volume forms*. This terminology is explained by the possibility of defining the "curvilinear integrals"

$$\int_V f(x_1, \dots, x_n) dx^1 \wedge \dots \wedge dx^n$$

over any measurable subregion of U. In the case $f = 1$, this integral is the Euclidean volume of the region V, whose properties we described in §5 of Chapter 2.

For $p < n$ it is possible to define the integral of any form $\omega \in \Lambda^p(\Omega^1)$ over "p-dimensional differentiable hypersurfaces". All modules of exterior forms are related by the remarkable "exterior differential" operators $d^p : \Lambda^p \to \Lambda^{p+1}$, which in terms of coordinates are defined by the formula

$$d^p\left(\sum f_{i_1 \ldots i_p} dx^{i_1} \wedge \ldots \wedge dx^{i_p}\right) = \sum \frac{\partial f}{\partial x^{i_{p+1}}} dx^{i_{p+1}} \wedge dx^{i_1} \wedge \ldots \wedge dx^{i_p}.$$

These operators satisfy the condition $d^{p+1} \circ d^p = 0$ and appear in the formulation of the generalized Stokes theorem, which relates the integral over a p-dimensional hypersurface with the boundary ∂V:

$$\int_{V^p} d\omega^{p-1} = \int_{\partial V^p} \omega^{p-1}.$$

The exterior 2-forms ω^2 , satisfying the condition $d\omega^2 = 0$, play a special role. The apparatus of Hamiltonian mechanics is formulated in an invariant manner in terms of these forms.

§9. Tensor Products in Quantum Mechanics

9.1. Unification of systems. The role of tensor products in quantum mechanics is explained by the following fundamental proposition, which continues the series of postulates formulated in §6.8 and §§9.1–9.6 of Chapter 2.

Let $\mathcal{H}_1, \ldots, \mathcal{H}_n$ be the state spaces of several quantum systems. Then the state space of the system obtained by unifying them is a subspace $\mathcal{H} \subset \mathcal{H}_1 \otimes \ldots \otimes \mathcal{H}_n$.

Strictly speaking, in the infinite-dimensional case the tensor product on the right side should be replaced by the completed tensor product of Hilbert spaces, but we shall disregard this refinement, and work, as usual, with finite-dimensional methods.

The subspace of $\mathcal{H}_1 \otimes \ldots \otimes \mathcal{H}_n$ which corresponds to the unified system must be determined on the basis of further rules, which we shall consider below. Here, however, we consider the case $\mathcal{H} = \mathcal{H}_1 = \mathcal{H}_1 \otimes \ldots \otimes \mathcal{H}_n$ and we attempt to explain how the first postulate of quantum mechanics – *the principle of superposition* – already leads to completely non-classical couplings between systems. To this end, let us clarify the possible states of the unified system. Let $\psi_i \in \mathcal{H}_i$ be some states of the subsystems. Then the factorizable tensor $\psi_1 \otimes \ldots \otimes \psi_n$ is one of the possible states of the unified system, and we can assume that it corresponds to the case when each of the subsystems is in its state ψ_i. But such factorizable states by no means exhaust all vectors in $\mathcal{H}_1 \otimes \ldots \otimes \mathcal{H}_n$: arbitrary linear combinations of them are admissible. When the unified system is in one of the non-factorizable states, the idea of subsystems becomes meaningless, because they and their states cannot be uniquely distinguished. In other words, in most cases subsystems correspond only "virtually" to the unified system.

It is important to emphasize that this result in no way employs the idea of interaction of the subsystems in the classical sense of the word, presuming the exchange of energy between them. Einstein, Rosen and Podolsky proposed a thought experiment in which two subsystems of a unified system are spatially far apart from one another after the system decays and the act of observing one subsystem transfers the other one into a definite state, even though the classical interaction between the two subsystems requires a finite time. This consequence of the postulates of superposition and of the tensor product sharply contradicts the classical intuition. Nevertheless their adoption led to an enormous number of theoretical schemes which correctly explain reality, so that they must be trusted and a new intuition must be developed.

We note in passing that the description of interaction requires the introduction of the Hamiltonian of the unified system. In the simplest case it has the "free" form

$$H_1 \otimes \mathrm{id} \otimes \ldots \otimes \mathrm{id} + \mathrm{id} \otimes H_2 \otimes \ldots \otimes \mathrm{id} + \ldots + \mathrm{id} \otimes \ldots \otimes H_n,$$

where $H_i : \mathcal{H}_i \to \mathcal{H}_i$ is the Hamiltonian of the ith system and id are identity mappings. In this case it is said that the systems do not interact. By way of some explanation of this we remark that if a unified system has such a Hamiltonian and is initially in the factorizable state $\psi_1 \otimes \ldots \otimes \psi_n$, then at any time t it will be in the factorizable state $e^{-itH_1}(\psi_1) \otimes \ldots \otimes e^{-itH_n}(\psi_n)$, that is, its subsystems will evolve independently of one another. In the general case, the Hamiltonian is a sum of the free part and an operator which corresponds to the interaction.

9.2. Indistinguishability. There exist two fundamental cases when the state space of a unified system does not coincide with the complete space $\mathcal{H}_1 \otimes \ldots \otimes \mathcal{H}_n$. In both cases the systems being unified are identical or indistinguishable, for example, they are elementary particles of one type; in particular, $\mathcal{H}_1 = \ldots = \mathcal{H}_n = \mathcal{H}$.

a) *Bosons.* By definition, a system with the state space $\mathcal{H}$ is called a boson if the state space generated by the unification of n systems is the nth symmetric power $S^n(\mathcal{H})$.

According to experiment, photons and alpha particles (helium nuclei) are bosons.

b) *Fermions.* By definition a system with a state space $\mathcal{H}$ is called a fermion if the state space generated by the unification of n such systems is the nth exterior product $\Lambda^n(\mathcal{H})$.

According to experiment, electrons, protons, and neutrons are fermions.

9.3. Occupation numbers and the Pauli principle. Let $\{\psi_1, \ldots, \psi_m\}$ be a basis of the state of a boson or fermion system. Then physicists write the elements of the symmetrized (or antisymmetrized) tensor basis of $S^n(\mathcal{H})$ (or $\Lambda^n(\mathcal{H})$) in the form

$$|a_1,\ldots,a_m\rangle = \begin{cases} S(\underbrace{\psi_1\otimes\ldots\otimes\psi_1}_{a_1}\otimes\ldots\otimes\underbrace{\psi_m\otimes\ldots\otimes\psi_m}_{a_m}) & \text{in } S^n(\mathcal{H}), \\ A(\underbrace{\psi_1\otimes\cdots\otimes\psi_1}_{a_1}\otimes\ldots\otimes\underbrace{\psi_m\otimes\cdots\otimes\psi_m}_{a_m}) & \text{in } \Lambda^n(\mathcal{H}). \end{cases}$$

In both cases $a_1+\ldots+a_m=n$. However, for bosons the numbers a_i can assume any non-negative integral values, while for fermions they can only assume the values 0 or 1; otherwise the corresponding antisymmetrizations equal zero and do not determine a quantum state.

The numbers a_i are called the "*occupation numbers*" of the corresponding state. It is understood that in the state $|a_1,\ldots,a_m\rangle$ of the unified system a_i, subsystems are in the state ψ_i. Since, however, the unified system cannot, in general, be in a state described by the factorizable tensor $\psi_1\otimes\ldots\otimes\psi_m$, in either the fermion or boson case, except for the case when all ψ_i are the same (for bosons), this means that even in the basis states $|a_1,\ldots,a_m\rangle$ one cannot say "which" of the subsystems is, for example, in the state ψ_i. The systems are indistinguishable.

The condition $a_i=0$ or 1 in the fermion case is interpreted to mean that two subsystems cannot be in the same state. This is the famous Pauli exclusion principle.

When the number n is very large, many physically important assertions about the spaces $S^n(\mathcal{H})$ and $\Lambda^n(\mathcal{H})$ are made in terms of probabilities, for example, in terms of the number of the states $|a_1,\ldots,a_n\rangle$ under one or another set of conditions relative to the occupation numbers. For this reason it is often said that bosons and fermions obey different *statistics* — Bose-Einstein or Fermi statistics, respectively.

9.4. The case of variable numbers of particles. In the course of the evolution of a quantum system its constituent "elementary subsystems" or particles can be created or annihilated. To describe such effects, in the boson and fermion case, respectively, subspaces of the states $\bigoplus_{i=1}^{\infty} S^i(\mathcal{H})$ (more precisely, the completion of this space) or $\bigoplus_{i=0}^{\infty}\Lambda^i(\mathcal{H})$, that is, the fully symmetric or exterior algebra of the one-particle space $\mathcal{H}$, are used.

The operator multiplying vectors in $S^n(\mathcal{H})$ (and, correspondingly, in $\Lambda^n(\mathcal{H})$) by n ($n=0,1,2,3,\ldots$), is called the *particle number operator*. Its kernel — the subspace $C=S^0(\mathcal{H})$ or $\Lambda^0(\mathcal{H})$ – is called the *vacuum state*: there are no particles in it.

The special *particle creation and annihilation operators* also play an absolutely fundamental role. The operator $a^-(\psi_0)$ annihilating a boson in the state $\psi_0\in\mathcal{H}$ operates on the state $S(\psi_1\otimes\ldots\otimes\psi_n)$ according to the formula

$$a^-(\psi_0)S(\psi_1\otimes\ldots\otimes\psi_n)=\sqrt{n+1}\cdot\frac{1}{n!}\sum_{\sigma\in S_n}(\psi_0,\psi_{\sigma(1)})\otimes\psi_{\sigma(2)}\otimes\ldots\otimes\psi_{\sigma(n)},$$

where $(\psi_0, \psi_{\sigma(1)})$ is the inner product in $\mathcal{H}$. The operator $a^+(\psi_0)$ creating a boson in the state $\psi_0 \in \mathcal{H}$ is defined as the adjoint of the operator $a^-(\psi_0)$ in the sense of Hermitian geometry. Analogous formulas can be written down for the fermion case. The role of these standard operators of tensor algebra is explained by the fact that important observables, primarily Hamiltonians, can be formulated conveniently in terms of them.

EXERCISES

In the following series of exercises, we set out the basic facts of the theory of tensor rank, which is important for estimates of computational complexity. The foundations of this theory were laid by F. Strassen.

1. Let $L_1, \ldots, L_n$ be finite-dimensional linear spaces over the field $\mathcal{K}$, $t \in L_1 \otimes \ldots \otimes L_n$, $t \neq 0$. By the rank $\operatorname{rank} t$ of the tensor t we mean the smallest number r such that for suitable vectors $l_i^{(j)} \in L_i$, $j = 1, \ldots, r$,

$$t = \sum_{j=1}^{r} l_1^{(j)} \otimes \ldots \otimes l_n^{(j)}.$$

It is clear that when $n = 1$ we have $\operatorname{rank} t = 1$ for any $t \neq 0$.

Let $t \in L_1^* \otimes L_2 = \mathcal{L}(L_1, L_2)$ (see §2.5). Prove that

$$\operatorname{rank} t = \dim \operatorname{im} t, \ t : L_1 \to L_2.$$

Hence derive the following facts:

a) for $n = 2$, $\operatorname{rank} t$ remains invariant under an extension of the basic field;

b) for $n = 2$, the set $\{t \mid \operatorname{rank} t \leq r\}$ is defined by a finite system of equations $P_{j,r}(t^{i_1 \ldots i_n}) = 0$, where the $P_{j,r}$ are polynomials in the coordinates.

Neither of these facts remains true for the case $n = 3$, which is of fundamental interest in the theory of computational complexity; see Exercises 4–9 below.

2. Let $L = \bigoplus_{i,j=1}^{2} \mathbb{C} a_{ij}$ be the space of complex 2×2 matrices. Prove that

$$\operatorname{rank}\left(\sum_{i,j,k=1}^{2} a_{ij} \otimes a_{jk} \otimes a_{ki} \right) = 7.$$

(Hint: use Exercise 12 of §4 in Chapter 1. The same hint applies to the next exercise.)

3. Prove that

$$\operatorname{rank}\left(\sum_{i,j,k=1}^{N} a_{ij}\otimes a_{jk}\otimes a_{ki}\right)\le cN^{\log_2 7}$$

for a suitable constant c. (Here $L=\bigoplus_{i,j=1}^{N}\mathbf{C}a_{ij}$; the tensor we are concerned with is $\operatorname{Tr} A\otimes A\otimes A$, where $A=(a_{ij})$ is the general $N\times N$ matrix.)

4. Let L be a finite-dimensional $\mathcal{K}$-algebra, where $L\bigotimes_{\mathcal{K}} L\to L:\ a\otimes b\mapsto ab$ is its law of multiplication. This law of multiplication is regarded as a tensor $t\in$ $\in L^*\otimes L^*\otimes L$. (Its coordinates are the structure constants of the algebra.) Calculate $\operatorname{rank} t$ for the case $\mathcal{K}=\mathbf{R}$, $L=\mathbf{C}$.

5. Using the notation of the previous exercise, let $L=\mathcal{K}^n$, with the coordinatewise multiplication:

$$(a_1,\dots,a_n)(b_1,\dots,b_n)=(a_1b_1,\dots,a_nb_n).$$

Calculate $\operatorname{rank} t$.

6. Using the results of Exercises 4 and 5, verify that the rank of the tensor of the structure constants of the algebra $\mathbf{C}$ over $\mathbf{R}$ is lowered when the basic field is extended to $\mathbf{C}$.

(Hint: $\mathbf{C}\bigotimes_{\mathbf{R}}\mathbf{C}$ is isomorphic to $\mathbf{C}^2$ as a $\mathbf{C}$-algebra.)

7. Let $L=\mathbf{C}e_1\oplus\mathbf{C}e_2$. Prove that the tensor

$$t=e_1\otimes e_1\otimes e_1+e_1\otimes e_2\otimes e_2+e_2\otimes e_1\otimes e_2$$

has rank 3.

8. Prove that the tensor t of the preceding exercise is the limit of a sequence of tensors of rank 2.

(Hint:

$$t+\epsilon e_2\otimes e_2\otimes e_2=\frac{1}{\epsilon}[e_1\otimes e_1\otimes(-e_2+\epsilon e_1)+(e_1+\epsilon e_2)\otimes(e_1+\epsilon e_2)\otimes e_2].)$$

9. Deduce from Exercises 7 and 8 that the set of tensors of rank ≤ 2 in $L\otimes L\otimes L$ is not given by a system of equations of the form

$$P_j(t^{i_1i_2i_3})=0,$$

where the P_j are polynomials.

10. By the limiting rank $\mathrm{brk}(t)$ of the tensor t we mean the smallest s such that t can be expressed as the limit of a sequence of tensors of rank $\leq s$. Prove that for a general 3×3 matrix A,

$$\mathrm{brk}(\mathrm{Tr}\, A \otimes A \otimes A) \leq 21.$$

11. What is the value of

$$\mathrm{rank}(\mathrm{Tr}\, A \otimes A \otimes A),\ \mathrm{brk}(\mathrm{Tr}\, A \otimes A \otimes A),$$

where A is a general $N \times N$ matrix ? (At the time of writing these lines, the answer is not known even for $N = 3$.)

12. Let L be an n-dimensional linear space over the field $\mathcal{K}$, and $M \subset \Lambda^2(L)$ an arbitrary subspace. Suppose that there exists $w \in L$ with $0 \neq v \wedge w \in M$ for each $v \in L$, $v \neq 0$. Prove that a basis $e_1, \ldots, e_n$ can be chosen in L such that

$$M + \Lambda^2(L_i) = \Lambda^2(L),\ 1 \leq i \leq n,$$

where $L_i = \bigoplus_{j \neq i} \mathcal{K} e_j$.

For the case $\mathcal{K} = \mathbf{F}_p$ (the field of p elements), an extremely complicated combinatorial proof of this result is known (M.R. Vaughan-Lee, J. Algebra, 1974,**32**, 278–285), which admits a group-theoretic interpretation.

It would be good to find a more direct approach.

Subject Index